PLANT EVOLUTION AND THE ORIGIN OF CROP SPECIES

Plant Evolution and the Origin of Crop Species

Second Edition

James F. Hancock
Department of Horticulture
Michigan State University

CABI Publishing

CABI Publishing is a division of CAB International

CABI Publishing
CAB International
Wallingford
Oxon OX10 8DE
UK
Tel: +44 (0)1491 832111
Fax: +44 (0)1491 833508
E-mail: cabi@cabi.org
Website: www.cabi-publishing.org

CABI Publishing
875 Massachusetts Avenue
7th Floor
Cambridge, MA 02139
USA
Tel: +1 617 395 4056
Fax: +1 617 354 6875
E-mail: cabi-nao@cabi.org

A catalogue record for this book is available from the British Library, London, UK.

Library of Congress Cataloging-in-Publication Data
Hancock, James F.
Plant evolution and the origin of crop species / James F. Hancock.--
2nd ed.
p. cm.
Includes bibliographical references (p.).
ISBN 0-85199-685-X (alk. paper)
1. Crops--Evolution. 2. Crops--Origin. 3. Plants--Evolution. I.
Title.
SB106.O74H36 2003
633-dc21
2003006924

ISBN 0 85199 685 X

Artwork provided by Marlene Cameron.
Typeset in 10pt Souvenir by Columns Design Ltd, Reading.
Printed and bound in the UK by Biddles Ltd, Guildford and King's Lynn.

Contents

Preface **ix**
Acknowledgements ix

Introduction **1**

Part 1. Evolutionary Processes

1. Chromosome Structure and Genetic Variability **3**
Gene and Chromosomal Structure 4
Types of Mutation 5
Measurement of Variability 16
Construction of Genetic Maps and Genome Evolution 28

2. Assortment of Genetic Variability **32**
Random Mating and Hardy–Weinberg Equilibrium 32
Migration 35
Selection 42
Genetic Drift 52
Evolution in Organelles 54
Interaction between Forces 55

3. The Multifactorial Genome **58**
Intragenomic Interactions 60
Coadaptation 64
Canalization 72
Paradox of Coadaptation 73

4. Polyploidy and Gene Duplication **77**
Factors Enhancing the Establishment of Polyploids 78

Evolutionary Advantages of Polyploidy 80
Genetic Differentiation in Polyploids 93
Chromosomal Repatterning 95
Genome Amplification and Chance 98

5. Speciation 100
What is a Species? 100
Reproductive Isolating Barriers 103
Modes of Speciation 110
Genetic Differentiation during Speciation 117
Hybridization and Introgression 119
Hybridization and Extinction 124
Crop–Weed Hybridizations 125
Risk of Transgene Escape into the Environment 127

Part 2. Agricultural Origins and Crop Evolution

6. The Origins of Agriculture 129
Rise of our Food Crops 129
Emergence of *Homo* 131
Evolution of *Homo* 137
Appearance of Modern Humans 139
Spread of *H. sapiens* 141
Agricultural Origins 143
Early Crop Dispersals 145
Transcontinental Crop Distributions 148

7. The Dynamics of Plant Domestication 151
Evolution of Farming 153
Early Stages of Plant Domestication 154
Origins of Crops 157
Characteristics of Early Domesticants 159
Changes During the Domestication Process 161
Genetic Regulation of Domestication Syndromes 163
Evolution of Weeds 166
Genetic Diversity and Domestication 169
Domestication and Native Diversity Patterns 170

8. Cereal Grains 174
Barley 174
Maize 176
Millets 181
Oats 183
Rice 185
Rye 187
Sorghum 188
Wheat 190

9. Protein Plants **195**
Chickpea 195
Cowpea 196
Pea 198
Lentil 200
Phaseolus Beans 203
Faba Beans 206
Soybean 206

10. Starchy Staples and Sugars **209**
Banana 209
Cassava 212
Potato 214
Sugar Cane 217
Sugar Beet 219
Sweet Potato 220
Taro 222
Yam 223

11. Fruits, Vegetables, Oils and Fibres **226**
Fruits
Apples 226
Citrus 228
Grape 230
Peach 231
Strawberry 232
Vegetables
Cole Crops 234
Squash and Gourds 236
Chilli Peppers 238
Tomato 240
Fibres and Oils
Cotton 241
Groundnut 243
Sunflower 244

12. Postscript: Germ-plasm Resources **246**
Ex situ Conservation 248
In situ Conservation 249

References **251**

Index **307**

Preface

The first edition of this book was published in 1992 by Prentice-Hall. This second edition incorporates the wealth of new information that has emerged over the last decade on plant evolution. The advent of molecular markers has generated a cascade of new information on evolutionary processes, the structure of plant genomes and crop origins. Ideas about the evolutionary role of introgression, hybridization and polyploidy have been dramatically altered, and the species origins of many recalcitrant crops have been elucidated. In addition, the major loci associated with domestication have been mapped and it has been shown that crop genomes can be quite fluid. To my knowledge, no other book on plant evolution has attempted to combine the last decade of molecular information with conventionally acquired information.

In this edition, I have tried very hard to show how natural and crop evolution are intimately associated. Much more of the crop information is incorporated in the early evolutionary discussions, and I take greater pains to describe the evolutionary mechanisms associated with crop domestication. The previous discussion about prehuman plant and animal evolution has been greatly abbreviated so that more space can be devoted to variation patterns associated with crop domestication and dispersal. All in all, I think this edition does a better job of describing the continuum between natural and crop evolution.

Acknowledgements

Numerous people contributed directly and indirectly to the book: first and foremost, my wife, Ann, who has been unflagging in her support over the years and has served as a model of creativity and drive. Marlene Cameron added greatly to the text with her exceptional artwork. Norm Ellstrand and Paul Gepts made numerous helpful suggestions. I am also indebted to the students who have pushed me in my crop evolution class over the last 20 years, and the Plant Breeding and Genetics Journal Club, which keeps uncovering important publications that I have missed.

Several previous texts have had particularly strong influences on me and been tremendous resources: J.R. Harlan's *Crops and Man*, C.B. Heiser's *Seed to Civilization*, V. Grant's *Plant Speciation* and *Organismic Evolution*, J.D. Sauer's *Historical Geography of Crop Plants: a Select Roster*, J. Smartt and N.W. Simmond's *Evolution of Crop Plants,* B. Smith's *The Emergence of Agriculture*, and D. Zohary and M. Hopf's *Domestication of Plants in the Old World.*

Introduction

This book has been written for advanced undergraduates and graduate students in the biological sciences. It is meant primarily as a text for crop evolution courses, but should serve well in a wide range of plant evolution and systematics courses. It is also intended as a resource book on individual crop histories. I have worked hard to combine the recently emerging molecular data with archaeological, morphological and cytogenetic information.

The book is arranged in two sections: Chapters 1–5 cover the genetic mechanisms associated with plant evolution, and Chapters 6–12 deal with the domestication process and the origin of crop species. In the first half of the book, little effort is made to distinguish between natural and crop evolution, since both kinds of change occur through the reorganization of genetic variability. The processes of change are the same, regardless of whether we are dealing with wild or domesticated populations; only the selector differs. In the first five chapters, I rely heavily on the evolutionary literature, but try to incorporate relevant crop species where appropriate. The goal of these chapters is to describe the overall framework of species change and demonstrate the intimacy of nature and crop evolution.

In the second half of the book, I focus on when and where crops were domesticated and the types of changes associated with their domestication. Chapters 6 and 7 give an overview of the emergence and diffusion of agriculture, and the ways species were changed during domestication. The next four chapters deal with the evolution of individual crops, representing grains, legumes, starchy roots, fruits, vegetables and oils. Whenever possible, the genetic mechanisms described in the first five chapters are highlighted. The last chapter contains a brief discussion of germ-plasm resources and why they need to be maintained. Clearly, if we are to continue to feed the human population, we must develop greater respect for our natural populations and their ongoing evolution.

The overall goal of this book is to describe the processes of evolution in native and cultivated populations and to provide a blueprint for the systematic study of crop origins. It is hoped that when the student completes this book she or he will understand the factors involved in species change and will have a greater appreciation of the coadaptive nature of plants and people.

1 Chromosome Structure and Genetic Variability

Introduction

Evolution is the force that shapes our living world. Countless different kinds of plants and animals pack the earth and each species is itself composed of a wide range of morphologies and adaptations. These species are continually being modified as they face the realities of their particular environments.

In its simplest sense, evolution can be defined as a change in gene frequency over time. Genetic variability is produced by mutation and then that variability is shuffled and sorted by the various evolutionary forces. It does not matter whether the species are cultivated or wild, the basic evolutionary processes are the same. The way in which organisms evolve is dependent on their genetic characteristics and the type of environment they must face.

A broad spectrum of evolutionary forces act to alter species including migration, selection and random chance. In the next four chapters, we shall describe these parameters and how they interact; but, before we do this, we shall begin by discussing the different types of genetic variability that are found in plants. The primary requirement for evolutionary change is genetic variability and mutation generates these building-blocks. A wide range of mutations can occur at all levels of genetic organization from nucleotide sequence to chromosome structure. In this chapter, we shall discuss how plant genes are organized in chromosomes and then the kinds of genetic variability present and their measurement.

Gene and Chromosomal Structure

Both gene and chromosome structure is complex. Genes are composed of coding regions, called exons, and non-coding regions, called introns. Both the introns and exons are transcribed, but the introns are removed from the final RNA product before translation (Fig. 1.1). There are also short DNA sequences both near and far from the coding region that regulate transcription, but are not transcribed themselves. Promoter sequences are found immediately before the protein coding region and play a role in the initiation of transcription, while enhancer sequences are often located far from the coding region and regulate levels of transcription.

Each chromosome contains not only genes and regulatory sequences, but also a large amount of short, repetitive sequences. Some of these are concentrated near centromeres in the densely stained portions of chromosomes called heterochromatic regions, and may play a role in the homologous pairing of chromosomes and their separation. However, there are numerous other repeating units that are more freely dispersed over chromosomes and do not appear to have a functional role. These have been described by some as 'selfish' or 'parasitic' as their presence may stimulate further accumulation of similar sequences through transposition, a topic we shall discuss more fully later (Doolittle and Sapienza, 1980; Orgel and Crick, 1980).

The overall amount of DNA in nuclei can vary dramatically between taxonomic groups and even within species. The total DNA content of nuclei is commonly referred to as the genome. There is a 100-fold variation in genome size among all diploid angiosperms, and congeneric species vary commonly by threefold (Price *et al.*, 1986; Price, 1988). In some cases, genomic amplification occurs in a breeding population in response to environmental or developmental perturbations (Walbot and Cullis, 1985; Cullis, 1987). Most of these differences occur in the quantity of repetitive DNA and not unique sequences.

Except in the very small genome of *Arabidopsis thaliana* (Barakat *et al.*, 1998), it appears that genes are generally found near the ends of chromosomes in clusters between various kinds of repeated sequences (Schmidt and Heslop-Harrison, 1998; Heslop-Harrison, 2000). The amount of inter-

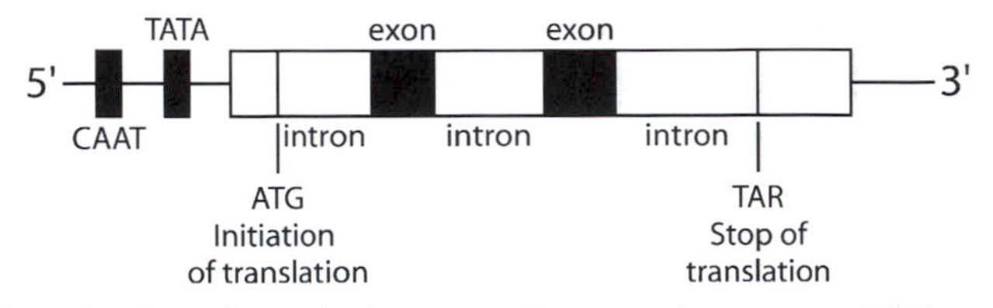

Fig. 1.1. Organization of a typical eucaryotic gene. A precursor RNA molecule is produced from which the introns are excised and the exons are spliced together before translation. The CAAT and TATA boxes play a role in transcription initiation and enhancement.

spersed repetitive DNAs can be considerable, making the physical distances between similar loci highly variable across species. However, the gene clusters may be 'hot spots' for recombination, making recombination-based genetic lengths much closer than physical distances (Dooner and Martinez-Ferez, 1997; Schmidt and Heslop-Harrison, 1998).

Types of Mutation

There are four major types of mutation: (i) point mutations; (ii) chromosomal sequence alterations; (iii) chromosomal additions and deletions; and (iv) chromosomal number changes. Point mutations arise when nucleotides are altered or substituted. For example, the base sequence CTT becomes GTT. Chromosomal sequence alterations occur when the order of nucleotides is changed within a chromosome. Chromosomal duplications and deletions are produced when portions of chromosomes are added or subtracted. Chromosomal numerical changes arise when the number of chromosomes changes.

Point mutations

Nucleotide changes occur spontaneously due to errors in replication and repair at an average rate of 1×10^{-6} to 10^{-7}. These estimates have come largely from unicellular organisms, such as bacteria and yeast, which are easy to manipulate and have tremendous population sizes, but good estimates have also been obtained in higher plants using enzymes (Kahler *et al.*, 1984) and a variety of seed traits (Table 1.1). Mutation rates can be increased by numerous environmental agents, such as ionizing radiation, chemical mutagens and thermal shock.

Table 1.1. Spontaneous mutation rates of several endosperm genes in *Zea mays* (from Stadler, 1942).

Gene	Character	No. of gametes tested	Mutation rate
R	Aleurone colour	554,786	0.00049
I	Colour inhibitor	265,391	0.00011
Pr	Purple colour	647,102	0.000011
Su	Sugary endosperm	1,678,736	0.000002
C	Aleurone colour	426,923	0.000002
Y	Yellow seeds	1,745,280	0.000002
Sh	Shrunken seeds	2,469,285	0.000001
Wx	Waxy starch	1,503,744	< 0.000001

Sequence alterations

Three types of DNA sequence alterations occur: translocations, inversions and transpositions. Small numbers of redundant nucleotide blocks may be involved or whole groups of genes. Translocations occur when nucleotide sequences are transferred from one chromosome to another. In homozygous individuals, nuclear translocations have no effect on fertility, but in heterozygous individuals only a portion of the gametes are viable, due to duplications and deficiencies (Fig. 1.2). Translocations are widespread in a number of plant genera, including *Arachis*, *Brassica*, *Campanula*, *Capsicum*, *Clarkia*, *Crepis*, *Datura*, *Elymus*, *Galeopsis*, *Gilia*, *Gossypium*, *Hordeum*, *Layia*, *Madia*, *Nicotiana*, *Paeonia*, *Secale*, *Trillium* and *Triticum* (Grant, 1975; Holsinger and Ellstrand, 1984; Konishi and Linde-Laursen, 1988; Stalker *et al.*, 1991; Livingstone *et al.*, 1999).

Populations are generally fixed for one chromosomal type, but not all. Translocation heterozygotes are common in *Paeonia brownii* (Grant, 1975), *Chrysanthemum carinatum* (Rana and Jain, 1965), *Isotoma petraea* (James, 1965) and numerous species of *Clarkia* (Snow, 1960). Probably the most extreme example of translocation heterozygosity is in *Oenothera biennis*, where all of its nuclear chromosomes contain translocations and a complete ring of chromosomes is formed at meiosis in heterozygous individuals (Cleland, 1972). In some cases, translocations have resulted in the fusion and fission of non-homologous chromosomes with short arms (Robertsonian translocations).

Inversions result when blocks of nucleotides rotate 180°. Nuclear inversions are called pericentric when the rotation includes the centromere and paracentric when the centromeric region remains unaffected (Figs 1.3 and 1.4). As with translocations, individuals that are heterozygous produce numerous unviable gametes but only if there is a crossover between chromatids; all the gametes of homozygotes are fertile regardless of crossovers.

Inversion polymorphisms have been described in a number of plant genera. One of the best documented cases of an inversion heterozygosity within a species is in *Paeonia californica*, where heterozygous plants are common throughout the northern range of the species (Walters, 1952). As we shall describe more fully in the chapter on speciation (Chapter 5), inversions on six chromosomes distinguish *Helianthus annuus*, *Helianthus petiolaris* and their hybrid derivative *Helianthus anomalus* (Rieseberg *et al.*, 1995). Tomato and potato vary by five inversions (Tanksley *et al.*, 1992), and pepper and tomato by 12 inversions (Livingstone *et al.*, 1999). The chloroplast genome of most angiosperm species has a large inverted repeat (Fig. 1.5), but its structure is highly conserved across families. Only a few species do not have the repeat and no intrapopulational variation has been described (Palmer, 1985).

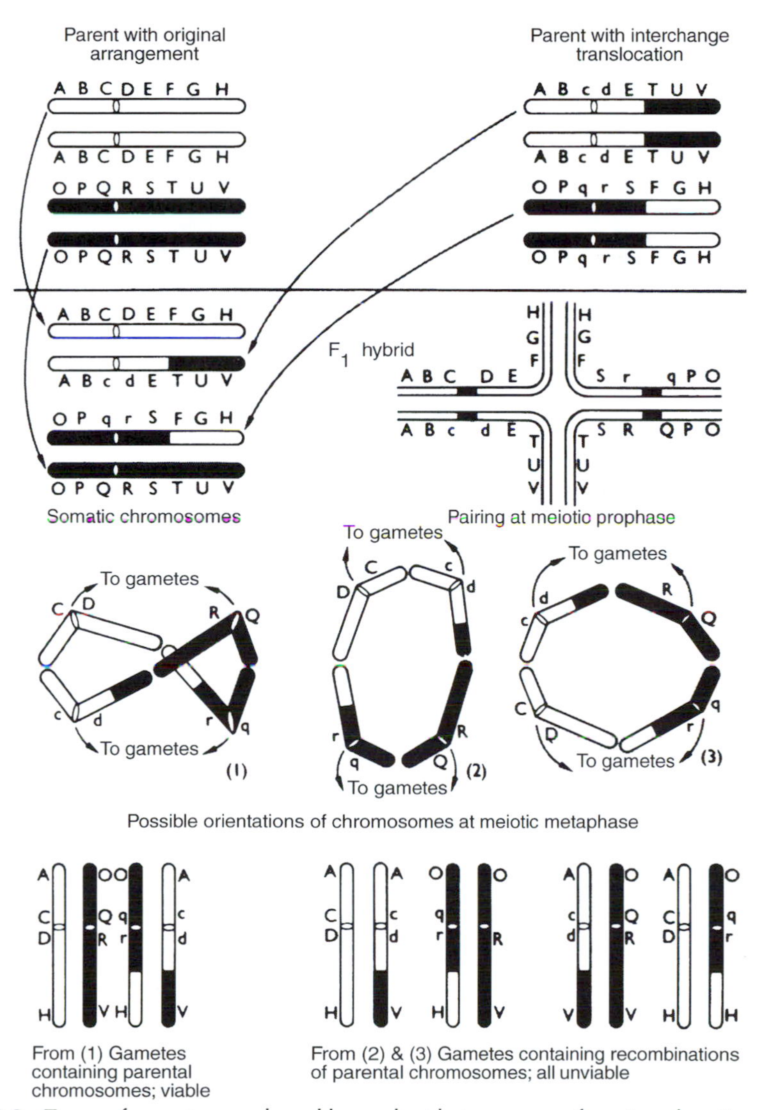

Fig. 1.2. Types of gametes produced by a plant heterozygous for a translocation. A ring of chromosomes is formed at meiosis and depending on how the chromosomes orient at metaphase and separate during anaphase, viable or unviable combinations of genes are produced. (Used with permission from T. Dobzhansky, © 1970, *Genetics of the Evolutionary Process,* Columbia University Press, New York.)

Transposition occurs when nucleotide blocks move from place to place in the genome (McClintock 1953, 1956; Bennetzen, 2000a; Fedoroff, 2000). There are two major classes of transposons: DNA and RNA transposable elements. The RNA transposable elements (retroelements) amplify via RNA intermediates, while the DNA transposons rely

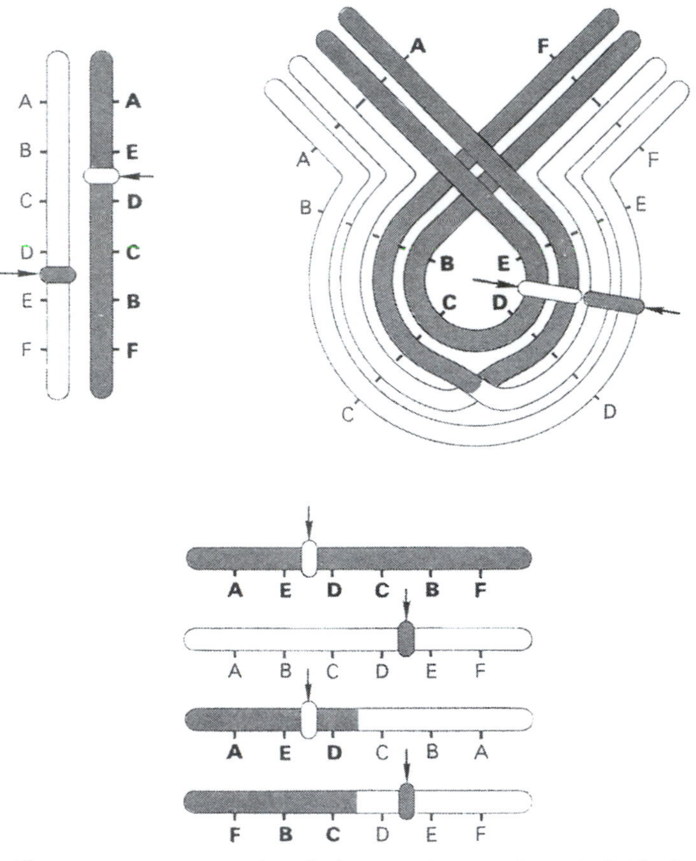

Fig. 1.3. Chromosome types produced after crossing over in an individual heterozygous for a pericentric inversion. Note the two abnormal chromatids, one with a duplication and the other with a deficiency. (Used with permission from T. Dobzhansky, © 1970, *Genetics of the Evolutionary Process,* Columbia University Press, New York.)

on actual excision and reinsertion. Both classes of transposition have been found in all plant species where detailed genetic analysis has been performed, and in many plant species, mobile elements actually make up the majority of the nuclear genome (SanMiguel and Bennetzen, 1998; Bennetzen, 2000a). Most of the transposons are inserted into non-coding regions, but sometimes they enter exons and, when they do, they can have extreme effects on phenotype. The wrinkled-seed character described by Mendel is caused by a transpose-like insertion into the gene encoding a starch-branching enzyme (Bhattacharyya *et al.*, 1990). Much of the flower colour variation observed in the morning glory is due to the insertion and deletion of transposable elements (Clegg and Durbin, 2000; Durbin *et al.*, 2001).

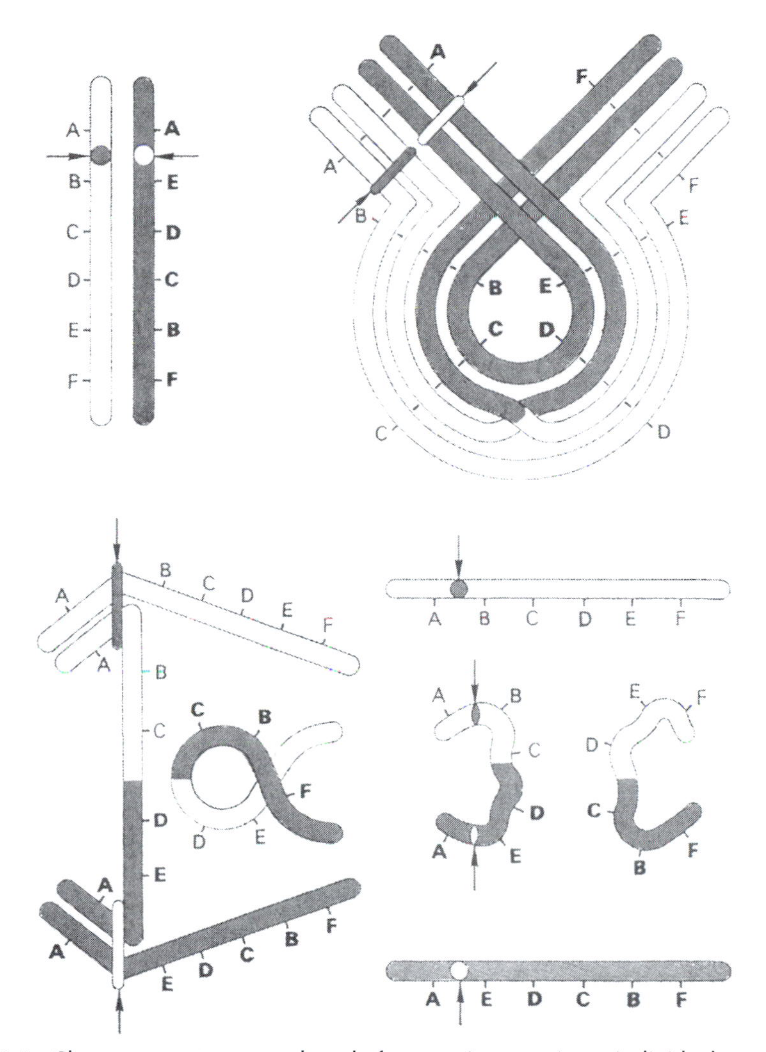

Fig. 1.4. Chromosome types produced after crossing over in an individual heterozygous for a paracentric inversion. Note the chromosomal bridge and the resulting chromatids with deletions. (Used with permission from T. Dobzhansky, © 1970, *Genetics of the Evolutionary Process,* Columbia University Press, New York.)

The DNA transposons range in size from a few hundred bases to 10 kb, and the most complex members are capable of autonomous excision, reattachment and alteration of gene expression. They all have short terminal inverted repeats (TIRs); the most complex ones encode an enzyme called transposase that recognizes the family's TIR and performs the excision and reattachment.

Retroelements (RNA transposons) are the most abundant class of transposons and they make up the majority of most large plant genomes. They transpose through reverse transcription of RNA intermediates, and as a result they do not excise when they transpose, resulting in amplification. The most abundant

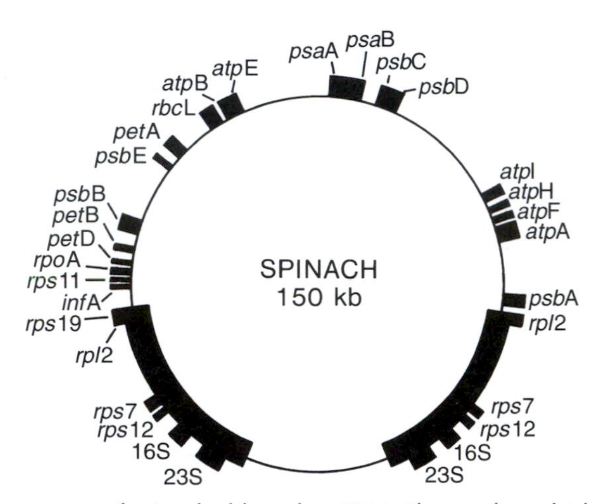

Fig. 1.5. The gene map of spinach chloroplast DNA. The two long thickenings in the lower half of the circle represent the inverted repeat. Gene designations: *rbc*L, the large subunit of ribulose bisphosphate carboxylase; *atp*A, *atp*B, *atp*E, *atp*F, *atp*H and *atp*I, the alpha, beta, epsilon, CF_0-I, CF_0-III and CF_0-IV subunits of chloroplast coupling factor, respectively; *psa*A and *psa*B, the P700 chlorophyll-*a* apoproteins of photosystem I; *psb*A, *psb*B, *psb*C, psbD and *psb*E, the Q-beta (32 kilodaltons (kDa), herbicide-binding), 51 kDa chlorophyll-*a*-binding, 44 kDa chlorophyll-*a*-binding, D2 and cytochrome-b-559 components of photosystem II; *pet*A, *pet*B and *pet*D, the genes for the cytochrome-f, cytochrome-b6 and subunit-4 components of the cytochrome-b6-f complex; *inf*A, initiation factor IF-1; *rpo*A, alpha subunit of RNA polymerase; *rpl*2, *rps*7, *rps*11, *rps*12, and *rps*19, the chloroplast ribosomal proteins homologous to *Escherichia coli* ribosomal proteins L2, S7, S11, S12 and S19, respectively; 16S and 23S, the 16S and 23S ribosomal RNAs, respectively. (Used with permission from J.D. Palmer, © 1987, *American Naturalist* 130, S6–S29, University of Chicago Press, Chicago.)

class of retroelements in plants are the LTR (long terminal repeat) retrotransposons, varying in size from a few hundred to over 10,000 nucleotides.

Duplications and deficiencies

Chromosomal deficiencies occur when nucleotide blocks are lost from within a chromosome, while duplications arise when nucleotide sequences are multiplied. These are caused by unequal crossing over at meiosis or translocation (Burnham, 1962). They can also occur when DNA strands mispair during replication of previously duplicated sequences (Levinson and Gutman, 1987). As previously mentioned, the genome is filled with high numbers of short, repeated sequences that vary greatly in length – so greatly, in fact, that some of them, such as single sequence repeats (SSRs), have proved valuable as molecular markers to distinguish species, populations and even individuals. Variations in gene copy number have been found in a very wide array of crop species. Gene amplifications are so common that Wendel (2000) has suggested that 'one generalization that has been confirmed and extended by the data emerging from the global thrust in genome sequencing and mapping is that most "single-copy" genes belong to larger gene families, even in putatively diploid organisms'. Using sequence data, the fraction of the genome represented by duplications has been estimated to be 72% in maize (Ahn and Tanksley, 1993; Gaut and Doebley, 1997) and 60% in *A. thaliana* (Blanc *et al.*, 2000).

Clusters of duplicated genes are often found scattered at multiple locations across the genome. When Blanc *et al.* (2000) compared the sequence of duplications on chromosomes 2 and 4 of *A. thaliana*, they identified 151 pairs of genes, of which 59 (39%) showed highly similar nucleotide sequences. The order of these genes was generally maintained on the two chromosomes, except for a small duplication and an inversion. When they compared the sequence of these duplicated regions to the rest of the genome, they found 70% of the genes to be present elsewhere. The genes were duplicated in 18 large translocations and several smaller ones (Fig. 1.6).

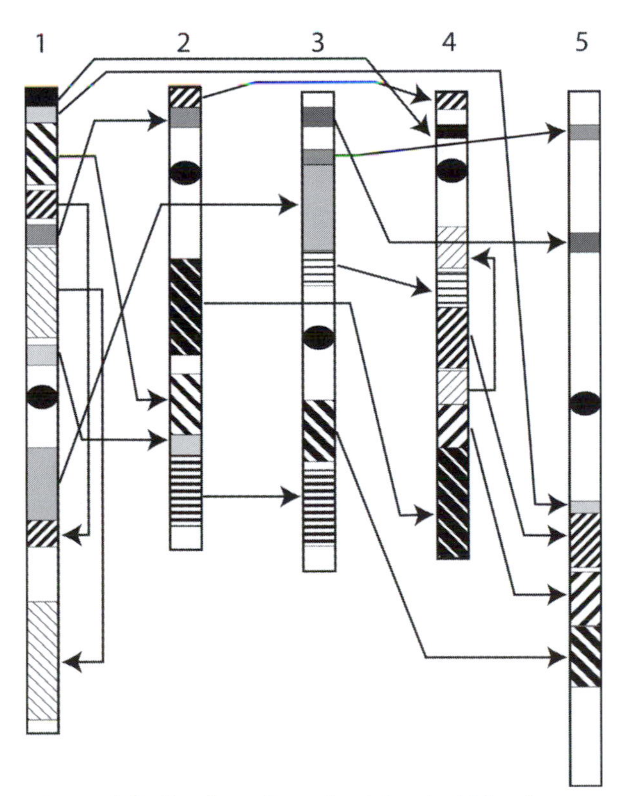

Fig. 1.6. Locations of duplications throughout the *Arabidopsis* genome. Similar blocks on different chromosomes are identified by similar coloured regions and arrows. (Modified from Blanc *et al.*, 2000.)

Chromosomal numerical changes

There are three primary types of numerical changes found in nuclear genomes: (i) aneuploidy; (ii) haploidy; and (iii) polyploidy. In aneuploidy, one or more chromosomes of the normal set are absent or present in excess. Haploids have half the normal chromosome set, while polyploids have more than two sets of homologous chromosomes.

Haploids are quite rare in nature, although they have been produced experimentally in many crops, including strawberries, lucerne and maize. They are generally unviable due to meiotic irregularities and low in vigour because lethal alleles are not buffered by heterozygosity.

Aneuploids are more common than haploids, although they are still relatively rare in natural populations. Most base numbers differ by a few chromosomes (*Citrus*, *Rubus*, *Poa*, *Nicotiana*, *Gossypium*, *Allium* and *Lycopersicon*), but extensive variations involving dozens of chromosomes occur in some groups (*Abelmoschus* and *Saccharum*). Aneuploids arise through the fusion and fission of chromosomes and when chromosomes migrate irregularly during meiosis. They are most common in polyploid species and hybrid populations resulting from trans-specific crosses.

Polyploidy is quite prevalent in higher plants; between 35 and 50% of all angiosperm species are polyploid (Grant, 1971). The number of polyploid crop species is even higher (78%) (Table 1.2) if we use the widely accepted assumption that chromosome numbers above $2n = 18$ represent polyploids. This is a conservative estimate, as many groups are thought to have haploid numbers lower than $x = 9$ (Grant, 1963). Most polyploids are thought to originate through the unification of unreduced gametes (Harlan and deWet, 1975; Bretagnolle and Thompson, 1995). Only occasionally do polyploids arise through somatic doubling and the generation of polyploid meristems, called 'sports'. Most commonly this is done artificially with the chemical colchicine.

Aneuploidy and polyploidy can have rather extreme effects on the physiology and morphology of individuals, since gene dosages are doubled. Cell sizes usually increase, developmental rates slow down and fertility is often reduced. The whole of Chapter 4 is devoted to the physiological and evolutionary ramifications of polyploidy.

There are two major types of polyploids: autopolyploids and allopolyploids (amphiploids) (Fig. 1.7). Allopolyploids are derived from two different ancestral species, whose chromosome sets cannot pair at meiosis. As a result, the chromosomes segregate and assort as in diploids, so inheritance of each individual duplicated loci of allopolyploids follows typical Mendelian patterns. In other words, allopolyploids display 'disomic inheritance', where two alleles segregate at a locus. The chromosome sets of autopolyploids are derived from a single or closely related ancestral species and they can pair at meiosis. The chromosomes associate together either as multivalents or random associations of bivalents. As a result,

Table 1.2. Chromosome numbers in selected crop species.

Diploids 2n ≤ 18		Polyploids 2n > 18	
Allium cepa (onion)	16	*Abelmoschus esculentus* (okra)	33–72
Avena sativa (oat)	14	*Aleurites fordii* (tung)	22
Beta vulgaris (sugar beet)	18	*Ananas comosus* (pineapple)	50
Beta oleracea (kale)	18	*Arachis hypogaea* (groundnut)	40
Carica papaya (pawpaw)	18	*Artocarpus altilis* (breadfruit)	56, 84
Cicer arietinum (chickpea)	16	*Avena abyssinica* (oat)	28
Citrus spp. (citrus)	18	*Avena nuda* (oat)	42
Cucumis sativus (cucumber)	14	*Avena sativa* (oat)	42
Daucus carota (carrot)	18	*Brassica campestris* (turnip)	20
Hordeum vulgare (barley)	14	*Brassica napus* (rape)	38
Lactuca sativa (lettuce)	18	*Cajanus cajan* (pigeon-pea)	22
Lens culinaris (lentil)	14	*Camellia sinensis* (tea)	30
Pennisetum americanum (pearl millet)	14	*Cannabis sativa* (hemp)	20
		Capsicum spp. (peppers)	24
Pisum sativum (pea)	14	*Carthamus tinctorius* (safflower)	24
Prunus amygdalus (almond)	16	*Chenopodium quinoa* (quinoa)	36
Prunus armeniaca (apricot)	16	*Citrullus lanatus* (water melon)	22
Prunus avium (sweet cherry)	16	*Cocos nucifera* (coconut)	32
Prunus persica (peach)	16	*Coffee arabica* (coffee)	44
Raphanus sativus (radish)	18	*Colocasia esculenta* (taro)	28
Ribes nigrum (blackcurrants)	16	*Cucumis melo* (musk melon)	24
Ribes sativum (redcurrants)	16	*Cucurbita* spp. (squash)	20
Secale cereale (rye)	14	*Dioscorea alata* (yam)	30, 40, 50, 60, 70, 80
Triticum monococcum (einkorn)	14	*Dioscorea esculenta* (yam)	40
Vicia faba (bean)	12	*Dioscorea rotundata* (yam)	40
		Elaeis guineensis (oil palm)	32
		Eleusine coracana (finger millet)	36
		Ficus carica (fig)	26
		Fragaria × *ananassa* (strawberry)	56
		Glycine max (soybean)	40
		Gossypium arboreum (cotton)	26
		Gossypium barbadense (cotton)	26
		Gossypium herbaceum (cotton)	52
		Gossypium hirsutum (cotton)	52
		Helianthus annuus (sunflower)	34
		Helianthus tuberosus (Jerusalem artichoke)	102
		Hevea brasiliensis (rubber)	36
		Ipomoea batatas (sweet potato)	90
		Linum usitatissimum (linseed)	30, 32
		Lycopersicon esculentum (tomato)	24
		Malus × *domestica* (apple)	34, 51
		Mangifera indica (mango)	40
		Manihot esculenta (cassava)	36
		Medicago sativa (lucerne)	
		Musa spp. (bananas)	22, 33, 44
		Nicotiana tabacum (tobacco)	48
		Olea europaea (olive)	46
		Oryza sativa (rice)	24
		Oryza globerrima (rice)	24

Continued

Table 1.2. *Continued.*

Diploids 2n ≤ 18	Polyploids 2n > 18	
	Persea americana (avocado)	24
	Phaseolus acutifolius (tepary bean)	22
	Phaseolus coccineus (runner bean)	22
	Phaseolus lunatus (lima bean)	22
	Phaseolus vulgaris (common bean)	22
	Phoenix dactylifera (date)	36
	Piper nigrum (pepper)	48, 52, 104, 128
	Prunus cerasus (sour cherry)	32
	Pyrus spp. (pears)	34,51
	Ricinus comminis (castor)	20
	Saccharum spp. (sugar cane)	60–205
	Sesamum indicum (sesame)	26
	Solanum melongena (aubergine)	24
	Solanum tuberosum (potato)	24, 36, 48, 60
	Sorghum bicolor (sorghum)	20
	Triticum aestivum (bread wheat)	42
	Triticum timopheevii (wheat)	28
	Triticum turgidum (emmer)	28
	Vaccinium corymbosum (blueberry)	48
	Vigna unguiculata (cowpea)	22
	Vitis spp. (grape)	40
	Zea mays (maize)	20

inheritance in autopolyploids does not follow typical Mendelian patterns, since the chromosomes carrying the duplicated loci can pair. Autopolyploids have more than two alleles at a locus and display what is called 'polysomic inheritance'.

The two major types of polyploids can be further divided into the following groups (Grant, 1971):

I. Autopolyploids
   1. Strict autopolyploids (one progenitor, polysomic inheritance).
   2. Interracial autopolyploids (closely related progenitors, polysomic inheritance).

II. Amphiploids
   1. Segmental allopolyploids (partially divergent progenitors, mixed inheritance).
   2. Genomic allopolyploids (divergent progenitors, disomic inheritance).
   3. Autoallopolyploids (complex hybrid of related and divergent progenitors).

Strict autopolyploids are based on the doubling of one individual, while interracial autopolyploids arise after the hybridization of distinct individuals within the same or closely related species. They form multivalents at meiosis or there is a random association of homologues into bivalents, resulting in polysomic inheritance. The simplest case is tetrasomic inheritance (four alleles

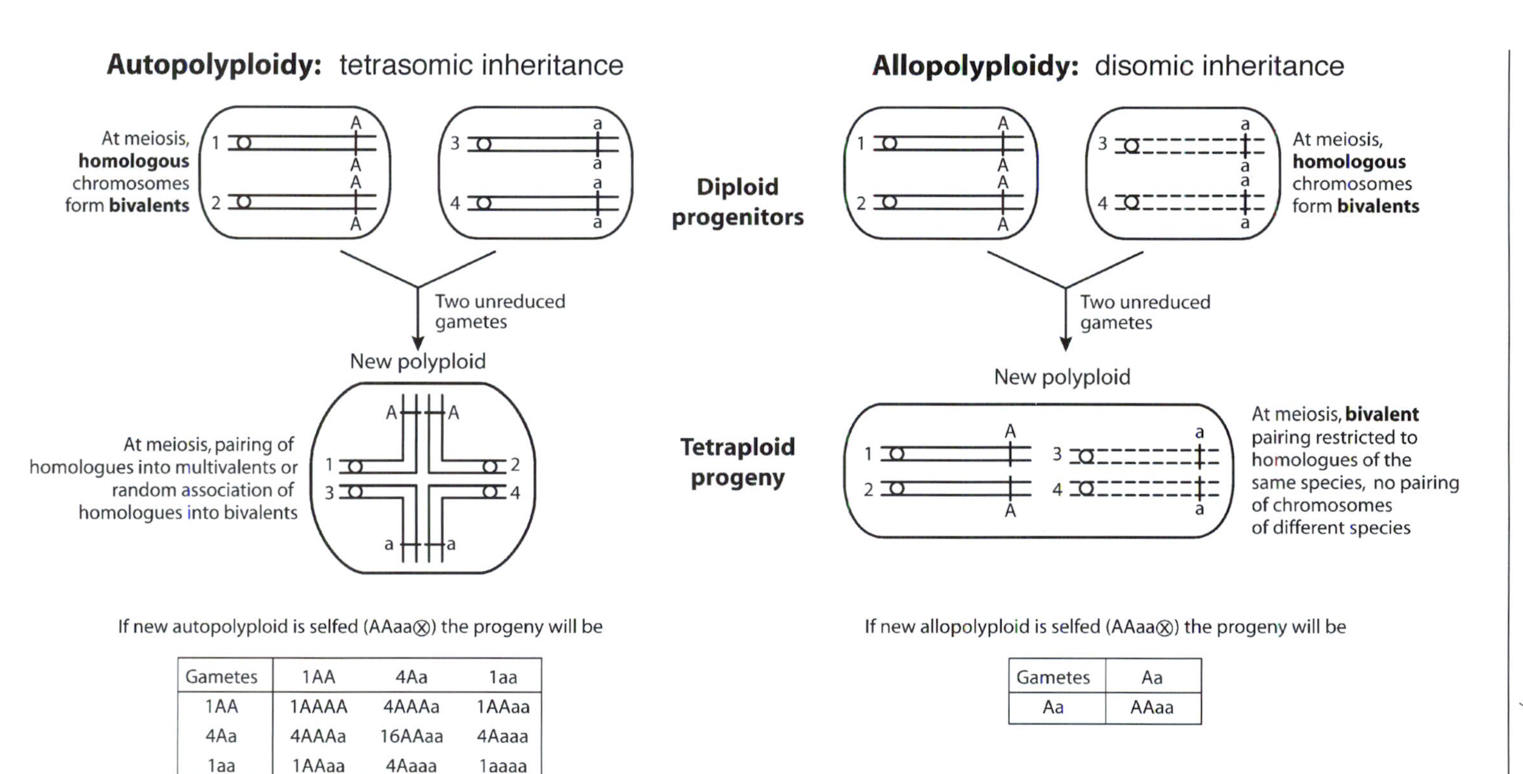

Gametes	1AA	4Aa	1aa
1AA	1AAAA	4AAAa	1AAaa
4Aa	4AAAa	16AAaa	4Aaaa
1aa	1AAaa	4Aaaa	1aaaa

Gametes	Aa
Aa	AAaa

Fig. 1.7. Inheritance patterns in autopolyploid and allopolyploid species.

at a locus), which is found in autotetraploids. In some cases, polysomic inheritance is observed in polyploids derived from what have been classified as separate species (Qu *et al.*, 1998). Because of their chromosomal behaviour, these should be considered bipartite autopolyploids.

Genomic allopolyploids originate from separate species with well differentiated genomes, form mostly bivalents at meiosis and display disomic inheritance. The progenitors of segmental allopolyploids are differentiated structurally to an intermediate degree and have varying levels of chromosomal associations, resulting in mixed inheritance. Autoallopolyploids have gone through multiple phases of doubling, involving similar and dissimilar species.

Nuclear autopolyploids have been traditionally considered rarer than allopolyploids in wild and cultivated species, but many of the classifications were based solely on morphological and cytogenetic data without information on inheritance patterns. Inheritance data are usually the only unequivocal means of distinguishing between auto- and allopolyploidy (Hutchinson *et al.*, 1983). Looking at metaphase I for bivalent pairing is insufficient, since chromosomal associations in autopolyploids can occur either as multivalents or as bivalents, as described above. In fact, Ramsey and Schemske (1998) discovered in a survey of the published literature that autopolyploids average fewer multivalents than expected due to random chiasmata formation, and allopolyploids average more. Tetrasomic inheritance has been documented in several bivalent pairing polyploids, including lucerne (Quiros, 1982), potato (Quiros and McHale, 1985), *Haplopappus* (Hauber, 1986), *Tolmiea* (Soltis and Soltis, 1988), *Heuchera* (Soltis and Soltis, 1989c) and *Vaccinium* (Krebs and Hancock, 1989; Qu *et al.*, 1998).

All plant organelle genomes are highly autopolyploid. There are 20–500 plastids per leaf cell and within each plastid are hundreds of identical plastid genomes (plastomes), depending on species, light levels and stage of development (Scott and Possingham, 1980; Boffey and Leech, 1982; Baumgartner *et al.*, 1989). There are also high numbers of mitochondria per cell, but there are few estimates of genome copy number per mitochondria. Lampa and Bendich (1984) reported 260 copies per leaf in mature pea leaves and 200–300 copies in etiolated hypocotyls of water melon, courgette and musk melon.

Measurement of Variability

Plant evolutionists typically assess genetic variability at several different organizational levels, from the actual DNA base sequence to quantitative morphological traits. In spite of the low mutation rates in plants, large amounts of genetic variation have accumulated at all levels of organization, and recent molecular technologies have uncovered an astonishing degree of genetic polymorphism in natural and domesticated plant populations.

Morphological variation has been described for both single and multiple gene systems by countless investigators. The first modern genetic analysis of

Mendel (1866) was based on allelic variation at several independent loci in cultivated peas. He looked at discrete loci controlling such things as pod colour, seed surface and leaf position. Since these seminal studies, the genetics of numerous monogenic traits have been described. A few examples are listed in Table 1.3.

In the early 1900s, geneticists began to wonder whether more continuous traits, such as plant height and seed weight, were inherited according to Mendelian laws. Johannsen (1903) showed that such variation was indeed influenced by genes, but that the environment also played a role – he was the first to distinguish between genotype and phenotype. Yule (1906) hypothesized that quantitative variation could be caused by several genes having small effects, and Nilsson-Ehle (1909) and East (1916) confirmed this suspicion using wheat and tobacco.

Table 1.3. Commonly studied traits regulated by one gene (sources: Hilu, 1983; Gottleib, 1984).

Structure	Trait	Example
Flower	Corolla shape	*Primula sinensis* (primrose)
	Gender	*Cucumis sativus* (cucumber)
	Male sterility	*Solanum tuberosum* (potato)
	Petal number	*Ipomoea nil* (morning glory)
	Pistil length	*Eschscholzia californica* (poppy)
	Self-incompatibility	*Brassica oleracea* (cabbage)
	White vs. coloured	*Viola tricolor* (violet)
	2- or 6-rowed inflorescences	*Hordeum vulgare* (wild barley)
Leaf	Angle	*Collinsia heterophylla* (innocence)
	Chlorophyll deficiency	*Hordeum sativum* (barley)
	Cyanogenic glucosides	*Lotus maizeiculatus* (bird's-foot trefoil)
	Leaflets vs. tendrils	*Pisum sativum* (garden pea)
	Margin	*Lactuca serriola* (lettuce)
	Rust resistance	*Triticum aestivum* (wheat)
Seeds and fruit	Fruit location	*Phaseolus vulgaris* (pea)
	Fruit pubescence	*Prunus persica* (peach)
	Fruit shape	*Capsicum annuum* (pepper)
	Fruit spiny	*Cucumis sativus* (cucumber)
	Fruit surface	*Spinachia oleracea* (spinach)
	Pod clockwise vs. anticlockwise	*Medicago truncatula* (wild lucerne)
	Rachis persistence	Cereal grasses
	Seeds winged vs. wingless	*Coreopsis tinctoria* (tick seed)
Physiological	Annual vs. biennial	*Meliotus alba* (clover)
	Determinate vs. indeterminate growth	*Lycopersicon esculentum* (tomato)
	Flowering photoperiod	*Fragaria* × *ananassa* (strawberry)
	Tall vs. short stature	*Oryza sativa* (rice)

The greater the number of gene loci that determine a trait, the more continuous the variation will be (Fig. 1.8). The genes that have cumulative effects on variation in quantitative traits are called polygenes or quantitative trait loci (QTL). As Johannsen originally described, the expression of quantitative traits is confounded by the environment, so that variation patterns are generally a combination of both genetic and environmental influences.

Several statistical techniques have been developed to partition the total variability within a population into its genetic and environmental components (for excellent reviews see Mather and Jinks, 1977; Falconer, 1981; Fehr, 1987). The overall relationship can be written as:

$$V^P = V^G + V^E + V^{GE}$$

where V^P represents the phenotypic or total variation within a population, V^G the genetic variation and V^E the environmental variation. V^{GE} represents the interaction between the environmental and genetic variance where the performance of the individuals is dependent on the particular environment they are placed in.

The genetic variance can be further broken down into additive (V^A), dominance (V^D) and epistatic or interaction variation (V^I). With additive genes, the substitution of a single allele at a locus results in a regular increase or decrease in a phenotypic value, e.g. aa = 4, aA = 5, AA = 6. Dominance effects occur when the heterozygote has the phenotype of one

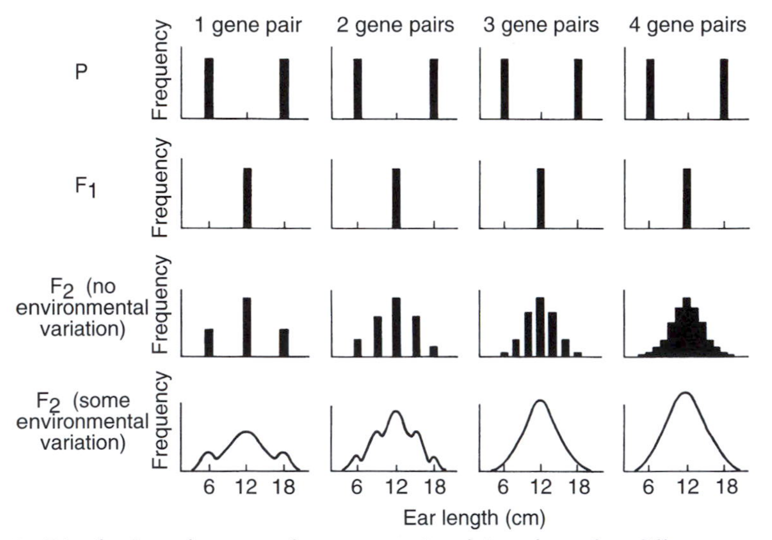

Fig. 1.8. Distribution of progeny from crosses involving plants that differ at one, two, three or six gene loci. The F_2 populations are shown without any environmental variation (row 3) and with 25% environmental variation (row 4). (Used with permission from Francisco Ayala, © 1982, *Population and Evolutionary Genetics: a Primer*, Benjamin/Cummings Publishing Company, Menlo Park, California.)

of the homozygotes, e.g. aa = 4, aA = 6, AA = 6. Epistatic effects occur when the influence of a gene at one locus is dependent on the genes at another locus, e.g. aA = 5 in the presence of bb, but aA = 6 in the presence of Bb. Thus, genotypic variation can be written as:

$$V^G = V^A + V^D + V^I$$

and the phenotypic variation can be written as:

$$V^P = V^A + V^D + V^I + V^E + V^{GE}$$

The most commonly employed measurement of quantitative variation is called heritability (h^2). Heritability is expressed in a broad or narrow sense, depending on which component of genetic variation is considered. Broad sense heritability is the ratio of the total genetic variance to the total phenotypic variance ($h^2 = V^G/V^P$). Heritability in the narrow sense is the ratio of just the additive genetic variation to the phenotypic variation – the effects of dominance and epistatic interactions are statistically removed ($h^2 = V^A/V^P$).

High heritabilities have been found in a plenitude of traits in both natural and cultivated populations (Table 1.4). Most measurement traits determining dimension, height and weight are quantitative, but numerous exceptions have been reported (Gottleib, 1984). As we shall discuss later, a large range in the contribution of QTL is often found. It is not unusual to find one gene that has a major effect on a trait and a number that modify its effects slightly.

Table 1.4. Broad sense heritability estimates for several traits in representative plant species.

Species	Trait	Heritability	Source
Solanum tuberosum (potato)	Tuber number	0.25	Tai (1976)
	Tuber weight	0.87	
	Yield	0.26	
Avena sativa (oats)	Days to heading	0.86	Sampson and Tarumoto (1976)
	Plant height	0.90	
	Stem diameter	0.76	
	Yield	0.70	
Fragaria × *ananassa* (strawberry)	Fruit size	0.20	Hansche *et al.* (1968)
	Firmness	0.46	
	Appearance	0.02	
	Yield	0.48	
Holeus lanatus (grass)	Tiller weight	0.24	Billington *et al.* (1988)
	Stolon number	0.17	
	Stolon weight	0.15	
	Leaf width	0.17	
	Flowering time	0.10	
	Inflorescence number	0.14	

Over the last couple of decades, plant evolutionists have been using several classes of biochemical compounds to further assess levels of genetic variability and calculate evolutionary relationships. Alkaloids and flavonoids have achieved some popularity (Harborne, 1982), but gel electrophoresis of enzymes has become the predominant mode of biochemical analysis. This technique takes advantage of the fact that proteins with different amino acid sequences often have different changes and physical conformations, and they migrate at different rates through a charged gel matrix (Fig. 1.9). The major advantage of this type of analysis is that many loci and individuals can be measured simultaneously and most alleles are codominant so that heterozygotes can be identified (Fig. 1.10).

Electrophoretic variability in plant and animal populations is described in a number of different ways. Different molecular forms of an enzyme that

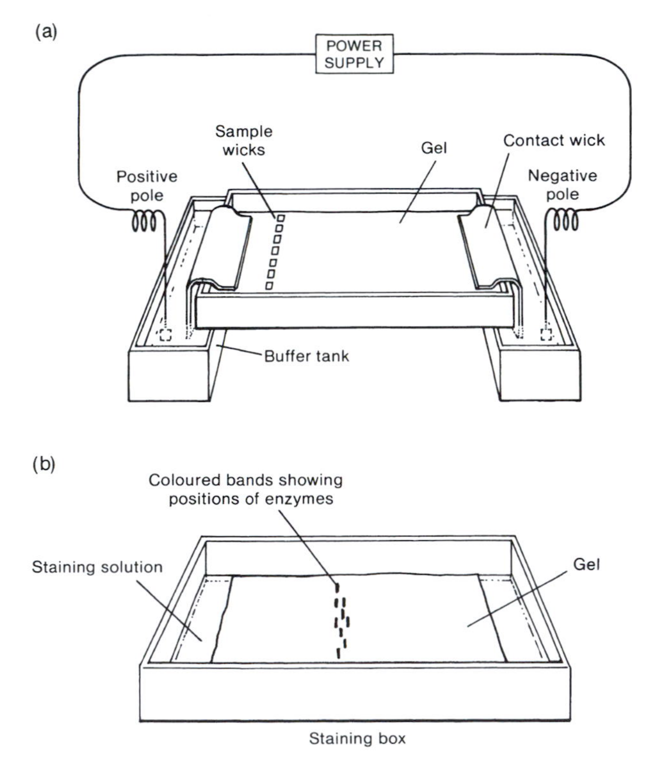

Fig. 1.9. Technique of starch gel electrophoresis. (a) Crude tissue homogenates are extracted in a buffer, loaded into paper wicks and placed in a gel, which is subjected to an electric current. (b) The enzymes migrate at different rates in the gel for several hours and their position is visualized by removing the gel from the electrophoresis unit and putting them in a box with protein-specific chemicals. The genotype of each individual can be determined from the spots (bands) which develop on each gel. (Used with permission from Francisco Ayala, © 1982, *Population and Evolutionary Genetics: a Primer,* Benjamin/Cummings Publishing Company, Menlo Park, California.)

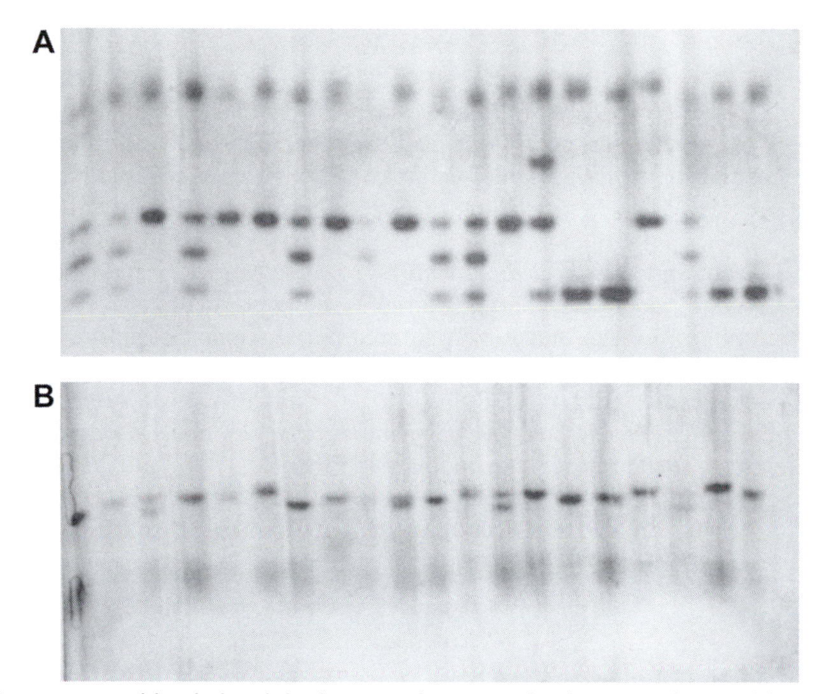

Fig. 1.10. A gel loaded with leaf extracts from 20 red oak trees and stained for either: (A) leucine aminopeptidase (LAP) , or (B) phosphoglucoisomerase (PGI). LAP is a monomeric enzyme (one subunit) and plants with one band are homozygous and those with two bands are heterozygotes. PGI is a dimeric enzyme (two subunits) and plants with only one band are also homozygous, but heterozygous individuals have three bands. (A gift from S. Hokanson.)

catalyse the same reaction are called isozymes if they are coded by more than one locus, and allozymes if they are produced by different alleles of the same locus. An individual with more than one allozyme at a locus is referred to as heterozygous. A population or species with more than one allozyme at a locus is called polymorphic. Populations are usually represented by their proportion of polymorphic loci (P) or the average frequency of heterozygous individuals per locus (H). Genetic distance or identity values can also be calculated between groups using allozyme frequency data. The most common measurement is that of Nei (1972). The identity of genes between two populations at the *j* locus is calculated as:

$$I_j = \frac{\Sigma x_i y_i}{(\Sigma x_i^2 \Sigma y_i^2)^{1/2}}$$

where x_i and y_i are the frequencies of the *i*th allele in populations *X* and *Y*. To represent all loci in a sample, the total genetic identity of *X* and *Y* is referred to as

$$I = \frac{J_{xy}}{(J_x J_y)^{1/2}}$$

where J_x, J_y and J_{xy} are the means over all loci of Σx_i^2, Σy_i^2 and $\Sigma x_i \Sigma y_i$. The genetic distance representing the divergence of two populations is estimated as

$$D = -\ln I$$

These values range from 0 to 1, with $I = 1$ representing populations with identical gene frequencies, and $I = 0$ representing populations with no alleles in common. Cluster analysis on the matrix of genetic distances can then be used to develop a dendrogram where the branches are expressed in units of genetic distance (Rohlf, 1998; Fig. 1.11). Numerous other identity measurements have been employed with different mathematical and biological assumptions. Good reviews can be found in Hedrick (1983) and Nei (1987).

Striking levels of polymorphism have been observed within most of the plant species examined for electrophoretic variation. An average of two to

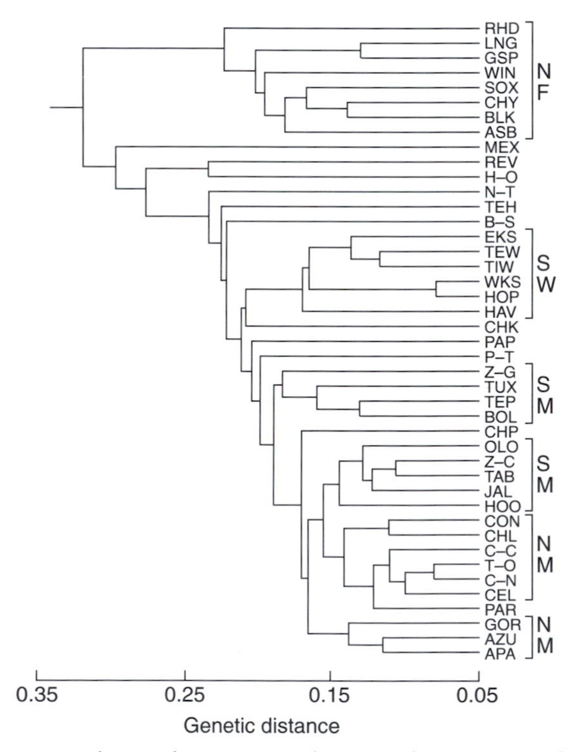

Fig. 1.11. Genetic similarity of maize races from North America and Mexico. Initials represent 18 populations of Northern Flints (NF) and several others from the south-western USA (SW), southern Mexico (SM) and northern Mexico (NM). (Used with permission from J. Doebley, M.G. Goodman and C.W. Stuber, © 1986, *American Journal of Botany* 73, 64–69.)

three alleles at a locus is the norm in natural and cultivated populations (Table 1.5), and plant breeders have even found sufficient variability in cultivated varieties to use isozymes in varietal patent applications (Bailey, 1983). Populations of the same species (conspecific) generally have genetic identity values in the range of 0.95–1.00, while identities between species range from 0.28 in *Clarkia* to 0.99 in *Gaura*, with an average 0.67 ± 0.04 (Gottleib, 1984). The genetic identity of crops and their wild progenitors is generally above 0.90 (Doebley, 1989).

While the use of electrophoresis uncovers more variability than morphological and physiological analyses, it still misses a substantial amount of the variability in the total DNA sequence. The DNA code is redundant, so mutations in many base pairs do not result in amino acid substitutions and not all amino acid substitutions result in large enzyme conformational or activity changes. Only about 28% of the nucleotide substitutions cause amino acid replacements that change electrophoretic mobility (Powell, 1975).

Table 1.5. Genetic variability uncovered by electrophoresis in selected plant species (sources: Gottleib, 1981; Doebley, 1989).

Species	Number		Average no.	Proportion
	Populations	Loci	of alleles	loci polymorphic
Capsicum annuum var. *annuum*[a]	–	26	1.4	0.31
Chenopodium pratericola	26	12	2.2	0.33
Clarkia rubicunda	4	13	2.3	0.46
Cucurbita pepo var. *ovifera*[a]	17	12	1.7	0.32
Gaura longifolia	3	18	3.2	0.33
Glycine max[a]	109	23	2.1	0.36
Lens culinaris subsp. *culinaris*[a]	31	15	1.6	0.33
Limnathes alba subsp. *alba*	6	15	2.8	0.80
Lycopersicon cheesmanii	54	14	2.6	0.57
Lycopersicon pimpinellifolium	43	11	4.5	1.00
Oenothera argillicola	10	20	2.2	0.20
Phlox cuspidata	43	20	2.4	0.25
Phlox drummondii	73	20	2.3	0.30
Picea abies	4	11	3.9	0.86
Pinus ponderosa	10	23	2.2	0.61
Raphanus sativus[a]	24	8	2.6	0.49
Stephenomeria exigua subsp. *carotifera*	11	14	4.8	0.57
Tragapogon dubius	6	21	2.5	0.24
Xanthium strumarium subsp. *chinense*	2	12	2.5	0.17
Zea mays subsp. *mays*[a]	94	23	7.1	0.50

[a]Crop species.

In the last 15 years, a number of DNA marker systems have been developed that more fully represent the molecular diversity in plants (Staub *et al.*, 1996; Jones *et al.*, 1997). These molecular markers measure variability directly at the DNA sequence level and thus uncover more polymorphisms than isozymes, and they are frequently more numerous, allowing for examination of a greater proportion of the genome. There are four major marker systems that have emerged: (i) restriction fragment length polymorphisms (RFLPs), where the DNA is cut into fragments using enzymes called restriction endonucleases and specific sequences are identified after electrophoresis by hybridizing them with known, labelled probes (Botstein *et al.*, 1980); (ii) randomly amplified polymorphic DNAs (RAPDs), where a process called the polymerase chain reaction (PCR) is used to amplify unknown sequences (Williams *et al.*, 1993; Welsh and McClelland, 1994); (iii) SSRs or microsatellites, where the PCR is used to amplify known repeated sequences (Tautz, 1989; Weber and May, 1989); and (iv) amplified fragment length polymorphisms (AFLPs), where the DNA is cut with restriction enzymes and the resulting fragments are amplified with PCR (Vos *et al.*, 1995).

In the RFLP analysis, DNA is digested by specific restriction enzymes, which recognize 4-, 5- or 6-base sequences. For example, the enzyme *Hin*dII recognizes and cleaves the nucleotide sequence CCGG. The enzyme cuts the DNA wherever this recognition site occurs in the molecule. The samples are then electrophoresed in agarose or acrylamide gels and the different fragments migrate at different rates due to their size differences (Fig. 1.12). In cases where the fragments are quite prevalent they can be seen directly under ultraviolet light after staining with ethidium bromide (Fig. 1.13). Because of the low complexity of the chloroplast genome and its high copy number, such restriction site analyses were widely used by molecular systematists to construct phylogenies. However, when the fragments are uncommon, as is usually the case with nuclear sequences, they are blotted from the gel on to a nitrocellulose filter and denatured into single-stranded DNA, and known pieces of labelled DNA are hybridized with the test samples (Southern blot). The fragments that light up are the RFLPs. These markers are codominant, meaning that both alleles can be recognized in heterozygous individuals.

In the RAPD analysis involving PCR, a reaction solution is set up which contains DNA, short DNA primers (usually ten oligonucleotides), the four nucleotide triphosphates found in DNA (dNTPs) and a special heat-stable enzyme, Taq polymerase. The DNA strands are separated by heating and then cooled, allowing the primers to hybridize (anneal) to complementary sequences on the DNA (Fig. 1.14). The polymerase enzyme then synthesizes the DNA strand next to the primers. The solution is then heated and cooled in numerous cycles, and the DNA is amplified over and over again by the same set of events. The resulting fragments are then separated by electrophoresis, as in the RFLP analysis, and visualized by staining them with ethidium bromide. RAPD polymorphisms originate from DNA sequence variation at primer binding sites (whether the primers

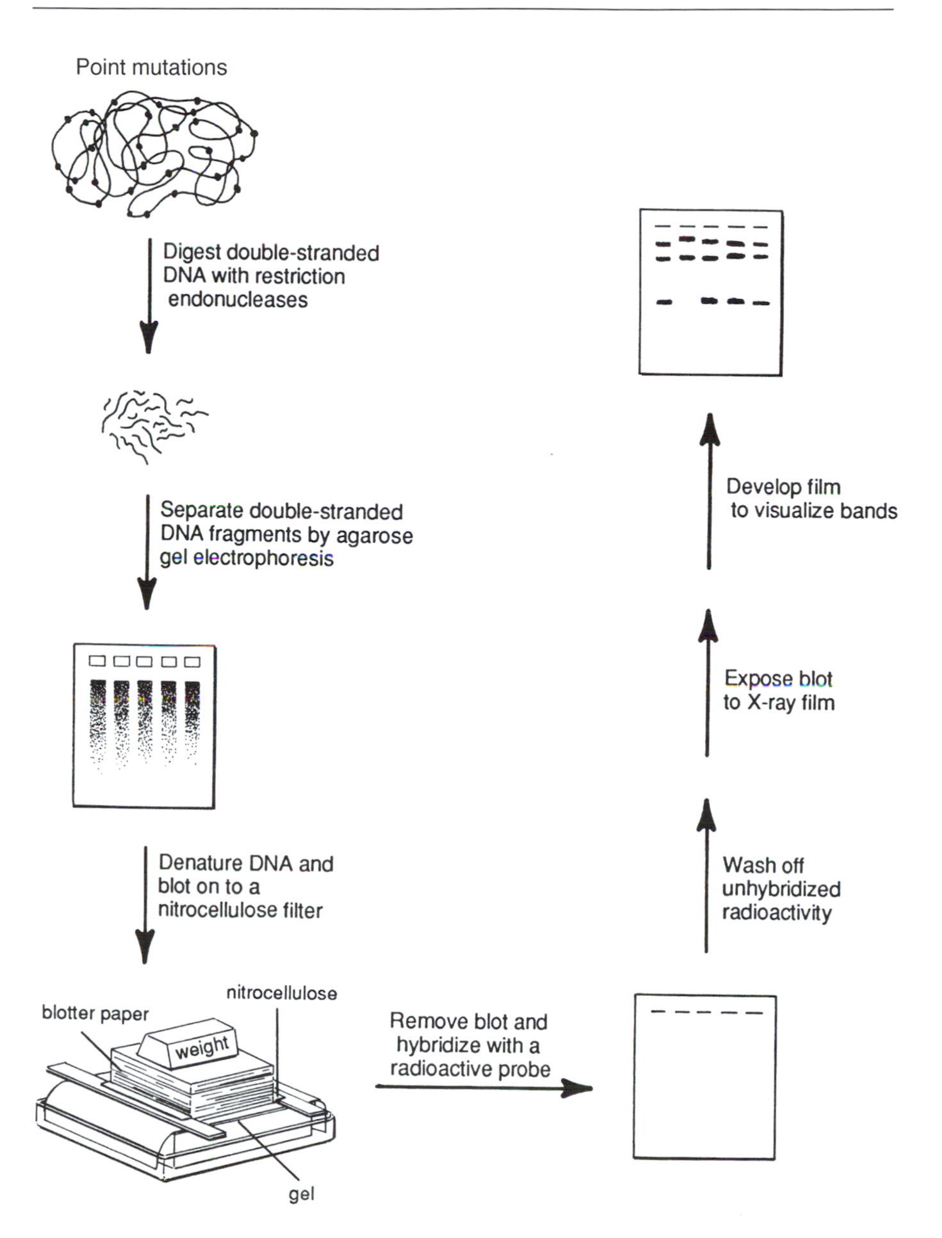

Fig. 1.12. Steps involved in DNA restriction fragment length analysis. DNA is cut into pieces using restriction endonucleases; the dots depicted above on the DNA strands represent cut sites. The different sized fragments are separated by electrophoresis in a gel made of agarose and are then transferred by blotting on to a nitrocellulose filter. The fragments are then denatured into single-stranded DNA and known pieces of radioactively labelled DNA are hybridized with them. The filters are then placed on X-ray film and bands 'light up' where successful hybridizations have occurred. (A gift from M. Khairallah.)

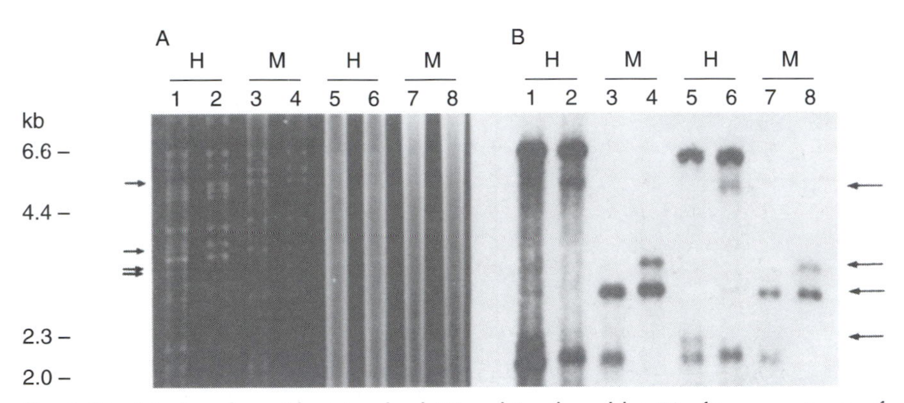

Fig. 1.13. Ethidium bromide stained gel (A) and Southern blot (B) of two genotypes of lucerne. Lanes 1–4 represent purified chloroplast DNA and lanes 5–8 contain total cell DNA. The DNA was digested with two enzymes, *Hin*dII (H) and *Msp*I (M), and hybridized with a piece of tomato plastid DNA in the Southern analysis. The total cell DNA is blurred in the ethidium bromide stained gel because it represents a much larger genome than the plastid DNA and has many more cut sites, which produce a continual array of fragment lengths. (From Schumann and Hancock, 1990.)

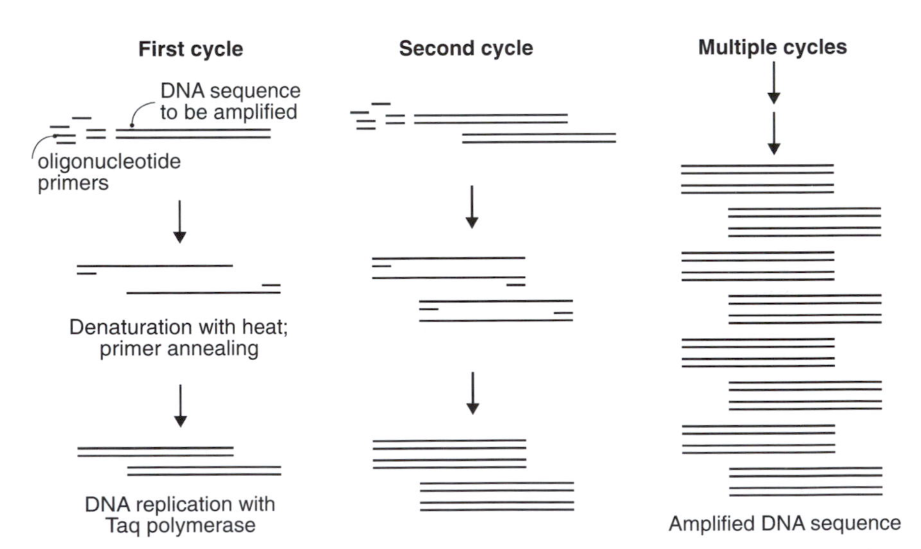

Fig. 1.14. Schematic drawing of the polymerase chain reaction (PCR). See text for details.

anneal or not) and DNA length differences between primer binding sites due to insertion or deletions of nucleotide sequences. RAPD fragments are dominant, being either present (dominant) or absent (recessive), and, as a result, heterozygotes cannot be distinguished from homozygous dominant individuals (Fig. 1.15).

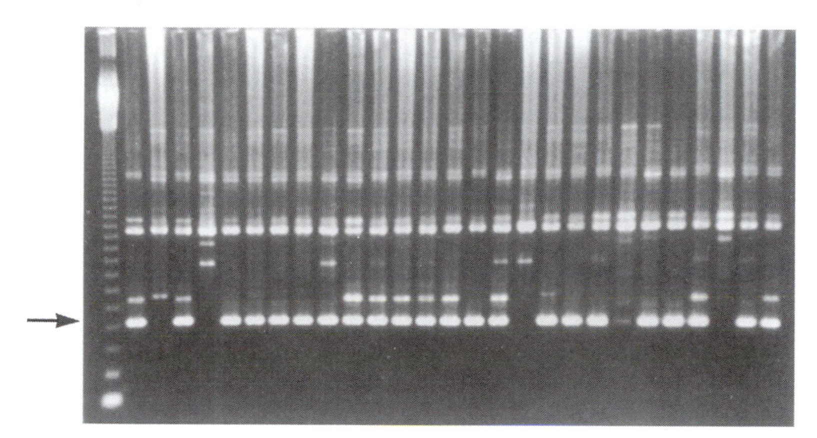

Fig. 1.15. A gel showing RAPD fragments of blueberries. (A gift from Luping Qu.)

The analysis of SSRs or microsatellites is similar to that of RAPDs, except that primers are used that flank known rather than unknown sequences. The recognized sequences are highly repeated units of 2–5 bases (for example,...GCAGCAGCA...). These repeat units are quite common in plants and can consist of hundreds of copies, which vary greatly in number across individuals. They are identified by cloning DNA fragments of individuals into bacterial carriers or bacterial artificial chromosomes (BACs), determining which BACs carry the repeat units by Southern hybridization and then sequencing the DNA flanking the microsatellite regions. Primers are then developed from the flanking sequences which will amplify the microsatellites using PCR. Polymorphisms in SSRs are observed when there are different numbers of tandem repeats in different individuals. The SSRs are codominant (heterozygotes can be recognized).

In AFLP analysis, the DNA from plants of interest is first cut with restriction enzymes. Specific primer sequences of DNA are then attached (ligated) to the resulting fragments and these are then amplified using PCR. The resulting fragments are separated by electrophoresis and visualized with DNA-specific stains. AFLP polymorphisms originate from DNA sequence variation (where the cut sites are located) and DNA length differences between primer binding sites. They are dominant markers like RAPDs, based on the presence or absence of a band (Fig. 1.16).

In comparing the various marker systems (Table 1.6), RFLPs have the advantage of being codominant and often represent known DNA sequences, but they are sometimes avoided because radioactivity is used in the process and the evaluation of segregating populations takes the greatest investment of time. RAPDs are probably the cheapest markers, but they suffer from problems with reproducibility and the fact that they are dominant. SSRs produce extremely high levels of allelic variability, but they are very expensive to generate. AFLPs produce the highest amount of variability per gel run of

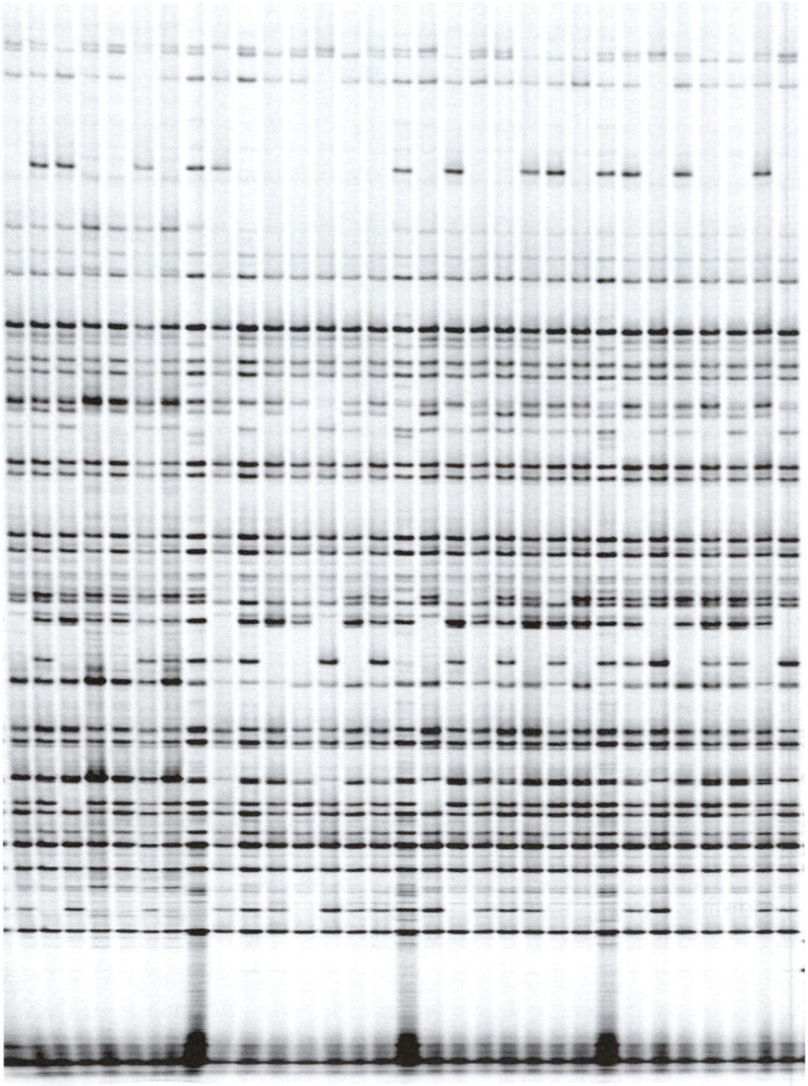

Fig. 1.16. A gel displaying amplified fragment length polymorphisms (AFLPs) of sugar beets. (A gift from Daniele Trebbi.)

any of the marker systems, but their dominance remains limiting. At present, SSRs appear to be the marker of choice, if the time and costs associated with their development are not prohibitive. AFLPs are probably the second choice when molecular markers are currently not available in a crop or high numbers of new markers are desired for systematic or mapping projects. RFLPs remain important in those crops where they have already been devel-

Table 1.6. A comparison of the various molecular marker systems.

	Marker systems			
Characteristic	AFLPs	RAPDs	RFLPs	SSRs
Cost of development	Low	Low	Intermediate	High
Cost per hybridization or PCR reaction	High	Low	High	Low
Cost per marker	Low	Low	High	Intermediate
Number of polymorphic loci generated per hybridization or PCR reaction[a]	Very high	High	Intermediate	Low
Number of alleles per locus generated per hybridization or PCR reaction[a]	Low	Low	Intermediate	High
Nature of gene action	Dominant	Dominant	Codominant	Codominant

[a]Data obtained from Powell *et al.*, 1996; Russell *et al.*, 1997; Pejic *et al.*, 1998.

oped and their utility has been established. RAPDs are only used where low cost and rapid development are deemed most important.

DNA markers have now been successfully used to estimate genetic distance in a wide range of plant species (Powell *et al.*, 1996; Staub *et al.*, 1996). For the codominant marker data (RFLPs and SSRs), genetic distances are often estimated using the equation of Nei and Li (1979). For dominant markers (RAPDs and AFLPs), a simple matching coefficient is generally employed to measure genetic distance (Jaccard, 1908). Dendrograms are then constructed with these types of values, just like those built with allozyme data. Too few comparisons have been made to date to generate firm conclusions, but the genetic similarity trees generated using RFLPs, SSRs and AFLPs appear to be generally correlated, while RAPD-based trees are commonly distinct from these. Part of this incongruity may have to do with a lack of reproducibility in RAPD markers because of mismatch pairing, but they may also be representing a different or more limited portion of the genome.

High levels of genetic variation have also been elucidated by the actual sequencing of the base pairs of a number of specific plant genes and building evolutionary trees based on sequence homologies (Ritland and Clegg, 1987; Soltis, D.E. and Soltis, 2000). During the 1990s, the mainstay of molecular phylogenetic studies was *rbcL* (large subunit of ribulose 1,5-bisphosphate carboxylase/oxygenase) and 18S ribosomal DNA (rDNA). These genes have proved most useful for inferring higher relationships, as their sequences are evolutionarily conserved. Several other chloroplast genes are now being used in phylogenetic studies, including rDNA internal transcribed spacer (ITS), *atpB* (a subunit of ATP synthase), *ndhF* (a subunit of nicotinamide adenine dinucleotide (NADH) dehydrogenase), *matK* (a maturase involved in splicing introns) and the *atpB–rbcL* intergenic region. Rates of evolution in *matK*, *ndhF* and the *atpB–rbcL* intergenic region appear to be high enough to resolve intergeneric and interspecific relationships.

While plant molecular systematists have relied primarily on genes from the chloroplast, the use of nuclear gene sequences has greatly expanded in the last few years. Most efforts have involved rDNA, although several other genes have gained favour, including *adh1* and *adh2*, *gapA* (glyceraldehyde 3-phosphate dehydrogenase) and *Pgi* (phosphoglucose isomerase). The rDNA gene is composed of several genes including the 5S, 5.8S, 18S and 26S rDNAs, and ITS. The 18S and 26S sequences are highly conserved and have been used to resolve higher order relationships, while the ITS regions are much more variable and have proved useful in distinguishing species. The 5S and 5.8S have been little used. The nuclear genes *Pgi*, *adh* and *GapA* are gaining in popularity because they are rapidly evolving and provide variability at the species and even population level.

Construction of Genetic Maps and Genome Evolution

Molecular markers are increasingly being used to construct genetic maps of species (Lander *et al.*, 1987; Tanksley *et al.*, 1988). These maps have greatly enhanced our knowledge of genome structure and evolution. By comparing maps of related species, we can evaluate the kinds of genomic changes that accompany speciation. Genetic maps are constructed by observing the frequency of recombination between markers in large segregating populations. It is assumed that the frequency of recombination is associated with the degree of linkage or the physical distance across chromosomes. A single map unit is called a centimorgan (cM) and is equal to 1% recombination.

The degree of linkage conservation or 'genome evolution' appears to vary greatly across plant species (Bennetzen, 2000b; Paterson *et al.*, 2000). Gene order is highly conserved among the *Fabaceae* lentil and pea (Weeden *et al.*, 1992), mung bean and cowpea (Menacio-Hautea *et al.*, 1993) and the solanaceous potato and tomato (Tanksley *et al.*, 1992). Linkage arrangements in the *Poaceae* species maize, rice and sorghum are so highly conserved that a circular map can be constructed that contains representatives of all three species (Devos and Gale, 2000). However, the *Gramineae* rye and wheat differ by at least 13 chromosomal rearrangements (Devos *et al.*, 1992a,b) and, in the *Solanaceae*, pepper is quite divergent from its relative the tomato (Livingstone *et al.*, 1999). Some of the diploid and amphiploid *Brassica* species have among the highest levels of divergence (Paterson *et al.*, 2000).

In spite of these genomic reshufflings, linkage relationships within gene blocks are often sufficiently conserved for numerous homologous regions to be identified across species, even in groups as widely diverged as *Arabidopsis* and tomato (Ku *et al.*, 2000), and *Arabidopsis*, common bean, cowpea and soybean (Lee *et al.*, 2001). The location of gene clusters is greatly variable across species, but the order of the genes within the clusters is highly conserved.

Molecular markers have also provided a means to dissect the genetic basis of the complex traits that are regulated by many individual QTL (Tanksley, 1993; Paterson *et al.*, 1998). The analysis is done by developing a genetic map of molecular markers, and searching for those markers whose presence is significantly associated with a phenotypic difference in a large segregating population, such as large vs. small seeds. A significant association indicates that the marker is genetically linked to one or more of the QTL. The number of genes regulating a particular trait can be determined through this process, along with an estimate of their relative importance. Plant breeders now regularly employ QTL analyses to identify the key genes regulating agronomic traits, and evolutionary biologists have found the technique to be valuable in identifying the genetics of important adaptive traits and reconstructing the speciation process (Rieseberg, 2000). These topics will be covered more fully in the chapters on plant domestication and speciation.

Summary

Most species have accumulated high levels of genetic variability through the constant slow pressure of mutation. Widespread variation has been observed in nucleotide sequence (point mutations), gene order on chromosomes (translocations, inversions and deficiencies), gene number (deficiencies and duplications) and genome number (polyploidy). Polyploidy is particularly important in plants, as up to half of all species probably originated in this manner. Much recent effort has concentrated on measuring molecular variation in populations and species. Levels of variation in most species are strikingly high, and a wide array of marker systems have now been successfully used to estimate genetic distance in numerous plant species. Genetic maps have also been used to compare rates of genome evolution. The degree of genome evolution appears to vary greatly across plant species, although blocks of genes can be found whose linkage relationships are conserved across many plant groups.

2

Assortment of Genetic Variability

Introduction

Evolution is the process by which the genetic constitution of populations is changed over time. Evolutionary change is a two-step process. First, mutation produces hereditary variation and then that variability is shuffled and sorted from one generation to another. The major forces of evolution are migration, selection and genetic drift. Migration is a movement of genes within and between plant populations via pollen or seeds. Selection is a directed change in a population's gene frequency due to differential survival and reproduction. Genetic drift is a non-directed change in a population's gene frequency due to random events.

This chapter is dedicated to the processes by which gene frequencies change. We shall first describe how populations behave with completely random mating and no extenuating circumstances, and then discuss how gene frequencies are influenced by migration, drift and selection. The various evolutionary forces will be described individually as if the others were not acting, and then in concert.

Random Mating and Hardy–Weinberg Equilibrium

In 1908, G.H. Hardy and W. Weinberg formulated what has become known as the Hardy–Weinberg Principle. What these two men discovered mathematically is that genotype frequencies will reach an equilibrium in one generation of random mating in the absence of any other evolutionary force (Hardy, 1908; Weinberg, 1908). The frequencies of different genotypes will then depend only upon the allele frequencies of the previous generation. If gene frequencies do not accurately predict genotype frequencies, then plants are crossing in a non-random fashion or some other evolutionary force is operating.

The general relationship between gene and genotype frequencies can be described in algebraic terms. If p is the frequency of one allele (A) in a population and q the frequency of another (a), then $p + q = 1$ when there are no other alleles. The equilibrium frequencies of the genotypes are given by the expansion of the binomial $(p + q)^2 = p^2 + 2pq + q^2$.

Suppose you have a population with the following composition:

55 AA 40 Aa 5 aa

Because each individual is diploid and carries two alleles at each locus, total allele frequencies are:

$$pA = \frac{2(55) + 40}{2(55 + 40 + 5)} = 0.75$$

$$qa = \frac{40 + 2(5)}{2(55 + 40 + 5)} = 0.25$$

Under Hardy–Weinberg assumptions the expected genotype frequencies would be the binomial expansion:

$AA = p^2 = (0.75)^2 = 0.562$
$Aa = 2pq = 2(0.75)(0.25) = 0.375$
$aa = q^2 = (0.25)^2 = 0.063$

and the expected number of individuals of each genotype would be:

expected gene frequency × number of individuals in population
AA = (0.562) (100) = 56.2
Aa = (0.375) (100) = 37.5
aa = (0.063) (100) = 6.3

A chi-square (χ^2) test is most commonly used to determine if a population varies significantly from Hardy–Weinberg expectations, although other comparisons, such as the G test, are sometimes used (Sokal and Rohlf, 1969). In the chi-square test:

$$\chi^2 = \Sigma \text{ (obs.} - \text{exp.)}^2/\text{ exp.}$$

Where obs. represents the observed number in a genotype class, exp. refers to the expected number in that same genotype class, and Σ indicates that the values are to be summed over all genotypic classes. A statistical table or graph is consulted to determine the probability (P) that a sampling error could have produced the deviation between the observed and expected values (Fig. 2.1). The degrees of freedom (d.f.) are equal to the number of genotype classes minus two. When the deviations are small, high probability values are produced, which indicate that the population is in Hardy–Weinberg equilibrium and the differences observed are probably

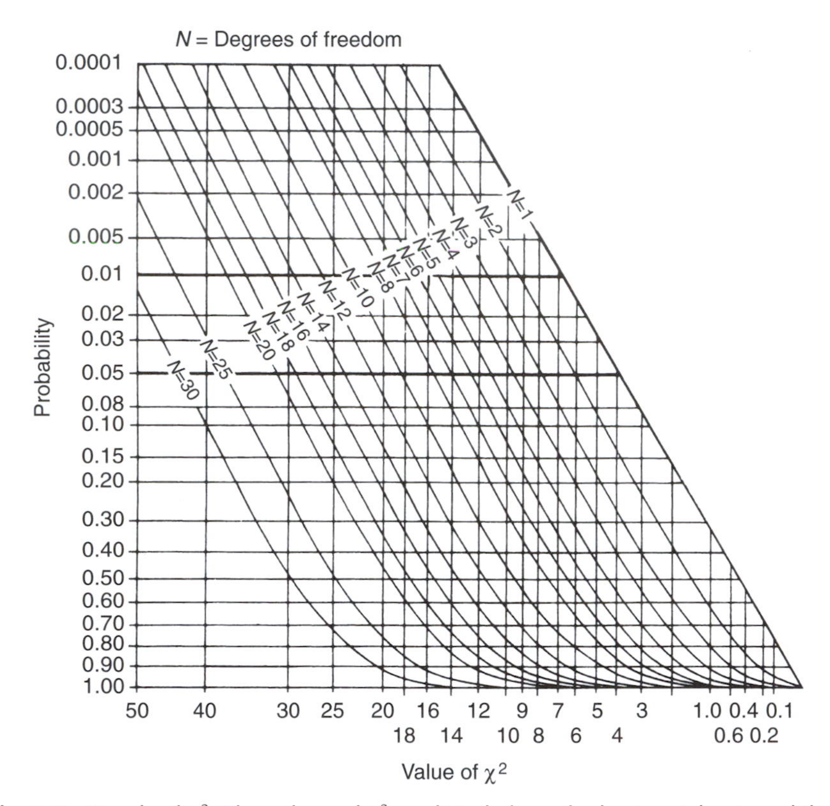

Fig. 2.1. Graph of χ^2. The values of χ^2 are listed along the horizontal axis and the associated values of p are listed along the vertical axis. N = degrees of freedom. (From Crow, 1945.)

due to minor sampling errors. Low probability values reflect deviations too great to be explained solely by chance and therefore some other evolutionary force could be operating on the populations. Usually, a population is considered in Hardy–Weinberg equilibrium unless its probability value is less than or equal to 0.05. Table 2.1 gives a chi-square test using the data generated in the above example. A χ^2 value of 0.47 with two degrees of freedom has a probability value of 0.80. Therefore, our population is very close to Hardy–Weinberg expectations and one of the evolutionary forces is probably not influencing the population, unless drift, selection and migration are acting in opposition and balancing each other out (Workman, 1969).

If more than two alleles exist in the populations, the Hardy–Weinberg formula takes the form of the polynomial square. For three alleles whose frequencies are p, q and r, then $p + q + r = 1$ and the equilibrium frequency is given by the trinomial square

$$(p + q + r)^2 = p^2 + 2pq + 2pr + q^2 + 2qr + r^2.$$

Table 2.1. Chi-square test.

	AA	Aa	aa	Total
Observed	55.0	40	5	100
Expected	56.2	37.5	6.3	100
Obs. – exp.	1.2	2.5	1.3	
Above number squared	1.44	6.25	1.69	
Above divided by expected	0.03	0.17	0.27	
	$\chi^2 = 0.03 + 0.17 + 0.27 = 0.47$, d.f. = 2			

While the Hardy–Weinberg formula is very useful in describing the completely randomized situation, it is a rare natural population that is in fact in Hardy–Weinberg equilibrium. More commonly, mating is not random and the populations are subjected to other evolutionary forces, such as migration, genetic drift and natural selection. The next section describes how differential levels of gene migration can influence variation patterns; later sections discuss how drift and natural selection affect gene frequencies. Linkage can also have a profound effect on gene frequencies and will be discussed in the next chapter.

Migration

Interpopulational

The influx of genes from another population can increase variability in the recipient population if the immigrants are unique. The factors limiting the relative importance of gene flow are the difference in gene frequency between populations and the rate of migration.

The rate of migration (m) varies from 0 to 1; $m = 0$ signifies no immigrant alleles and $m = 1$ represents a complete replacement by immigrants. Mathematically:

$$m = \text{number of immigrants per generation/total}$$

If the frequency of an allele among the natives is represented by q_o and among the immigrants by q_m, then the allele frequency in the mixed population will be:

$$q = m(q_m - q_o) + q_o$$

The change in gene frequency per generation (q) due to migration is:

$$\Delta q = m(q_m - q_o)$$

For example, if the native population originally contains $q_o = 0.3$, the immigrant group is $q_m = 0.5$ and the migration rate is $m = 0.001$, then the change in gene frequency due to one generation of immigration is:

$$\Delta q = 0.001\ (0.5 - 0.3) = 0.0002$$

Obviously, two populations must have different gene frequencies to be greatly affected by gene flow (unless an unequal sample of genes is transmitted), and the greater the input of novel genes, the greater the overall change.

Numerous methods of measuring migration or gene flow among plant populations have been developed (Moore, 1976; Handel, 1983b; Slatkin, 1985). In some, pollen and seeds are simply captured in traps and identified, while in others the movement of pollinators and dispersers is closely monitored. Pollen and seeds have also been tagged with dyes and other chemical markers. The most popular method is to find unique alleles such as allozymes and trace their movement via electrophoresis (Ellstrand and Marshall, 1985).

Small amounts of geographical separation have been found to greatly minimize levels of migration. The bulk of pollen from both wind- and insect-pollinated species travels 5–10 m at most (Fig. 2.2; Colwell, 1951; Levin and Kerster, 1969; Handel, 1983a; Golenberg, 1987). Seed dispersal is also generally leptokurtic and limited to several metres (Levin and Kerster, 1969). These observations suggest that most populations are isolated against high influxes of foreign genes.

This is not to say that long distance dispersal does not occur at all. Wind-carried pollen has been found hundreds of miles from its source (Ehrlich and Raven, 1969; Moore, 1976) and insects have been shown to carry viable pollen for several kilometres (Ellstrand and Marshall, 1985; Devlin and Ellstrand, 1990). Many seeds have long distance dispersal mechanisms such as wings to catch air patterns, stickers that attach to animal fur or hard seed-coats that survive trips through the digestive tracts of migrating animals. All these mechanisms ensure at least some long range dispersal of genes, although only a small percentage travel more than a few metres from a plant.

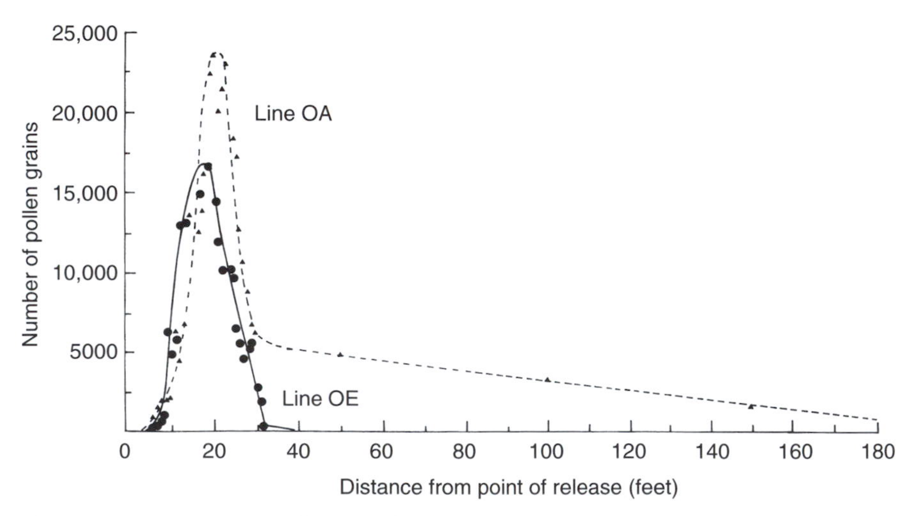

Fig. 2.2. Distance pine pollen travels after release. Line OA is directly downwind from a tree, while line OE is 45° from line OA. (From Colwell, 1951.)

Human beings have played an important role in plant evolution by planting cultigens next to wild species and by providing disrupted sites for population expansion (Anderson, 1948). Numerous crop species have been shown to hybridize with their wild progenitors such as barley, carrots, wheat, oats, sorghum and rye (Harlan, 1965; Wijnheijmer *et al.*, 1989). This sexual transfer of genes to weedy relatives has led to concerns about the escape of engineered genes into the natural environment (Ellstrand and Hoffman, 1990). The long term effects of interspecies hybridizations will be discussed in depth later (Chapter 5).

Not only has migration been important in the development of plant species, but it may also have played an important role in the spread of agriculture among human beings. There has long been a debate over whether agriculture spread from the Middle East to Europe via word of mouth or actual migration of agricultural peoples. Recent analyses of blood proteins in extant human populations indicate that agricultural peoples slowly spread from their Middle East origin and hybridized along the way with local peoples (Sokal *et al.*, 1991). This led to a slow diffusion of genes from east to west over a period of 4000 years (see Chapter 7).

Intrapopulational

The degree of gene movement affects variation patterns not only between populations but also within populations. Limited gene flow results in a highly substructured population where only adjacent plants have a high likelihood of mating. Several terms have been developed to describe the substructuring of plant populations due to limited gene flow. A deme or panmictic unit is the group within which random mating occurs (Wright, 1943). A neighbourhood is defined as the number of individuals in a circle with a radius twice the standard deviation of the migration distance or the number of individuals with which a parent has a 99% certainty of crossing (Wright, 1946). Neighbourhood sizes have been estimated to vary from four individuals in *Lithospermum caroliniense* (Kerster and Levin, 1968) to 282 in *Phlox pilosa* (Levin and Kerster, 1967).

The size of a neighbourhood has a strong effect on the variation patterns observed in polymorphic populations. In substructured populations with many small neighbourhoods, genes will move very slowly across the total population resulting in patchy distributions, while in large, less structured populations patterns will appear much more regular (Fig. 2.3). Expected levels of heterozygosity in the population as a whole can also be reduced by small neighbourhood size if the gene frequencies within individual subpopulations vary significantly from the populational mean (Wahlund, 1928). This occurs because the expected number of heterozygotes calculated from average gene frequencies is not always the same as the average of the subpopulations. For example, in the substructured population in Table 2.2, the expected number of heterozygotes

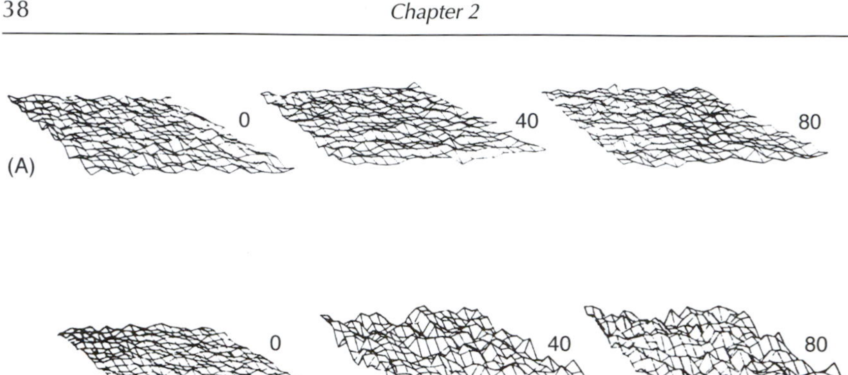

Fig. 2.3. A computer simulation of the influence of neighbourhood size on gene frequency variation. Random changes in gene frequency are simulated for a group of 10,000 evenly spaced individuals. In series A, all individuals have an equal probability of mating, while, in series B, each individual mates within a neighbourhood of nine individuals. Local gene frequency differences are represented by the heights of peaks above the plane after 0, 40 and 80 generations. (Used with permission from F.J. Rohlf and G.D. Schnell, © 1971, *American Naturalist*, University of Chicago Press, Chicago.)

Table 2.2. Expected frequency of heterozygosity in substructured population.

Neighbourhood	Gene frequency p	Gene frequency q	Expected frequency of heterozygosity ($2pq$)
A	0.2	0.8	0.32
B	0.5	0.5	0.50
C	0.8	0.2	0.32
D	0.3	0.7	0.42
E	0.7	0.3	0.42
Average	0.5	0.5	0.40

calculated using the average gene frequency across all populations is $H = 2pq = 2(0.5)(0.5) = 0.5$, but the actual average across neighbourhoods is $H = 0.40$. Thus, subpopulations can be in Hardy–Weinberg equilibria even though the population as a whole is not.

Another factor that has a strong influence on variation patterns is the breeding system of the species. Largely self-pollinated species tend to be much more homozygous than outcrossed ones, since heterozygote percentages are reduced by 50% with each generation of selfing. For instance, if we start with a population of only heterozygotes (Aa) and self-pollinate the progeny each year, the ratios shown in Table 2.3 will be produced over time. The reduction in heterozygosity occurs because half of each heterozygote's progeny will be homozygous in each generation due to Mendelian segregation, while the homozygotes continue to produce only homozygotes.

Table 2.3. Heterozygotes and self-pollination.

Generation	Genotypes AA	Aa	aa	Proportion of heterozygotes relative to those in the initial population
0	0	1	0	1
1	1/4	1/2	1/4	1/2
2	3/8	1/4	3/8	1/4
3	7/16	2/16	7/16	1/8
4	15/32	2/32	15/32	1/16
∞	1/2	0	1/2	0

Any level of inbreeding will reduce the number of heterozygotes expected from random mating. The inbreeding coefficient F is commonly used to measure levels of inbreeding and is defined as the probability that two genes in a zygote are identical by descent (Wright, 1922). Mathematically, the chance of a newly arisen homozygote with identical alleles is $1/2N$ for any generation where N is the number of breeding diploid individuals. The probability that the remaining zygotes, $1-(1/2N)$ have identical genes is the previous generation's inbreeding coefficient. Thus, for succeeding generations:

$$F_0 = 0$$
$$F_1 = 1/2N$$
$$F_2 = 1/2N + (1 - 1/2N)F_1$$
$$F_3 = 1/2N + (1 - 1/2N)F_2$$

and the inbreeding coefficient for any generation n:

$$F_n = 1/2N + (1 - 1/2N)F_{n-1}$$

The increase in the inbreeding coefficient F with different types of matings is depicted in Fig. 2.4.

Plant species range widely in their breeding systems, from obligate outcrossers to completely selfed, but most fall somewhere in the middle (Table 2.4). It is not unusual to find some degree of cross-fertilization (1–5%) in even some of the strongest self-pollinating crops, such as barley, oats and broad bean (Allard and Kahler, 1971; Martin and Adams, 1989). Rick and Fobes (1977) found outcrossing rates in some populations of native self-pollinated tomatoes to range from 0 to 40% across South America.

When selfing does occur, it is generally the result of proximity – anthers and pistils are often very close to each other in hermaphroditic flowers and this close alignment encourages self-pollination. The presence of multiple flowers on the same plant also encourages self-pollination. Probably the most extreme mechanism ensuring selfing is the rare phenomenon called cleistogamy, where flowers never open at all (groundnuts and violets). There are also apomictic species, which do not undergo any sexual reproduction.

Table 2.4. Reproductive systems of selected crops (sources: Fryxell, 1957; Allard, 1960).

Predominantly self-pollinated	Predominantly cross-pollinated	
Apricot	Almond	Papaya
Aubergine	Apple	Pear
Barley	Asparagus	Pecan
Broad bean	Banana[a]	Pistachio
Chickpea	Beet	Plum
Citrus[a]	Blueberry	Radish
Common bean	Broccoli	Raspberry
Cotton	Brussels sprouts	Rhubarb
Cowpea	Cabbage	Rye
Date	Carrot	Ryegrass
Fig[a]	Cauliflower	Safflower
Groundnut	Celery	Spinach
Lettuce	Cherry	Squash
Lima bean	Clover	Strawberry
Mung bean	Cucumber	Sunflower
Pea	Date	Sweet potato
Peach	Fig	Turnip
Rice	Filbert	Walnut
Sorghum	Grape	Water melon
Soybean	Hemp	
Spinach	Lucerne	
Tomato	Maize	
Wheat	Mango	
	Musk melon	
	Olive	
	Onion	

[a] Some parthenocarpic types.

Several mechanisms encourage outcrossing, including dioecism, dichogamy and self-incompatibility systems. Dioecism is found in a small percentage of species and is represented by separate pistillate (female) and staminate (male) plants. Dioecism has originated independently in many families, but there are no common underlying mechanisms (Ainsworth, 2000). In a few rare cases, even sex chromosomes have evolved. Some important dioecious crops are hops, date-palm, asparagus, spinach, strawberry and hemp. Dichogamy is also quite rare and occurs when the pollen of hermaphroditic plants is shed at times when the stigmata are not receptive. Representative species with this mating system are maples, oaks and sugar cane.

Self-incompatibility (SI) systems are very common and fall into two broad classes – gametophytic and sporophytic (Fig. 2.5). In gametophytic systems, pollen germination and tube growth are dependent on the genotype of the pollen, while in sporophytic systems, pollen performance is based on the genotype of the male parent. SI evolved independently in numerous lineages of plants; gametophytic systems are the most common and have been found in 60 families of angiosperms (Lewis, 1979; de Nettancourt, 2001). Modes of SI differ widely across dicotyledons, but they are all regulated by tightly linked multigene complexes referred to as haplotypes (McCubbin and

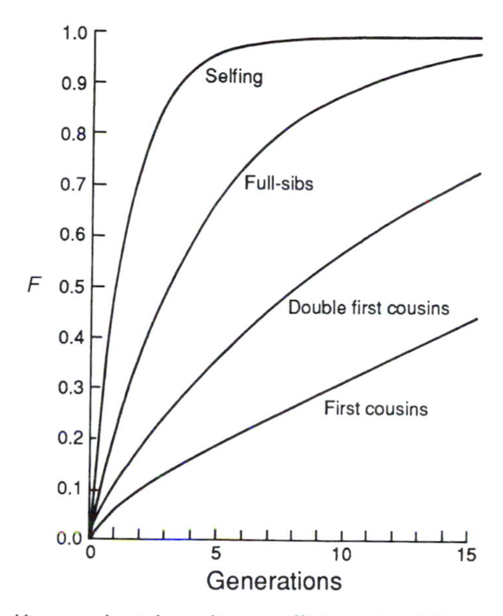

Fig. 2.4. Effect of selfing on the inbreeding coefficient *F* in different kinds of inbred populations (used with permission from F.J. Ayala, © 1982, *Population and Evolutionary Genetics: a Primer*, Benjamin/Cummings Publishing Company, Menlo Park, California).

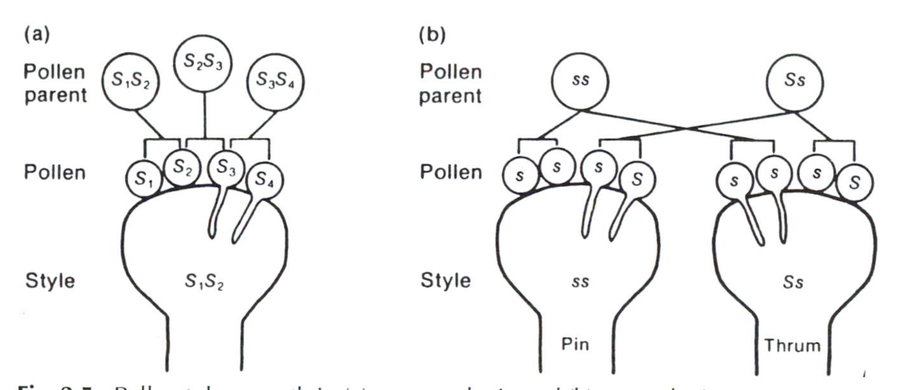

Fig. 2.5. Pollen tube growth in (a) gametophytic and (b) sporophytic incompatibility. The letters (*S*) denote incompatibility genes. Note that in the gametophytic system the genotype of the pollen itself determines whether it germinates or not, while in the sporophytic system it is the genotype of the pollen parent that is important. In gametophytic incompatibility, fertilization only occurs from pollen with a different allele from that of the female; thus, an S_1S_2 stigma cannot be fertilized by S_1 or S_2 pollen and only S_3 and S_4 pollen can germinate. In sporophytic incompatibility, the pollen can germinate if it contains the same allele as the stigmata, as long as its parent had a different genotype; thus, *s* pollen can grow on an *ss* stigma if it comes from an *Ss* parent. (Used with permission from K. Mather, © 1973, *Genetic Structure of Populations*, Chapman & Hall, London.)

Kao, 2000). Information on the genetics of SI in monocotyledonous grasses is just beginning to emerge, but it appears that at least two independent loci are involved in self-recognition (Bauman *et al.*, 2000).

SI can act both before or after fertilization. Prezygotic SI results when there is an interaction between maternal tissue and the male gametophyte that prevents self pollen tube development (Lewis, 1979; Williams *et al.*, 1994). Postzygotic reductions in self-fertility can arise due to late acting or ovarian self-incompatibility (OSI) where there is synchronous embryo failure (Seavey and Bawa, 1986; Sage *et al.*, 1994, 1999).

The modes of action of SI vary widely (McCubbin and Kao, 2000; Silva and Goring, 2001; Nasrallah, 2002). In the *Brassicaceae* (crucifers) with sporophytic incompatibility, self pollen germination is inhibited through the interaction of two pollen proteins (SP11/SCR), a stylar receptor kinase (SKR), a glycoprotein (SLG; S-locus receptor kinase gene) and several other proteins. The exact mechanism of rejection is not known, but may involve hydration of the pollen grains. In the *Papaveraceae* (poppy) with gametophytic SI, self pollen tube growth is rejected at the stigmatic surface by the interaction of a stigmatic S protein, a pollen S receptor (SBP), a calcium-dependent protein kinase (CDPK) and a glycoprotein. Pollen tube growth is arrested by a cascade of events associated with an increase in cytosolic calcium and a disruption of the cytoskeleton. In the gametophytic *Solanaceae* (tobacco, tomato, petunia and potato) and *Rosaceae* (apple, cherry and pear), self pollen tube growth is inhibited within styles by a pistil-released ribonuclease (S-RNase) that selectively destroys the RNA of self pollen tubes. Pollen S proteins are thought either to selectively allow self S-RNAs to enter the pollen tube or to inhibit non-self S-RNAs.

Self-sterility in some species is also caused by early acting inbreeding depression, where embryos abort as deleterious alleles are expressed during seed development or the outcrossed progeny display a hybrid vigour that enables them to outcompete the selfed ones (Charlesworth and Charlesworth, 1987; Waser, 1993). Evidence for this system is growing, particularly in long-lived, outcrossing species (Weins *et al.*, 1987; Seavey and Carter, 1994; Husband and Schemske, 1996; Carr and Dudash, 1996, 1997). Lucerne, blueberries and coffee have such a system (Busbice, 1968; Crowe, 1971; Krebs and Hancock, 1991). Such species are often misclassified as self-incompatible due to poor selfed seed set, but they are distinct from SI in that they commonly display the following: (i) a range in self-fertility among different genotypes; (ii) a significant positive correlation between self and outcross fertility; (iii) a significant correlation between the per cent aborted ovules and the inbreeding coefficient; and (iv) embryos abort at different stages of development (Hokanson and Hancock, 2000).

Selection

Selection can be defined as a change in a population's gene frequency due to differential survival and reproduction. Many factors influence the persistence of a gene, including germination rate, seedling survival, adult mortality, fertility and fecundity.

The fitness of a genotype is mathematically defined as the mean number of offspring left by that genotype relative to the mean number of progeny from other, competing genotypes. It is commonly designated by the letter *W*. Since it is a relative measure, the genotype with the highest fitness (most fit) is assigned a value of 1. Other fitness values are decimal fractions ranging between 1 and 0. The selection coefficient is the proportional reduction in each genotype's fitness due to selection.

Suppose a population had the following genotype frequencies before and after selection:

	AA	Aa	aa
Frequency before selection	0.25	0.50	0.25
Frequency after selection	0.35	0.48	0.17

The relative reproductive contribution of each genotype would be:

	Reproductive contribution
AA	0.35/0.25 = 1.40
Aa	0.48/0.50 = 0.96
aa	0.17/0.25 = 0.68

To assign the most fit a value of 1, we must divide each genotype by the reproductive contribution of the most successful genotype:

	Fitness *(W)*
AA	1.4/1.4 = 1
Aa	0.96/1.4 = 0.7
aa	0.68/1.4 = 0.4

The selection coefficient is $1 - W$; so:

AA $1.0 - 1.0 = 0$
Aa $1.0 - 0.7 = 0.3$
aa $1.0 - 0.4 = 0.6$

Jain and Bradshaw (1966) found selection coefficients in nature to range from 0.001 to 0.5. Humans probably used even more extreme values in the domestication of crop species (Ladizinsky, 1985).

The response of quantitative traits to selection can be predicted by the equation, $R = h^2S$, where h^2 is the estimate of heritability and S is the selective differential or the difference between the mean of the selected parents (μ_s) and the mean of all individuals in the parental population (μ) (Falconer and Mackay, 1996). The truncation point (T) represents the point at which individuals are either selected for or against (Fig. 2.6). For example, if we are

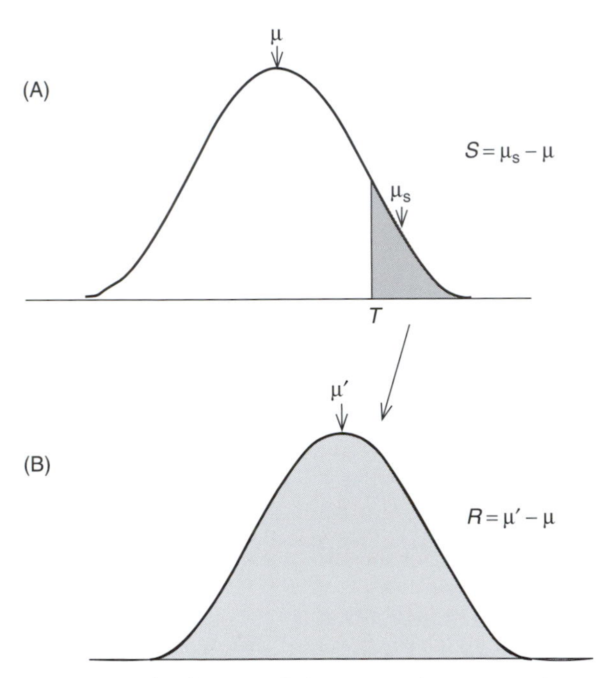

Fig. 2.6. (A) Quantitative distribution of phenotypes in a parental population with the mean μ. Only those individuals with phenotypes above the truncation point (*T*) produced progeny in the next generation. The selected parents are denoted by shading and their mean phenotype by μ_s. (B) Distribution of phenotypes in the next generation from offspring derived from the selected parents. The mean phenotype is denoted μ′. *S* represents the selection differential, and *R* is called the response to selection. (Used with permission from D.L. Hartl, © 1980, *Principles of Population Genetics*, Sinauer Associates, Inc., Sunderland, Massachusetts.)

interested in a trait such as pod length with a high heritability of $h^2 = 0.60$ and our total parental population has a mean of 12.30 cm and our selected population has a mean of 15.20 cm, then $R = 0.60$ (15.20 cm − 12.30 cm) = 1.80 cm. This means that our progeny population will have pods 1.80 cm longer than the original population.

Types of selection

There are three primary types of selection: (i) directional; (ii) stabilizing; and (iii) disruptive or diversifying (Fig. 2.7). In directional selection, one side of a distribution is selected against, resulting in a directional change. Under stabilizing selection, both extremes are selected against and the intermediate type becomes more prevalent. In disruptive selection, the intermediate types are selected against, resulting in the increase of divergent types. The amount of genetic variability present in a population and the strength of the selection coefficient determine how fast and how much a population will change.

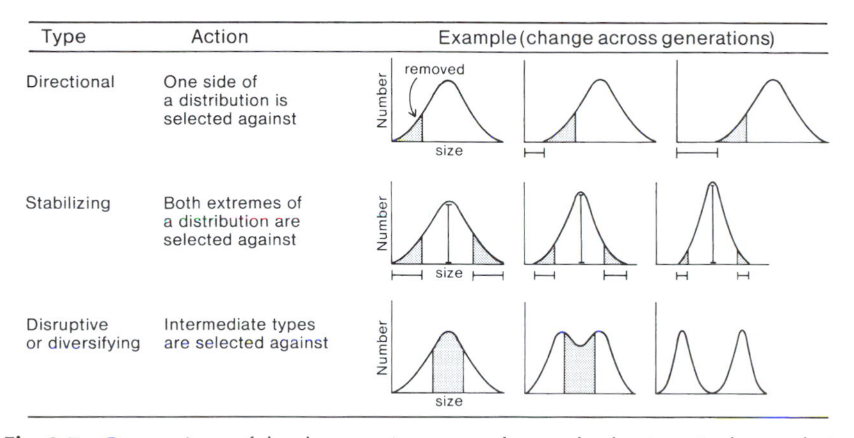

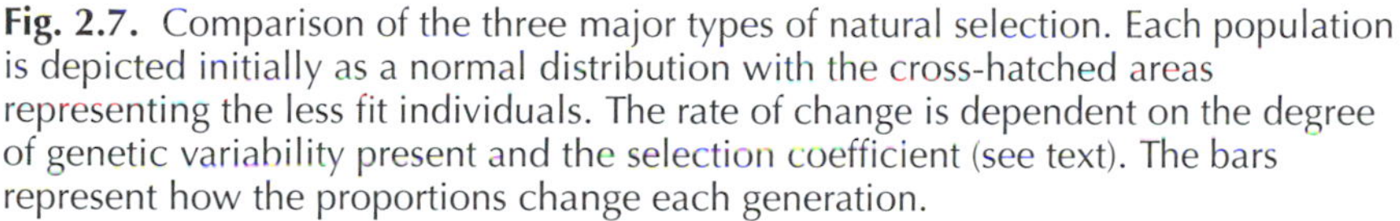

Fig. 2.7. Comparison of the three major types of natural selection. Each population is depicted initially as a normal distribution with the cross-hatched areas representing the less fit individuals. The rate of change is dependent on the degree of genetic variability present and the selection coefficient (see text). The bars represent how the proportions change each generation.

Numerous examples of directional selection have been presented in the evolutionary literature. Perhaps the most striking and repeatable is the evolution of heavy-metal tolerance in those plant species that come in contact with mine borders (Jain and Bradshaw, 1966; Antonovics, 1971). In less than 200 years, pasture species such as *Anthoxanthum odoratum* and *Agrostis tenuis* evolved distinct morphologies and means of coping with high levels of copper, lead and zinc in the face of substantial gene flow. As is shown in Fig. 2.8, a distinct change in metal tolerance can be observed across only a few metres. The domestication of maize also offers a dramatic example of directional selection. Under human guidance, the size of the ear was increased between 7000 and 3500 before present (BP) from about 2.5 cm to well over 15 cm (Fig. 2.9).

The fact that we recognize most species as definable entities argues that stabilizing selection is common in nature or else species lines would be much more blurred than they are. Huether (1969) provided some of the strongest evidence of stabilizing selection when he showed that *Linanthus androsaceus* almost always have five lobes per flower even though their environments vary greatly and there is genetic variation for the trait at every site. Plant taxonomists often find floral traits to be the most dependable species markers, because they are generally less variable than other traits.

Documentation of disruptive selection has come most frequently by correlating gene frequency with environmental variation – one of the most commonly cited examples is the work of Allard and co-workers (Allard and Kahler, 1971; Allard *et al.*, 1972) where they found reproducible correlations between allozyme frequencies in the oat *Avena barbata* and the relative moisture content of soils in California. Distinct allozyme frequencies are found on the wettest (mesic) and driest (xeric) sites (Table 2.5). These allozymes may not have been the specific traits selected, but they were at least associated with the selected loci through genetic linkage and/or partial selfing (Hedrick, 1980; Allard, 1988).

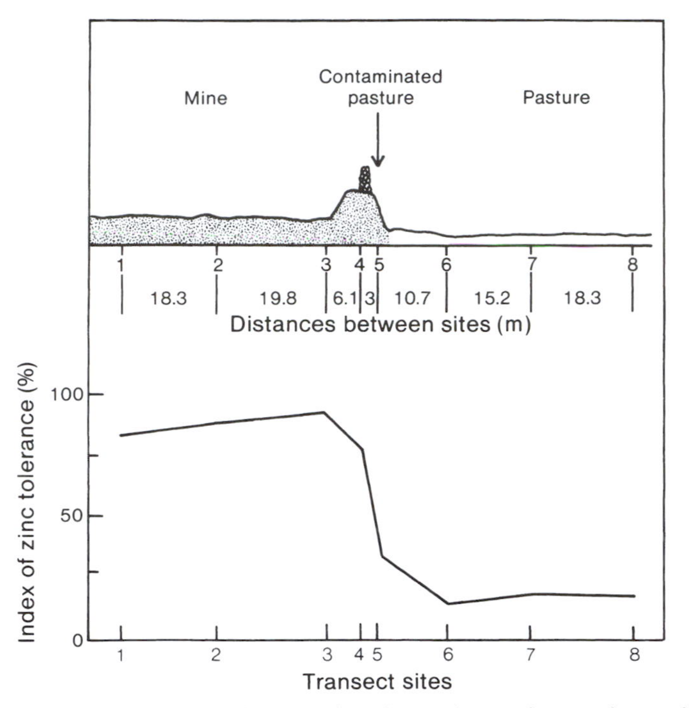

Fig. 2.8. Zinc tolerance in populations of *Anthoxanthum odoratum* located at different distances from a mine border (redrawn with permission from T. McNeilly and J. Antonovics, 1968, *Heredity* 23, 205–218, and J. Antonovics and A.D. Bradshaw, 1970, *Heredity* 27, 349–362).

Many unique races of crops were developed by human beings through disruptive selection as they gathered and grew populations from distinct regions and habitats (see Chapter 7). Perhaps the most distinct alteration came in *Brassica oleracea*, where several distinct crops were developed, including kale, broccoli, cabbage, kohlrabi and Brussels sprouts (see Fig. 11.3). Diverse races of rice, chickpea and chilli peppers also emerged in response to isolation and differential selection pressure from humans.

Plant breeders have on numerous occasions employed diversifying selection to produce differences in harvest dates and quality factors. For example, dramatic changes in the protein content of maize were produced over 60 generations by selecting repeatedly for high and low values (Fig. 2.10).

While directional selection can eliminate variability in a population, stabilizing and disruptive selection often act to maintain it. Dobzhansky (1970) called this maintenance of genetic polymorphism balancing selection. Disruptive selection maintains polymorphism when different genotypes are favoured in different environments. Stabilizing selection acts to maintain polymorphisms if heterozygotes produce the favoured phenotype. Numerous examples have been provided where superior vegetative or reproductive vigour have been associated with heterozygosity (Mitton and Grant, 1984; Strauss and Libby, 1987).

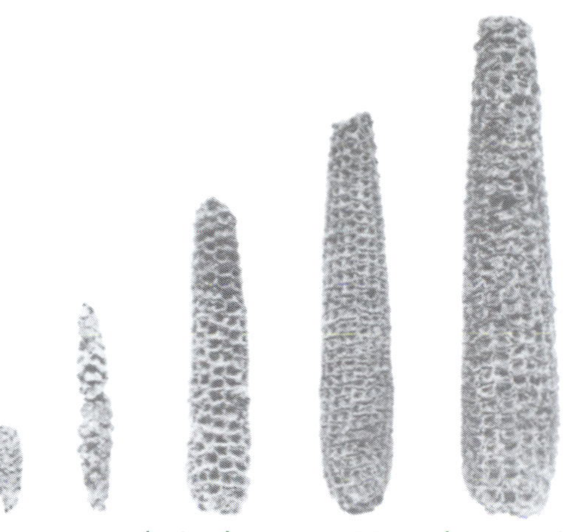

Fig. 2.9. Increase in maize cob size between 7000 and 3500 BP in Tehuacan Valley, Mexico. The cob on the left was about 1 inch long. (Used with permission from D.S. Byers (ed.), © 1967, *The Prehistory of the Tehuacan Valley*, Vol. 1. Andover Foundation for Archaeological Research, University of Texas, Austin.)

Table 2.5. Genotype frequencies in two populations of oats (*Avena barbata*) found in California. Three loci of esterase (E_1, E_4 and E_{10}) and one each of phosphatase (P_5) and anodal peroxidase (APX_5) are represented. (From Marshall and Allard, 1970.)

		Frequency	
Locus	Genotype	Mesic population	Xeric population
E_1	11	0.76	0.00
	12	0.07	0.00
	22	0.17	1.00
E_4	11	1.00	0.30
	12	0.00	0.11
	22	0.00	0.59
E_{10}	11	1.00	0.46
	12	0.00	0.13
	22	0.00	0.41
P_5	11	0.00	0.40
	12	0.00	0.15
	22	1.00	0.45
APX_5	11	0.86	0.48
	22	0.00	0.41
	33	0.09	0.00
	12	0.00	0.11
	13	0.05	0.00
	23	0.00	0.00

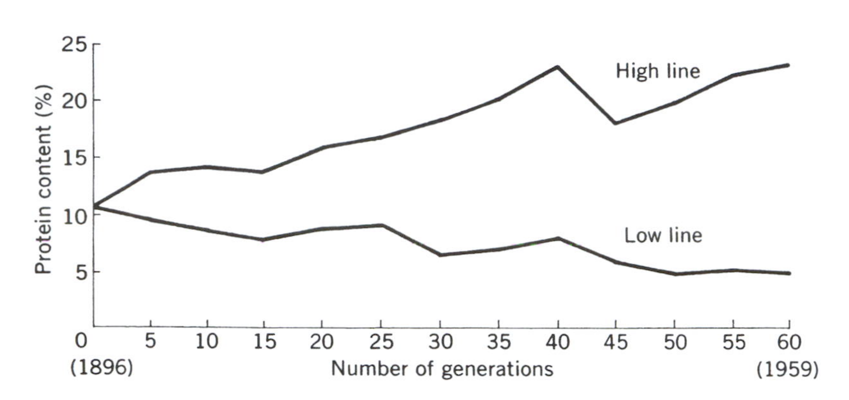

Fig. 2.10. Change in protein content of maize during 60 generations of artificial selection by E. Leng and Associates at the Illinois State Experiment Station. (Used with permission from V. Grant, © 1977, *Organismic Evolution,* W.H. Freeman and Company, San Francisco.)

Disruptive selection occurs not only across spatially heterogeneous environments, but also across time. Polymorphisms can be maintained across temporally varying environments, different life stages and variant sexes. There are also examples of polymorphisms being maintained by differing population frequencies and genetic backgrounds. Some alleles, such as the S alleles associated with SI, are even favoured when they are in low frequency, but disfavoured at high frequency.

Factors limiting the effect of selection

Several important factors limit the influence of selection on gene frequencies: (i) amount of genetic variability present; (ii) generation time; (iii) strength of the selective coefficient; (iv) degree of dominance; (v) initial frequency of the advantageous allele; and (vi) intragenomic interactions.

The only way a population can change is if there is genetic variability present. Many a breeding programme has stalled or species become extinct when the genetic variability for a particular trait was extinguished. This is true both for single gene traits and for quantitative traits where more than one locus is involved. In 1930, Fisher outlined his fundamental theorem of natural selection, which stated: 'The rate of increase in fitness of any organism at any time is equal to its genetic variance at that time.'

The mating behaviour of a species has important long term evolutionary ramifications. Outcrossed species often have high levels of heterozygosity and gene combinations are continually shuffled during reproduction through crossing over and independent assortment. Selfing species are frequently very homozygous and specific gene combinations are rarely disrupted by sexual recombination. An outcrossed, heterozygous species has a considerable amount of genic flexibility with which to meet environmental change, but the selfing, homozygous ones may be better adapted to specific, unchanging conditions. An intermediate strategy with variable outcrossing rates may yield the greatest long term success in an unpredictable world; this may be why few extreme selfers or outcrossers are found (Allard, 1988). Gornall (1983) has suggested that 'those colonizing species which had extraordinary flexible recombination systems, which allowed them both to store and release large amounts of variability, were in a sense pre-adapted to successful cultivation'.

The strength of selection will critically influence the rate of change in a population, as the greater the proportion of individuals being selected each year, the more rapid the change. Likewise, the generation time of an organism can have a strong effect on the rate of evolution simply because the longer the period between reproductive episodes, the longer the separation between meiotic reassortments of genes and seedling selection. Generation times in higher plants vary from a few weeks in *Arabidopsis* to dozens of years in some tree species.

The degree of dominance has an influence on rates of change because recessive alleles in the heterozygous state are masked from selection. Frequencies of a newly arisen recessive allele will therefore change very slowly in a population until its frequency is quite high (Fig. 2.11). Frequencies of codominant alleles or those with an intermediate effect in a heterozygote will change more rapidly than those of recessives, since they have a partial influence on the phenotype of heterozygotes. Frequencies of dominants will initially change more abruptly than those of either recessives or intermediates, but the rate of change will eventually slow as the frequency of recessives becomes so low that most are 'hidden' in heterozygotes. In all cases, populations with intermediate gene frequencies change more rapidly than those with low frequencies of deleterious or advantageous alleles (Fig. 2.11).

The number of selective deaths involved in one complete allelic substitution has been called the genetic cost (Haldane, 1957, 1960). By similar logic, the number of individuals that will die each generation due to selection has been called the genetic load (Wallace, 1970). The cost factor is high when the favoured allele is at a low frequency, and decreases as the initial frequency increases. Obviously, a population can tolerate only a limited number of genetic deaths each generation if it is to avoid extinction.

The adaptiveness of an allele is influenced not only by the external environment, but also by the other genes in the genome. In most of our discussion so far, we have treated genes as if they are independent entities; however, the genome is in reality a tightly interacting unit. This concept will be discussed at length in the next chapter on the complexity of plant genomes.

Coevolution

Most of this chapter has concerned selection within a single species. However, there are numerous examples of what is called coevolution, where species have evolved together rather than independently. The most commonly cited examples concern mutualism, character displacement and host–pathogen evolution.

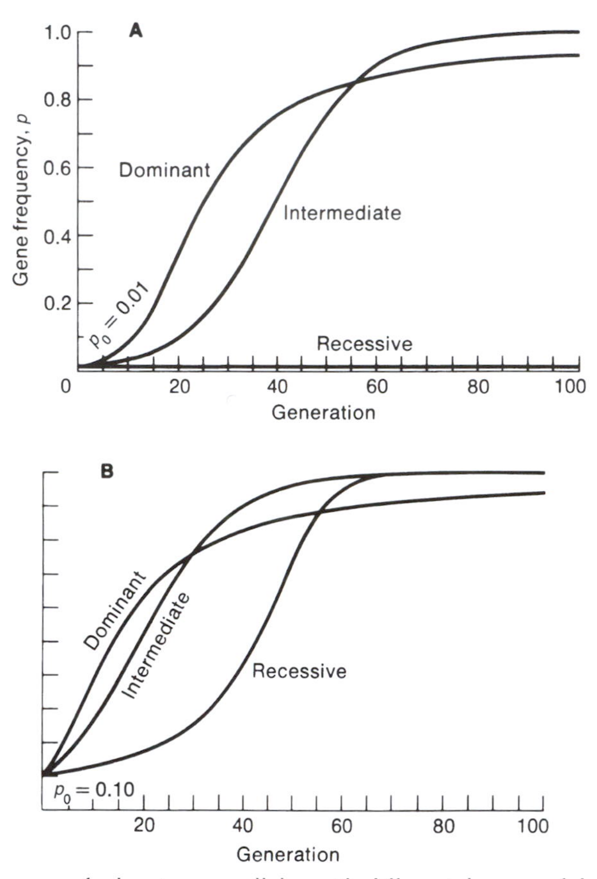

Fig. 2.11. Increase of advantageous alleles with different degrees of dominance. (A) New mutation with a very low frequency ($p = 0.01$). (B) Existing polymorphism at $p = 0.10$. (Used with permission from D.J. Futuyma, © 1979, *Evolutionary Biology*, Sinauer Associates, Sunderland, Massachusetts.)

In mutualism, there is a symbiotic reaction in which two species benefit by their interaction. Two classic examples in plants are soil mycorrhizae and nitrogen fixation bacteria. In both cases, microbes have become intimately associated with plant roots such that the plant gains mineral nutrients and the fungi or bacteria receive a carbon source. An even more dramatic example is found between the tree *Acacia* and the ant *Pseudomyrmex*, where the ant receives nourishment from the tree's nectaries and the ants defend the tree against herbivores and competing vegetation (Janzen, 1966).

When two species have overlapping ranges, one species is either selectively eliminated or the two undergo character displacements that minimize competition (Roughgarden, 1976). For example, where the ranges of *Phlox pilosa* and *Phlox glaberrima* do not overlap, they both have pink flowers, but, in areas of overlap, *P. pilosa* frequently has white flowers. Interspecific hybridizations are minimized because insects have a tendency to carry pollen from white to white flowers rather than from pink to white or vice versa (Levin and Kerster, 1967). Ecological races of many species also have different flowering times, which reduce the chance of hybridization. In some cases, these displacements are thought to be an early step in speciation (Chapter 5).

There are also numerous examples of host–parasite evolution, where there is widespread matching of host alleles conferring resistance to specific parasite strains (Flor, 1954; Vanderplank, 1978; Allard, 1990). Most of the best studied examples come from agricultural systems where constant battles are fought to find new sources of resistance to rapidly evolving pest populations (Wahl and Segal, 1986; Allard, 1990). For example, dominant alleles have been identified at more than five loci in wheat that confer resistance to the Hessian fly, but alleles exist for each locus in the fly that confers counter-resistance (Hatchett and Gallum, 1970). Dozens of examples of 'gene-for-gene' relations have been described in pathogen/host systems where there is a gene for susceptibility or resistance in the crop to match each gene for virulence or avirulence in the pathogen (Table 2.6). Numerous studies have been devoted to determining the nature of these resistances (Martin and Ellingboe, 1976; Leath and Pederson, 1986; Hautea *et al.*, 1987; Pederson and Leath, 1988).

Table 2.6. The reaction of 16 genotypes of wheat to three Canadian races of *Puccinia graminis tritici* (from Vanderplank, 1978).

Resistance genes	Race		
	C10	C33	C35
Sr 5	S	S	S
Sr 6	R	R	S
Sr 7a	R	S	S
Sr 8	R	S	S
Sr 9a	S	R	S
Sr 9b	S	R	S
Sr 9d	S	S	R
Sr 9e	S	S	R
Sr 10	S	S	R
Sr 11	S	S	R
Sr 13	S	R	R
Sr 14	S	S	S
Sr 15	S	R	S
Sr 17	S	R	R
Sr 22	R	R	R
Sr T2	S	R	R

S, susceptible; R, resistant.

Genetic Drift

Many people stress the importance of selection in shaping natural populations, but non-directional forces such as genetic drift can also play an important role. The common inclination is to assume that nature is ordered and that all is optimized. The simple truth is that variation patterns are regulated to a large extent by luck. A highly adapted genotype will not predominate in a population if most of its seeds fall on a rocky outcrop where they cannot germinate or there is no pollinator activity due to cool conditions when its pollen is dehisced. To proliferate, a genotype must be well adapted and its progeny must reach maturity before some accident eliminates it.

While it is easy to document selection by observing directional changes over time or genotype/environment correlations, drift is difficult to measure because it is non-directional and often masked by other forces (Schemske and Bierzychudek, 2001). Most demonstrations of drift have come by the process of elimination; no apparent associations were observed between any environmental parameter and an allele's frequency. For example, on a hillside of *Liatris cylindracea* in Illinois, Schaal (1975) found a substantial amount of allozyme variation in several loci that she could not associate with any environmental parameter, including moisture, soil pH, nutrient content or per cent organic matter (Fig. 2.12).

Sewell Wright (1931, 1969) provided numerous models that show how drift might operate. Drift will ultimately result in the fixation of one allele in a population like directional selection, but the approach will be much more variable (Fig. 2.13). Two primary factors influence the effects of drift: allele

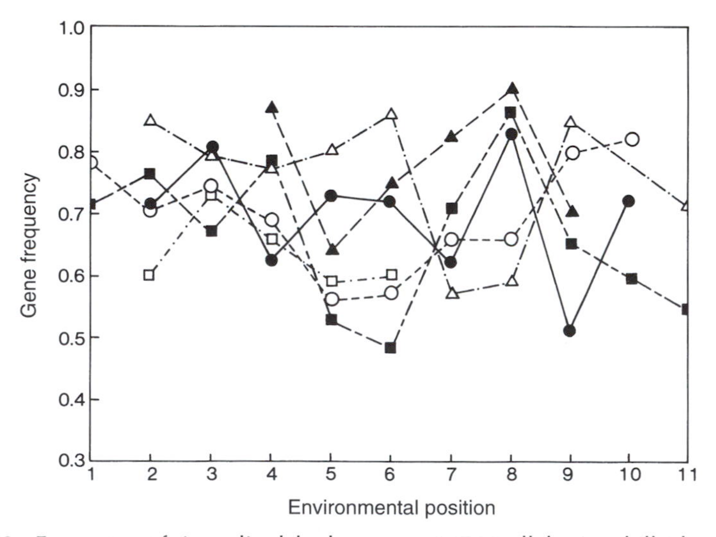

Fig. 2.12. Frequency of six malic dehydrogenase (MDH) alleles in a hillside population of *Liatris cylindracea*. The lines represent a transect down the hill. Point 1 is at the top and point 11 is at the bottom. (Used with permission from B.A. Schaal, © 1975, *American Naturalist* 109, 511–528, University of Chicago Press, Chicago.)

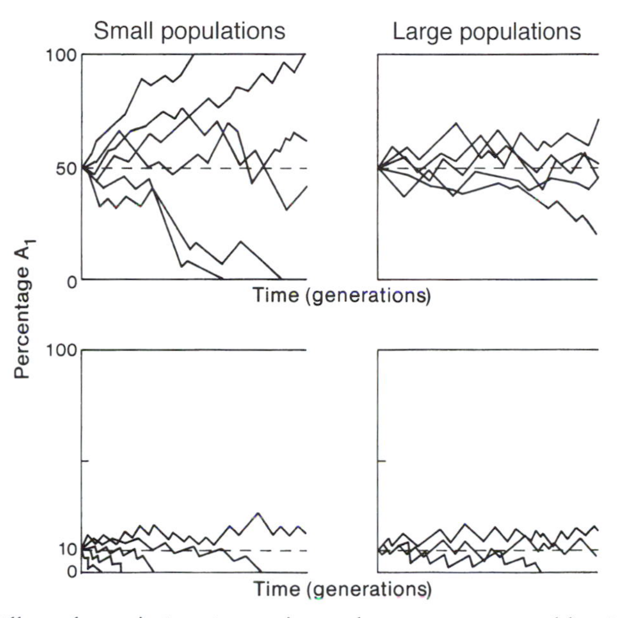

Fig. 2.13. Effect of population size and gene frequency on rate of fixation due to drift. Each line represents a different population.

frequency and population size. The more frequent an allele, the greater its chances of being fixed and the smaller the population, the faster it will stumble towards fixation. The probability that an allele will be fixed is equal to its frequency in a population.

In the 1980s, there was considerable debate over the relative importance of drift and selection in natural populations, particularly as they related to allozyme variation (Hedrick, 1983; Nei, 1987). The 'selectionists' felt that most variability was maintained by selection, while the 'neutralists' felt that drift was the most important force. While this debate continues, a reasonable compromise is to assume that both drift and selection are important and that the relative strengths of the two forces are dictated by each individual set of circumstances. The critical point to realize is that forces other than selection can shape variation patterns. This is particularly true in plant populations where limited migration results in most populations being effectively small.

Extreme cases of random genetic drift occur when a new population is initiated by only a few individuals; this is called the founder effect (Mayr, 1942). Gene frequencies may be quite different in a few colonizers compared with the population from which they originated because of sampling errors. Chance variations in allelic frequencies can also occur when populations undergo drastic reductions in numbers (bottlenecks) after environmental catastrophes. The early stages of most crop domestications were probably influenced by founder effects, as the bulk of our domesticated strains are based on relatively few genotypes (Ladizinsky, 1985). In general, the genetic identities (I) of crop plants and their nearest wild relatives are > 0.90 (Doebley, 1989).

Evolution in Organelles

Drift may also play an important role in the evolution of plastid and mitochondrial genomes. Plastids and mitochondria replicate independently from the nucleus and their assortment into daughter cells is generally random (Michaelis, 1954; Birky, 1983). As we have already discussed, these organelles are polyploid and exist in multiple copies in mature leaf cells, but in meristematic cells their numbers are greatly reduced (Butterfass, 1979). Both organelles are also inherited maternally due to the unequal contribution of egg cells at fertilization, with only a few exceptions (Sears, 1980; Schumann and Hancock, 1990). Any mutations that arise in an organelle genome can become fixed in cells during cell division due to random sorting and ultimately establish sexual tissues that produce gametes with a unique cytoplasm (Fig. 2.14; Michaelis, 1954, 1967; Birky, 1983).

As with nuclear genes, the ultimate evolutionary consequence of organelle replacement is dependent on its physiological ramifications. Many changes have little effect on the overall phenotype and are completely subject to drift, while others are dramatic and are greatly affected by selection.

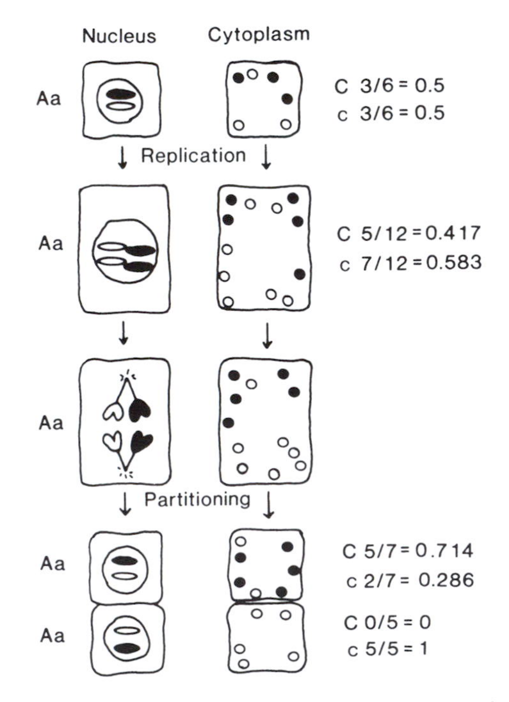

Fig. 2.14. Replication and partitioning of nuclear chromosomes and organelles. During cell division the nuclear chromosomes are partitioned regularly so that each daughter cell gets both homologues; organelles are partitioned randomly and as a result cells can become fixed for one type. (Used with permission from C.W. Birky, © 1983, *Science* 222, 468–475, American Association for the Advancement of Science.)

In the non-photosynthetic parasitic angiosperm *Epifogus virginiana*, de Pamphilis and Palmer (1990) found the photosynthetic genes to be largely deleted, while those involved in gene expression were still present. This indicates that, once photosynthesis was unnecessary, the plastid genes associated with it were selectively eliminated, but the plastid 'housekeeping' genes necessary for replication were maintained.

Interaction between Forces

As we mentioned at the outset of this chapter, the evolutionary forces of migration, drift and selection rarely act alone. The relative importance of these forces depends on the particular circumstances at hand.

Selection is directional, while drift is not. The relative influence of these two forces is tempered by the strength of the selection coefficient and the population size (Fig. 2.15). Where drift is the predominant force, alleles will be fixed essentially at random; where selection is the predominant force, particular alleles will be favoured. It is difficult to predict what will happen in the intermediate zones where neither drift nor selection strongly dominates.

Migration can act to supply variability to a population but it can also act in opposition to drift and selection. In general, very small amounts of migration can block the effects of genetic drift. These relationships can be seen by substituting migration rate for the selection coefficient (s) in Fig. 2.15. Likewise, if the migration rate is greater than or equal to the selection coefficient, then the importance of selection for population differentiation becomes almost insignificant (Ellstrand and Marshall, 1985).

Mutation can act like migration in tempering the effects of selection and drift. However, its rate of occurrence is generally quite low ($< 10^{-6}$) and therefore its greatest importance lies as a generator of new variability. Such rare variants can be established much faster in small populations than in large ones even if selection coefficients are quite high.

Sewell Wright (1931) combined migration, drift and selection into a unified process of evolution called the shifting balance theory. In it, he viewed most populations as being substructured into small periodically drifting groups and he envisioned fitness as being due to many interacting genes rather than a few key ones (see Chapter 3).

Wright described evolution as occurring essentially in five stages: (i) a population is 'trapped' at a certain fitness level where only a limited number of genotypes are well adapted; (ii) because of drift, a less adapted genotype becomes established by chance in a subpopulation; (iii) subsequent evolution reassorts the available genes in a new type that is more fit than the original; (iv) the subpopulation grows and begins to disperse; and (v) nearby neighbourhoods receive the new genotype and the whole population reaches an adaptive peak.

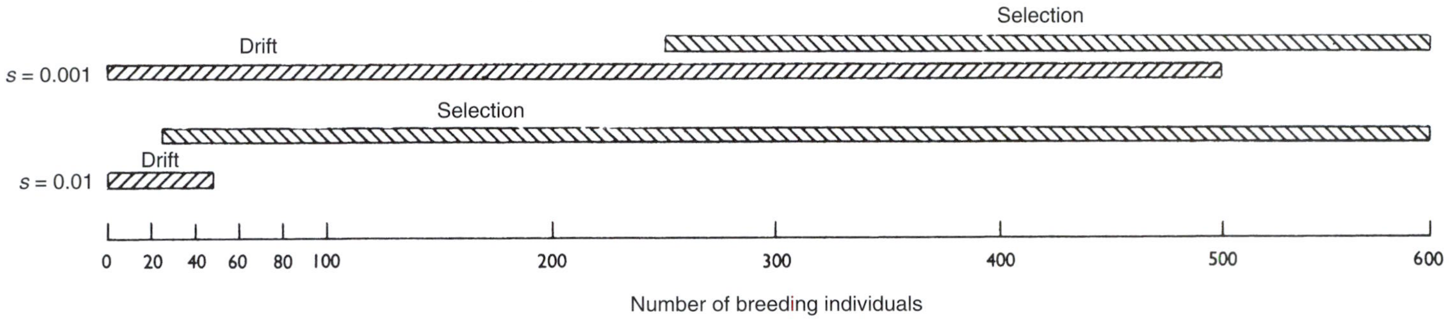

Fig. 2.15. Range of population sizes and selection coefficients where drift or selection will prevail. (Used with permission from V. Grant, © 1963, *The Origin of Adaptations*, Columbia University Press, New York.)

It is not known how important the shifting balance model is in nature. A considerable amount of recent debate has surrounded the frequency with which Wright's phases act in concert and whether simple models of mass selection explain most evolutionary change just as well (Coyne *et al.*, 1997; Peck *et al.*, 1998; Coyne *et al.*, 2000; Goodnight and Wade, 2000). The shifting balance model is almost impossible to prove in its entirety, but most of the individual aspects have been documented in natural populations (as we have discussed). At the very least, Wright's model is a stimulating attempt to describe the potential complexity of evolution in natural populations.

Summary

Numerous evolutionary forces shape plant populations, including migration, genetic drift and natural selection. Selection causes populations to change in several directed fashions, while genetic drift is synonymous with random change. Many factors influence the rate of change in populations. Of primary importance are the amount of variability present and the effective population size. Other important parameters are generation time, the strength of the selection coefficient and the degrees of dominance and epistasis. Selection is generally thought to play the leading role in shaping plant populations, but the critical influence of drift and migration cannot be excluded. Gene flow is generally limited in plant populations and as a result, effective population sizes are often quite small. This substructuring of populations can lead, even in the absence of selection, to patchy distributions of genes and the generation of unique allelic combinations. Both random and non-random forces have probably been important in plant evolution.

The Multifactorial Genome 3

Introduction

Up to this point, we have been describing evolution at primarily the single gene level. It is important to realize, however, that natural selection operates on the whole phenotype of an individual and not only on the product of a single gene. The fitness of an organism is dependent on the interaction of the complete genetic complement. This is alluded to in Wright's shifting balance theory.

The expression of a trait is regulated by both the genetic background of an individual and the environment that surrounds it (Chapter 1). Some genes have predictable, stable influences on phenotype, while others do not. There are genes with *incomplete penetrance* such that only a proportion of the individuals carrying it express the phenotype, and others with *variable expressivity* where the trait is expressed to various degrees in different individuals. The precise reasons for these differential gene effects are rarely known, but they are usually thought to relate to the particular environment of the individual and/or to interactions of the gene with others in the genotype.

The effect of a single locus is often lost in a myriad of genic and cellular interactions. Biochemists have provided us with numerous examples of lengthy, intercoordinated pathways that are filled with feedback loops and internal regulation (Fig. 3.1). In some cases, genes act independently of other genes, so that the phenotype is the sum of the contributions of the individual loci (additive effects), but, more frequently, the whole is not equal to the sum of its parts and subtle changes in the product of one locus can have a cascading effect on the overall phenotype.

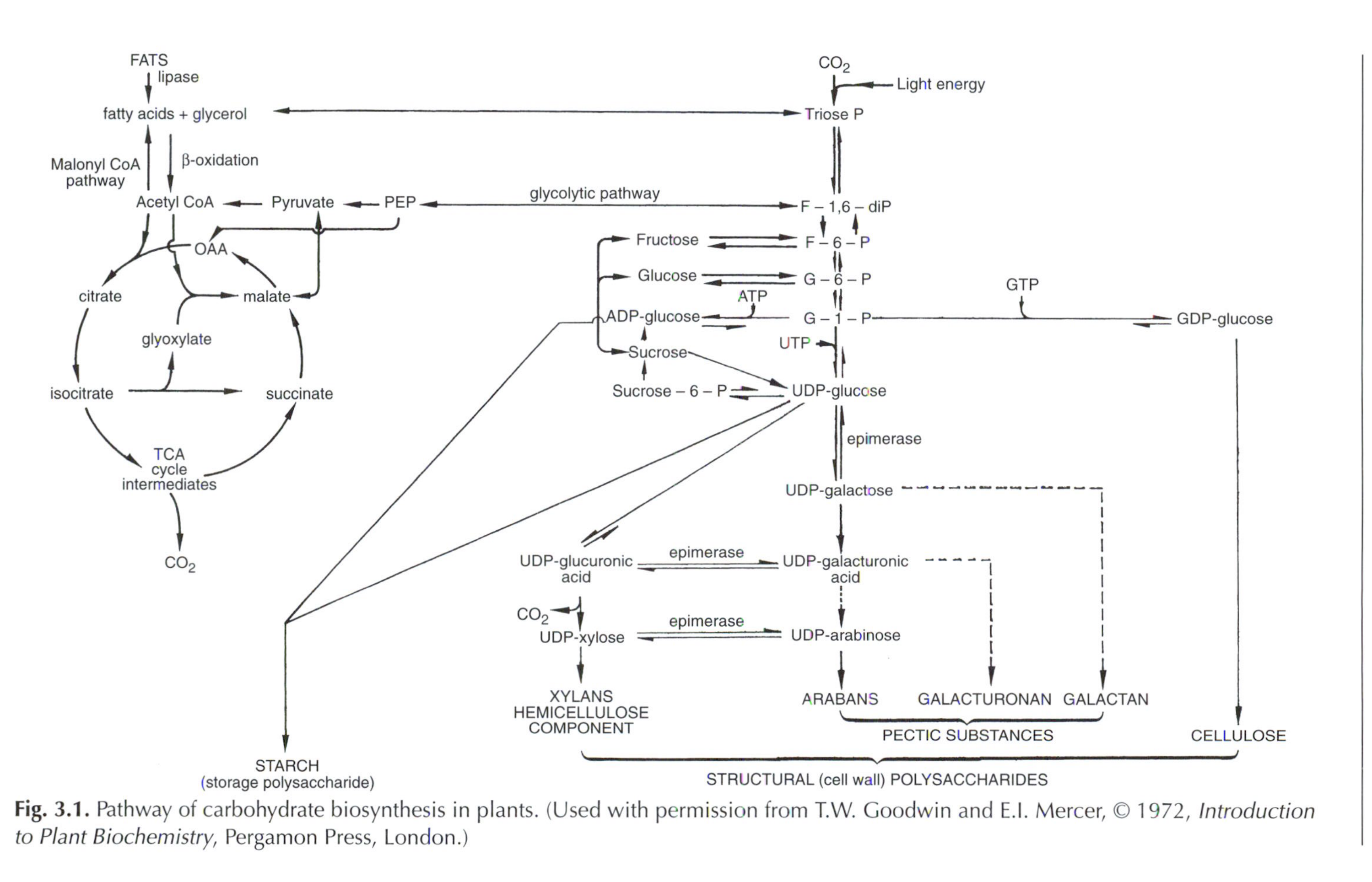

Fig. 3.1. Pathway of carbohydrate biosynthesis in plants. (Used with permission from T.W. Goodwin and E.I. Mercer, © 1972, *Introduction to Plant Biochemistry*, Pergamon Press, London.)

The evolutionary ramification of this genomic complexity is that alleles are selected at two levels: (i) how well they function in the environment; and (ii) how well they function together. There is selection for harmoniously coordinated alleles across loci. The situation can be compared to an orchestra: the great ones not only have outstanding individual players, but also perform as a finely tuned, integrated unit.

There is considerable evidence that the genome is cohesive and inheritance is generally multifactorial. Genetic phenomena such as pleiotropy, epistasis and specific combining ability are best explained as intragenomic interactions, and instances of coadaptation have been observed where groups of genes or whole genomes appear to be selected together. We shall begin our discussion of genomic complexity by first describing the various types of multigenic interactions that exist, and then evaluating the reported examples of coadaptation.

Intragenomic Interactions

Pleiotropy occurs when one allele has more than one phenotypic effect. Some clear examples of this phenomenon are the gene *S* in *Nicotiana tabacum*, which affects several structures, including the shapes of leaves, flowers and capsules (Fig. 3.2), and the colour gene *C* in onions, which not only regulates colour but also determines resistance to the smudge fungus through the production of specific phenolic acids. The allele *dl* in the tomato (*Lycopersicon esculentum*) reduces the number of hairs on stems, peduncles and stamens. It also produces separated anthers, which diminish levels of self-pollination (Rick, 1947). The fact that these single gene changes influence so many traits argues strongly that the genome must be interconnected.

As we shall discuss more fully in Chapter 6, a number of pleiotropic genes were important in the early domestication of beans and maize. The *fin* gene in dry beans conditions the earliness of flowering, and has significant effects on node number on stems, pod number and the number of days from flowering to fruiting (Koinange *et al.*, 1996). Several pleiotropic quantitative trait loci (QTL) have been identified in *Zea*, including: (i) *teosinte glume architecture 1* (*tb1*), which affects internode lengths, inflorescence sex and structure; (ii) *teosinte branched 1* (*te1*), which also affects internode lengths and numbers, and inflorescence sex; and (iii) *suppressor of sessile spikelets 1* (*sos1*), which affects branching in the inflorescence and the presence of single vs. paired spikelets in the ear (Doebley *et al.*, 1995).

In epistasis, the allelic constitution at one locus affects the level of expression of alleles at another locus, again illustrating the interactive nature of the genome. For example, the presence of prussic acid in clover requires a dominant allele at both of two loci. Bulb colour in onion is also regulated by alleles at two loci; one locus determines whether the bulb will be coloured at all and a second locus determines whether it will be red or yellow. The pres-

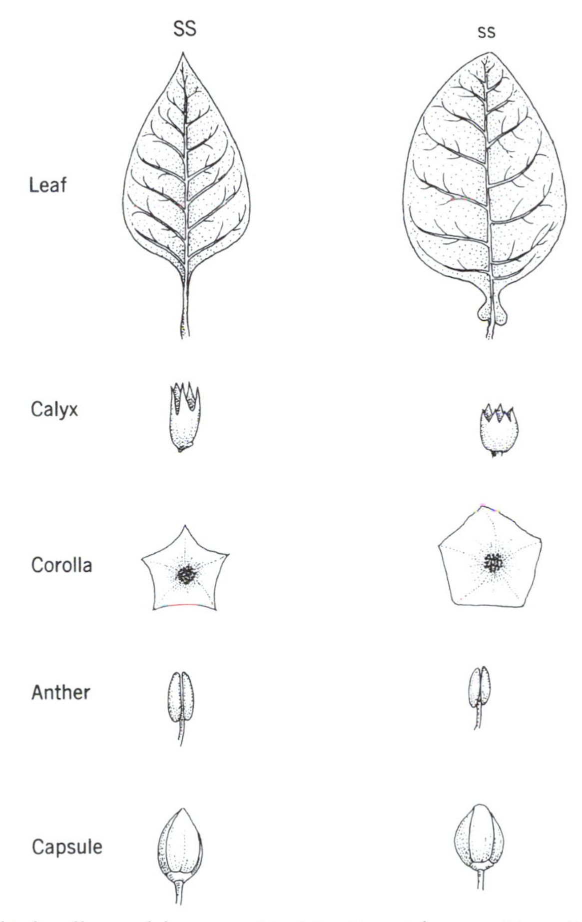

Fig. 3.2. Multiple effects of the gene *S* in *Nicotiana tabacum*. (From Stebbins, 1959.)

ence of the pungent chemical capsaicin in hot peppers is determined by a dominant allele at one locus, while the degree of heat is regulated by a series of modifiers at several other loci.

Quantitative traits are commonly influenced by epistatic interactions. These can be identified by plotting the trait values associated with each genotype (Fig. 3.3). Suppose you have two loci regulating plant height, A and B, with two alleles each and you plot the values of the genotypes of BB, Bb and bb for each allelic substitution at the A locus. If there is no epistasis and no dominance, the relative values for each genotype will rely solely on the additive combination of alleles at each locus, and the slopes of all the lines will equal 1. If there is no epistasis but there is complete dominance, the values for heterozygotes will equal those of one of the homozygotes and the trajectories of each line will level off at the same point. If there is epistasis, the alleles will interact in a more complex fashion, and the trajectories of each line will differ.

A. Additive interactions:

Genotype	AA	Aa	aa
BB	7	6	5
Bb	5	4	3
bb	3	2	1

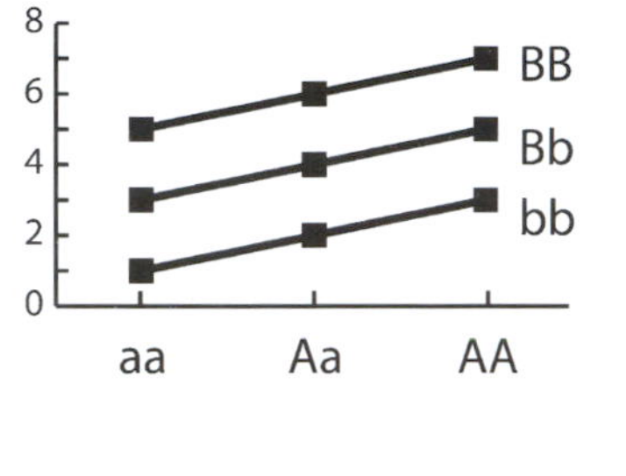

B. Dominance interactions:

Genotype	AA	Aa	aa
BB	4	4	2
Bb	4	4	2
bb	3	3	1

4 3 2 1 0
BB, Bb
bb
aa Aa AA

C. Epistatic interactions:

Genotype	AA	Aa	aa
BB	3	3	1
Bb	3	3	1
bb	1	1	1

3 2 1 0
BB, Bb
bb
aa Aa AA

Fig. 3.3. A demonstration of the effects of additive, dominant and epistatic interactions on a quantitative trait.

Statistical analyses have been developed to partition levels of quantitative variation into additive, dominance and epistatic interactions, using analysis of variance techniques (Falconer and Mackay, 1996). In one of the simplest analyses, breeders measure complex genomic interactions by calculating general and specific combining ability (Griffing, 1956). In this type of analysis, the breeder makes a series of crosses between a group of parents and compares their mean performance with that of their progeny. The mean performance of each line in crosses with all the other lines is called the general combining ability (GCA) and represents the additive component of variance. The deviation of a particular individual cross from the average GCA of the two lines is called the specific combining ability (SCA) and represents intra- and interlocus interactions (Fig. 3.4). Standard analysis of variance techniques are used to calculate the relative importance of GCA and SCA (Gilbert, 1967). More complex crossing and statistical approaches are required to separate dominance and epistatic interactions.

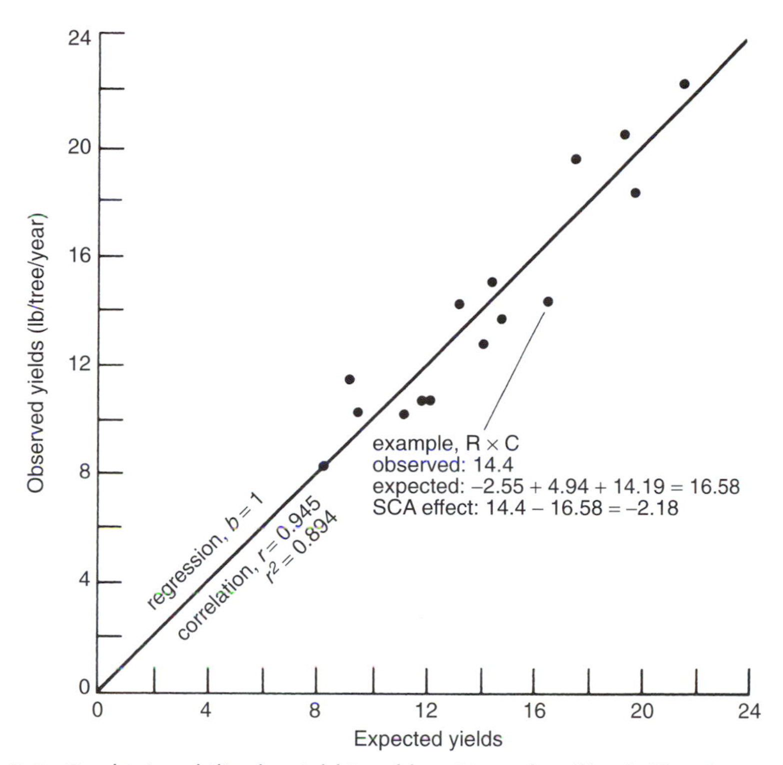

Fig. 3.4. Combining ability for yield in rubber, *Hevea brasiliensis*. The observed yield of progenies is plotted against the expected yields or general combining ability (GCA) of each line. The GCA is calculated as the mean performance of each line in crosses with all other lines. When a regression line is drawn through the various parental values, the deviations above and below the line represent specific combining ability (SCA). (Used with permission from N.W. Simmonds, © 1985, *Principles of Crop Improvement*, Longman, London.)

Classical statistical studies of quantitative variation have uncovered considerable evidence of epistasis (Falconer and Mackay, 1996), but the documentation of intergenic interactions in QTL mapping studies has proved more elusive (Tanksley, 1993; Paterson, 1995; Kim and Rieseberg, 2001). A large part of the difficulty in identifying epistatic interactions with molecular markers deals with the statistical power represented by smaller population sizes, and limitations in the analysis of variance technique itself (Wade, 1992; Doebley *et al.*, 1995). In spite of these limitations, several plant studies have clearly documented epistasis. Yamamoto *et al.* (2000) found a significant interaction between two QTL (*Hd2* and *Hd6*) involved in photoperiod sensitivity in rice. Yu *et al.* (1997) identified 32 QTL associated with four yield traits in rice, and found almost all of them to interact significantly with at least one other QTL. Doebley's group has discovered a significant epistatic interaction in maize between a QTL on chromosome arm 1L (QTL-1L) and another on chromosome arm 3L (QTL-3L), which both influence the number and length of the internodes in both the primary lateral branch and the inflorescence (Doebley *et al.*, 1995; Lukens and Doebley, 1999). Alleles at these loci derived from either maize or teosinte had the strongest phenotypic effect in their own species background, further signalling the complexity of genomic interactions in *Zea mays*. QTL-1L was determined to be the locus for *tb1*, and QTL-3L could be *te1*, which we have already described as being highly pleiotropic. Again, selection at these two loci would have had dramatic phenotypic effects during the domestication process.

Coadaptation

If we accept the notion that the genome is highly interactive, then we realize that the whole genotype is the unit of selection and not the individual genes themselves. The simultaneous selection of large blocks of genes has already been referred to as coadaptation. In most cases, it is very difficult to obtain direct evidence of coadaptation since the relationship between most gene products and the phenotype is hazy, but a great deal of circumstantial evidence has been accumulated. Phenomena considered to represent coadaptation include: (i) hybrid breakdowns; (ii) supergenes; and (iii) gametic disequilibria.

Hybrid breakdowns

Hybrids within populations of animal species are usually completely normal, but hybrids between populations are sometimes weak or unviable (Wallace, 1968). Such hybrid breakdowns are thought to arise because the gene pools within populations have been selected over time for their harmonious interaction. When individuals are mated from variant populations, their genes may not be well integrated and therefore produce poorly adapted offspring.

The majority of the most graphic examples of hybrid breakdowns have been demonstrated in animals. For example, Gordon and Gordon (1957) found that platyfish from different Amazonian river basins had characteristic patterns of small, dorsal spots. When fish from variant origins were mated, progeny were produced with gross, distorted collections of pigment. Backcross individuals proved to have even more unsightly patches.

Few examples of hybrid breakdowns have been described between populations of plant species, perhaps due to the lower developmental complexity of plants (Gottleib, 1984). However, crosses between related plant species often result in the production of weak or sterile progeny. The cross of *Gilia ochroleuca* × *Gilia latifolia* produces hybrids that have mainly abortive pollen grains and their ovules do not develop normally under any conditions (Grant and Grant, 1960a). Hybrids of *Gossypium hirsutum* × *Gossypium barbadense* have inrolled leaves, corky stems and bushy growth and are mostly sterile (Stephens, 1946). In some cases, unfavourable gene combinations do not appear until the F_2 generation and more complete reassortment occurs. Crosses of *Zauschneria cana* × *Zauschneria septentrionalis* (Clausen et al., 1940) and *Layia gaillardioides* × *Layia hieracioides* (Clausen, 1951) produce vigorous, semifertile hybrids but most of the F_2 individuals are weak and dwarfish. We shall describe these types of relationships more fully in the chapter on speciation.

Probably the most frequently mentioned case of hybrid breakdown within a plant species involves populations of the bean *Phaseolus vulgaris* (Gepts, 1988, 1998). There are large- and small-seeded races from South America and Mexico that when crossed produce high percentages of weak, semi-dwarf progeny. This reduction in hybrid fertility and vigour is associated primarily with two independent loci, DL_1 and DL_2, although many other differences exist in the gene pools of the geographical races, including distinct phaseolin seed proteins, electrophoretic alleles, flowering times and floral structures (Gepts and Bliss, 1985; Shii *et al.*, 1981).

Proper development and function depend not only on nuclear interactions but also on nuclear–organelle cooperation. Dramatic differences are found between reciprocal crosses of *Epilobium hirsutum* and *Epilobium luteum* (Michaelis, 1954). *E. hirsutum* × *E. luteum* yields hybrids with stunted growth, narrow yellow-mottled leaves and sterile anthers, while *E. luteum* × *E. hirsutum* produces normal-looking plants and fertile anthers. These differences are presumably the result of the egg providing most of the cytoplasm for the fertilized egg. As with most angiosperms, the egg cell of *Epilobium* is rich with organelles, but the sperm cells have little cytoplasm.

Stubbe (1960, 1964) found the plastids of diploid *Oenothera* in the section *Euoenothera* to vary in their functionality in different nuclear backgrounds. He identified five plastome types that differed in their nuclear compatibility (Fig. 3.5). Although the chloroplasts can often survive in the nuclear background of another species, they show varying degrees of bleaching. This dysfunction suggests that the nuclear and plastid genomes coevolved to produce a finely tuned cooperative relationship.

Multigene complexes

There are a number of examples in plants and animals where several genes with related functions are found in close proximity on a chromosome and may represent coadaptation (Ford, 1975). Darlington and Mather (1949) coined the term supergene to describe the case where a series of genes rarely undergo recombination due to tight linkage on a chromosome or association within an inversion. They felt that genes which were originally separate might occasionally migrate together to unify coadapted complexes as selection 'capitalized' on cytological aberrations.

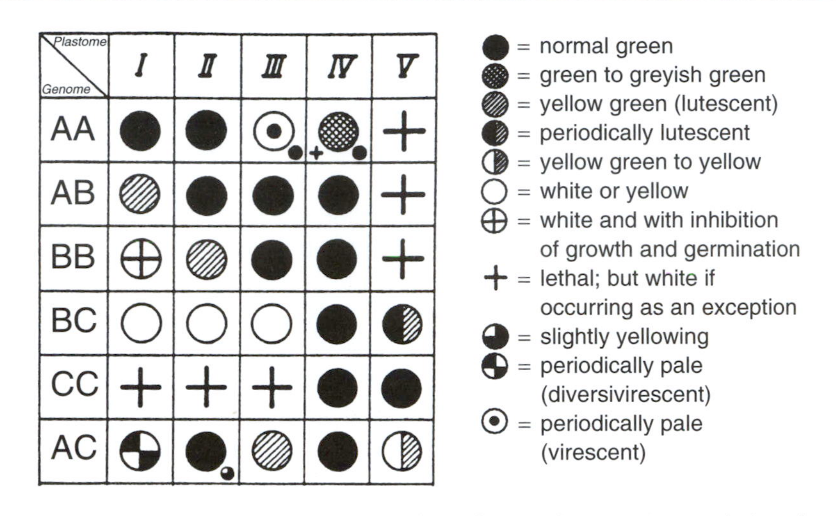

Fig. 3.5. Compatibility between different diploid *Oenothera* nuclear and plastid genomes. Small symbols represent a less frequent occurrence in the A genomes. (From Stubbe, 1964.)

Heterostyly in primroses is a clear example of such supergenes (Dowrick, 1956; Crowe, 1964). Successful pollination occurs only between individuals with their stigmata and anthers in the same position (Fig. 3.6) and numerous other characteristics are linked to these traits, including size and ornamentation of pollen grains and the size of stigmatic papillae. Other classic examples of heterostyly are found in the *Polygonaceae*, *Linoceae*, *Lythraceae*, *Oxalidaceae* and *Boraginaceae* (Grant, 1975).

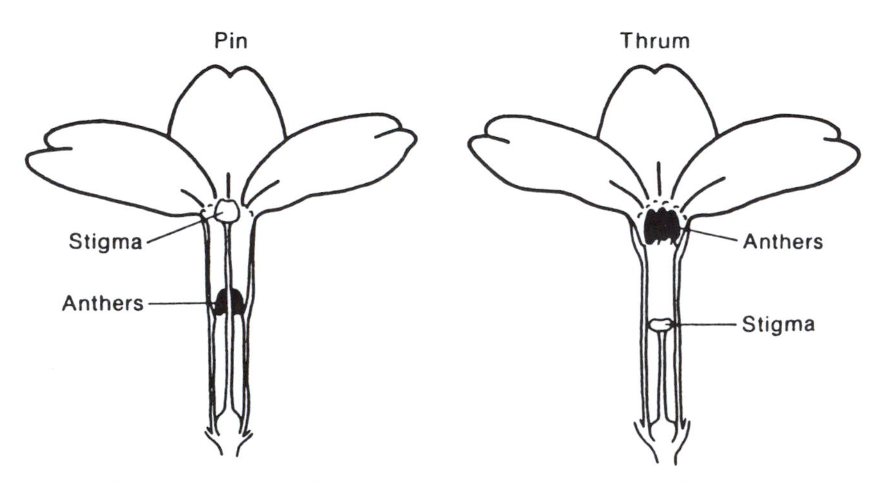

Fig. 3.6. Relative positions of the stigma and anthers in pin and thrum flowers of the primrose. (From P.W. Hedrick, © 1983, *Genetics of Populations*, Jones and Bartlett Publishers, Boston, Figure 5.3, p.173. Reprinted with permission.)

The speltoid mutant in hexaploid wheat is another example of a supergene in higher plants that may even represent a reversion to primitive characteristics. The cultivated bread wheat *Triticum aestivum* normally has a tough rachis (main fruiting axis) with loose glumes (bracts) that allow the grain to fall out easily. The speltoid mutant mimics the primitive wheat *Triticum spelta,* in which the rachis is brittle and the glumes are tightly attached. *Triticum aestivum* yields mostly naked seeds when harvested, while the rachides of the speltoid mutants shatter and the seeds remain attached (Frankel and Munday, 1962). The genes involved in this syndrome of traits are clustered together on one chromosome (IX) and all speltoid mutants have a deficiency for a segment called Q (Sears, 1944; MacKey, 1954).

Chromosomal inversion heterozygosities have also been used to argue for coadaptation. Probably the most complete story has been developed in the fruit fly *Drosophila* (Dobzhansky and Pavlovsky, 1953; Vetukhiv, 1956). Numerous chromosomal arrangements exist in natural populations of this genus and many of these chromosomal types are maintained in relatively stable frequencies. Dobzhansky and Pavlovsky suggested that these polymorphisms represented coadapted blocks of well integrated genes and that the most fit individuals were those heterozygous for chromosomal rearrangements. They tested their hypothesis by comparing the fitness of homozygous and heterozygous genotypes of *Drosophila willistoni* and *Drosophila pseudobscura* from different localities. When experimental populations were initiated with individuals heterozygous for two inversions, high frequencies of inversion heterozygotes were maintained if the populations were begun with individuals from the same locality, but frequencies of inversion heterozygotes diminished in populations initiated with strains of different localities. They suggested that the genes in the inversions from similar populations were well integrated, while those from distinct populations were not. The gross structure of the chromosomal arrangements found in two populations may have been similar, but the alleles they carried were different. This divergence in allelic frequency was later documented by Prakash and Lewontin (1968, 1971) using allozymes.

Such dramatic examples of inversion polymorphism are unusual in plants, but the maintenance of translocation heterozygotes in species like *Oenothera* are thought to reflect coadapted complexes (Chapter 1). Permanent translocation heterozygotes are maintained through a system of balanced lethals where homozygotes are either weak or not produced due to gametic unviabilities. For example, in *Oenothera lamarchiana* the chromosomes form a ring of 14 at meiosis and are oriented in such a way that alternate chromosomes pass to each pole (Fig. 3.7). This results in only two types of gametes being produced, one with chromosomes from only the pollen parent and one with chromosomes from only the mother. A system of balanced lethals then operates in *Oenothera* to allow only heterozygous zygotes to survive. In one case, only one set of chromosomes produces viable pollen and one set viable eggs (gametophytic lethals), while, in other instances, both types of gametes are formed, but only heterozygous zygotes survive (zygotic lethals).

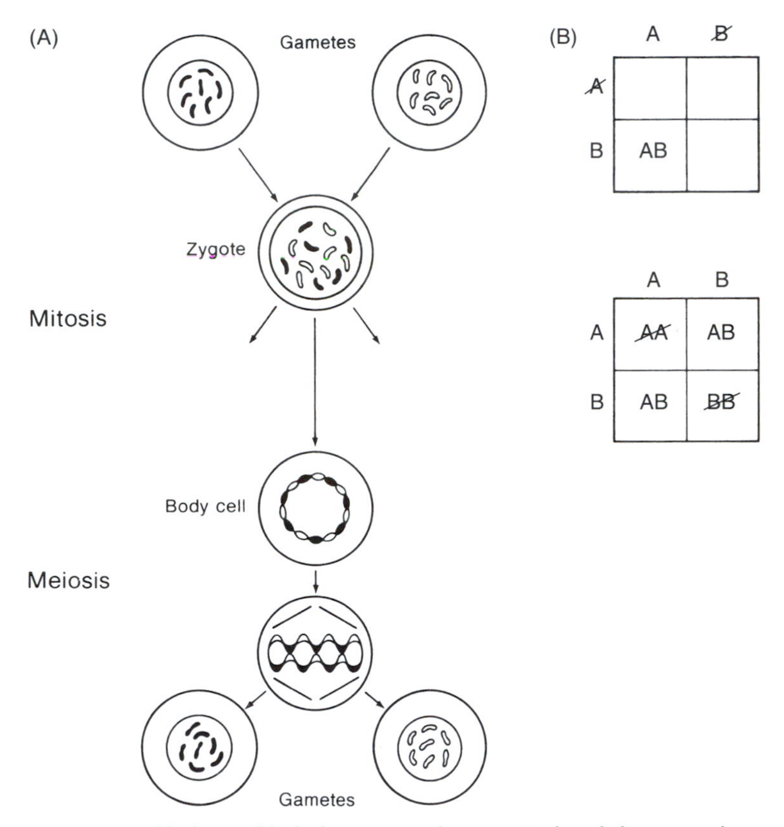

Fig. 3.7. System of balanced lethals in *Oenothera*. (a) Stylized diagram of chromosome segregation in *O. lamarchiana*. The chromosomes form a ring of 14 at meiosis, oriented in such a way that alternate chromosomes pass to each pole. Only two types of gametes are produced, one with all its chromosomes derived from the pollen parent, and the other with a full set from the mother. (b) Two types of balanced lethal systems found in *Oenothera* where only heterozygous zygotes result. Top, gametophytic lethals where only one chromosome set produces viable eggs and pollen. Bottom, zygotic lethals where both types of gametes are formed, but only heterozygous ones survive. The slashes represent mortality or death. (Used with permission from T. Dobzhansky, F.J. Ayala, G.L. Stebbins and J.W. Valentine, © 1977, *Evolution*, W.H. Freeman and Company, San Francisco.)

It is generally agreed that this complex system of structural hybridity could have arisen as a mechanism to preserve coadapted complexes. To test this possibility, Levy and colleagues examined electrophoretic variation among chromosomal complexes of several *Oenothera* species which carried the gametophytic lethal system (Levy and Levin, 1975; Levy *et al.*, 1975; Levy and Winternheimer, 1977). He discovered that the species had very few alleles at individual polymorphic loci, but that individual strains had unique combinations of the whole allelic array. The egg and sperm lines of most strains differed significantly in allelic frequencies at a number of loci and intergenomic linkage disequilibrium accounted for 97.5% of the observed heterozygosity. Thus, particular genic arrays were indeed being maintained in each population – an observation at least consistent with the coadaptation hypothesis.

Gametic disequilibria

In electrophoretic examinations of plant and animal populations, alleles at groups of linked and unlinked genes are often found to be out of Hardy–Weinberg equilibrium. The non-random association of alleles at different loci in gametes is referred to as gametic phase disequilibrium or the shortened form, gametic disequilibrium (Lewontin and Kojima, 1960; Crow and Kimura, 1970).

As we have already noted, genetic analyses involving multiple loci are quite complex. To measure gametic phase disequilibria, the frequencies of all possible allelic combinations are calculated and then compared with each other. Excesses of one gametic type that do not equilibrate after several generations of mating are thought by many population biologists to represent coadapted complexes.

For example, using the simplest case of two loci with the alleles Aa and Bb:

Gametes carrying AB and ab are said to be coupling gametes.
Gametes carrying Ab and aB are said to be repulsion gametes.
Gametic phase disequilibrium (D) = product of coupling genotype frequencies − product of repulsion genotype frequencies

Mathematically, if:

Locus 1	Locus 2
$A = p_1$	$B = p_2$
$a = q_1$	$b = q_2$

then:

$$D = (p_1p_2)(q_1q_2) - (p_1q_2)(q_1p_2)$$

and to weight by gene frequency:

$$D' = D/pq$$
$$= \text{relative gametic phase disequilibrium}$$

D' varies from $+1$ to -1, and:

if $D' = 0$, there is no disequilibrium
if D' is negative, there are more repulsion than coupling gametes
if D' is positive, there are more coupling than repulsion gametes

Populations with positive and negative values of D have a surplus of one or the other gametic type and are therefore in gametic phase disequilibrium. This means that alleles at each of the loci are not segregating independently and distinct genic assemblages are being maintained.

The most commonly cited examples of gametic phase disequilibria in plants are those of Allard and his co-workers (Allard *et al.*, 1972; Allard, 1988), although others have been documented (Golenberg, 1989). We described previously the association between allelic frequencies at single loci and moisture levels in oats (Chapter 2). These alleles were further clustered into a few non-random associations across loci. The mesic environments were almost monomorphic for a set of alleles at five loci (denoted 21112), while on the more xeric sites only two sets of alleles predominated (12221, 12211) (Table 3.1). Three of these loci were on the same chromosome, while the other two segregated independently (Clegg *et al.*, 1972). When the D′ values were averaged across all of the possible two-locus pairings, values ranged from 0.36 to 0.71 in each of the various subdivisions. The direct adaptive benefit of these alleles was not measured, but the most common assemblages were thought to represent coadapted complexes since they were not in the expected frequencies.

Allard and co-workers (Clegg *et al.*, 1972; Weir *et al.*, 1972) also found associations between alleles at different electrophoretic loci in cultivated populations of barley (*Hordeum vulgare*). They examined seed samples of two highly heterozygous experimental populations (CCII and CCV), which were initially generated by hybridizing 30 parents from the major barley production regions of the world. These populations were propagated in large plots under agricultural conditions for decades without conscious human-directed selection. It was discovered that some allelic combinations had

Table 3.1. Changes in five-locus gametic percentages in wild oats along a moisture gradient. Subdivisions A, B, C and D represent a change from mesic to xeric conditions. (From Allard *et al.*, 1972.)

Gametic type[a]	Subdivision				Location total
	A	B	C	D	
21112	91.5	39.9	16.9	1.9	56.7
12221	1.7	4.1	27.9	31.9	11.3
12211	0.1	1.5	3.2	30.5	4.0
11112	0.7	4.8	2.7	0.6	1.8
21121	0.7	5.8	8.8	5.7	4.0
21221	0.0	4.6	5.6	1.6	2.2
12212	0.4	10.7	1.5	0.3	2.2
22221	0.2	3.9	6.5	2.0	2.5
D′	0.71	0.36	0.52	0.39	0.64

[a]Two alleles (1 and 2) at five loci – esterase loci E_4, E_9, E_{10}, phosphatase locus P_5 and anodal peroxidase locus APX_5.

increased over time, while others had declined (Table 3.2). The Allard group hypothesized that natural selection was structuring 'the genetic resources of these populations into sets of highly interacting, coadapted gene complexes'.

While these gametic disequilibria could indeed be the result of coadaptation, the possibility cannot be excluded that the changes are the result of the 'hitchhiking' of allozyme loci with other major adaptive genes (Hedrick and Holden, 1979; Hedrick, 1980). The loci examined by the Allard group may be selectively neutral but linked to other genes that are selectively important. Since both oats and barley are highly selfed, gametic disequilibria could arise without strong epistatic interactions.

A similar argument can also be made about the electrophoretic variation observed in the chromosomal inversion types of *Drosophila* and *Oenothera*. Over time, neutral allozyme loci within the different chromosomal types might have gradually diverged due to random forces, since the chromosomal types would be operating as separate gene pools due to the unviability of heterogenetic crossovers (Nei and Li, 1975). Still, the persistence of chromosome heterozygosities in natural populations argues that at least some adaptively important genes are found on the variant chromosomal types.

Complex gene interactions in polyploids

Poor performance in selfed autopolyploids is usually attributed to the loss of higher order allelic interactions in what is known as the overdominance

Table 3.2. Most common tri-allelic four-locus gametic types (percentages) found in complex hybrid populations of barley (CCII and CCV) over several generations. Only the most common genotypes are shown. (From Clegg *et al.*, 1972.)

	CCII Generation		
Gamete	7	18	41
1221	7.1	6.2	5.2
2112	10.9	2.1	49.7
2113	11.5	11.5	24.8
	CCV Generation		
Gamete	5	17	26
1221	0.9	3.0	17.3
2111	12.9	17.0	17.3
2112	3.8	6.3	8.5

model of inbreeding depression (ID) (Bever and Felber, 1992). Stated in another way, vigour in autopolyploids is positively correlated with the number of different alleles at each locus.

Evidence for the existence of complex interallelic interactions in autopolyploids has come from the observation that inbreeding depression in autopolyploid lucerne (Busbice and Wilsie, 1966; Busbice, 1968) and potato (Mendoza and Haynes, 1974; Mendiburu and Peloquin, 1977) is much greater than would be predicted by the coefficient of inbreeding in a two-allele model. Busbice and Wilsie (1966) suggested that this rapid loss of vigour is associated with the theoretical rate at which loci with three and four alleles are lost (tri- and tetra-allelic loci). Correlative data have come from comparisons of autotetraploids with different genetic structures. Bingham and his group (Dunbier and Bingham, 1975; Bingham, 1980) produced diploids from natural tetraploids by haploidy, generated diploid hybrids and then doubled the diploid hybrids using colchicine treatments to obtain defined two-allele duplexes (di-allelic loci). These were then crossed to produce double hybrids with presumed tetra-allelic interactions. When the performance of these different structured populations was compared, 'progressive heterosis' was observed as the diploid hybrids had higher herbage yield (540 g/plant) than their diploid parents (237 g/plant), and the double hybrids had the highest yield of all (684 g/plant).

Canalization

Individuals of plant and animal species generally maintain a recognizable identity even though they are highly polymorphic and suffer a wide range of environmental extremes. This suggests that their developmental patterns are buffered against a broad amplitude of environmental and genetic conditions. Waddington (1942) and Mather (1943) originally suggested that organisms become resistant to both genetic and environmental perturbations through an extreme type of coadaptation called canalization. They evolve to the point where most allelic substitutions and environmental disturbances do not substantially alter developmental patterns. In a sense, canalization is the end result of a complex form of stabilizing selection.

The phenotype is constructed by successive interactions of the genotype with the environment in which development occurs. Due to highly complex interallelic networks, there is often a wide range of environments that produce the same phenotype and it frequently takes extreme conditions to drastically disrupt development. This has been referred to as environmental canalization. Likewise, numerous allelic substitutions can occur at most loci without severely affecting development, due to the counterbalancing influences of alleles at other loci. The only alleles that persist in a population are those that integrate well under a large range of environments to produce a normal phenotype. This has been referred to as genetic canalization.

While the concept of canalization is intellectually satisfying, it has been difficult to prove and evolutionary theory surrounding the process of canalization has been slow to emerge (Gibson and Wagner, 2000). Some of the earliest experiments documenting canalization involved *Drosophila melanogaster* (Waddington, 1953, 1957). Waddington took a population of wild type individuals and exposed them to a heat shock. Most were phenotypically normal, but a few showed a cross-veinless trait on their wings that did not appear under normal conditions. After a number of generations of heat shock and selection for cross-veinlessness, the trait began to appear in flies without treatment. Subsequent genetic analysis discovered that the trait was being caused by a group of polygenes with small cumulative effects. Waddington suggested that cross-veins on wings were canalized and only unusual events such as heat shock and strong selection could lead to the appearance of the abnormal type and the subsequent alteration of the normal phenotype.

As previously described, Huether (1968, 1969) found that most individuals of *Linanthus androsaceus* had five-lobed flowers in nature (> 95%). However, he was able to increase the frequency of abnormal types through very strong disruptive selection and by maintaining the plants under various environmental stresses, including long days, high temperatures and decapitation. The combination of strong selection and environmental shock produced the highest frequency of anomalous individuals after five generations (Fig. 3.8). Thus, petal number was highly canalized and it took extreme environments and/or strong selection to alter development.

Obviously, all phenotypic traits are to some extent modifiable or we should not see so much variation in nature. However, the modifiability of a character depends upon the importance of the character to the general fitness of the individual. For example, gross physical or developmental abnormalities in reproductive structures could severely limit fitness by reducing attractiveness to pollinators. Such traits would remain constant unless the population were subjected to shifting balance processes or a drastic change in the environment beyond the normal threshold of the trait.

Paradox of Coadaptation

Until a trait or process is canalized, the cost of maintaining coadapted complexes can be extremely high due to recombination (recombinational load (Wallace, 1970)). If we consider a complex composed of only five loci with two alleles each, there are 243 possible diploid genotypes generated through recombination (Table 3.3). If only a few of these are well adapted, a lot of individuals will expire before maturity.

Several genetic systems limit free recombination and therefore help maintain coadapted complexes. We have already suggested that chromosomal rearrangements and tight linkage can hold coadapted genes together by

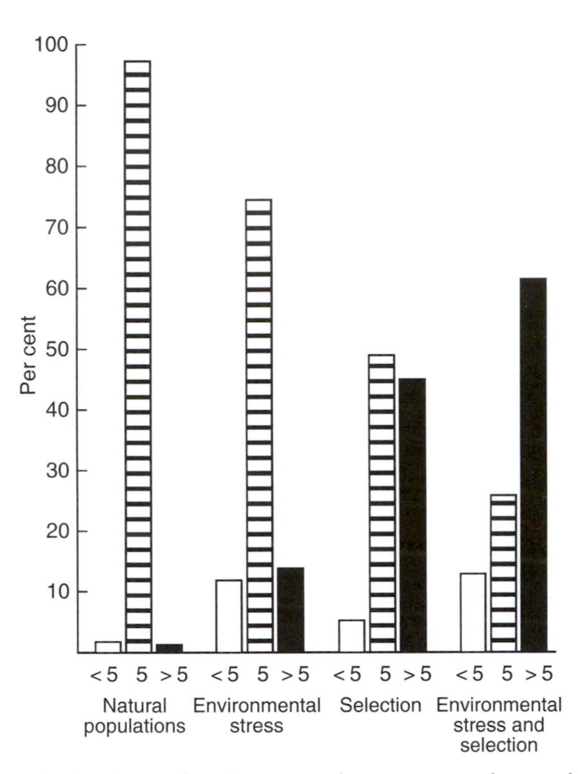

Fig. 3.8. Effect of selection and environmental extremes on the number of *Linanthus androsaceus* corolla lobes after five generations. Abnormal types are rare in natural populations, but they increase in proportion after strong selection or environmental shock. (From Huether, 1968.)

limiting recombination. Inbreeding and asexual reproduction (apomixis) can also have the same effect, although these systems fix whole genomes rather than blocks of genes.

Simple linkage reduces recombination since the closer two genes are on a chromosome, the less likely there will be a chiasma between them at meiosis. Inversions and translocations act as crossover suppressors because crossovers in heterozygotes result in the production of unviable gametes (see previous section on mutation). The mortality in this system is high, but it is thought to be much lower than if free recombination occurred among coadapted genes.

Selfing limits recombination by promoting homozygosity: homozygous individuals have no variability to reassort. Some species of plants rely on asexual reproduction for proliferation and as a result do not undergo sexual recombination at all. This phenomenon, called apomixis, can occur through agamospermy where viable seeds arise without fertilization, and vegetative propagation via structures such as stolons, tubers, rhizomes or suckers. Here, gene assemblages are maintained forever except when spontaneous mutations arise. Numerous genera contain apomicts, including *Taraxicum*, *Aster*,

Table 3.3. The number of possible diploid genotypes that can be produced by recombination among various numbers of separate loci, each of which possesses various numbers of alleles (from Grant, 1963).

Number of alleles of each gene	Number of genes				
	2	3	4	5	n
2	9	27	81	243	3^n
3	36	216	1,296	7,776	
4	100	1,000	10,000	100,000	
5	225	3,375	50,625	759,375	
6	441	9,261	194,481	4,084,101	
7	784	21,952	614,656	17,210,368	
8	1,296	46,656	1,679,616	60,466,176	
9	2,025	91,125	4,100,625	184,528,125	
10	3,025	166,375	9,150,625	503,284,375	
r	$\frac{r(r+1)^2}{2}$	$\frac{r(r+1)^3}{2}$	$\frac{r(r+1)^4}{2}$	$\frac{r(r+1)^5}{2}$	$\frac{r(r+1)^n}{2}$

Erigeron, Rudbeckia, Poa, Crepis, Musa, Manihot, Malus, Rubus, Potentilla, Citrus, Allium and *Tulipa* (Grant, 1981). Some of our most important cultivated crops are asexually reproduced naturally through tubers (white potatoes, yams, sweet potatoes and arrowroot) or artificially via cuttings (banana and cassava).

Apomictic plants cannot undergo free recombination, but a surprising number have collected sufficient variation to possess races with variant ecological requirements (Babcock and Stebbins, 1938; Lyman and Ellstrand, 1984). Most cultivars of banana, manihot and yam arose as distinct apomictic variants (Chapter 4). Solbrig and Simpson (1977) have even described asexual biotypes of dandelion (*Taraxicum officinale*) that are differentially adapted to mowed and unmowed lawns. Such variation arose either through mutation or previous hybridizations between sexual progenitors.

Many plant breeders have exploited clonal reproduction in producing cultivated varieties. Crops like strawberry and blueberry can be sexually crossed to produce variable progeny but when highly productive, coadapted genotypes arise they can be propagated asexually through runners or cuttings. This eliminates the need to develop homozygous, balanced parental lines to regenerate desirable heterozygous varieties.

Summary

The overall fitness of an individual is generally dependent not on just one gene, but rather on the interactive relationships of all the genes making up the genome. Pleiotropy, epistasis and SCA represent these complex intragenomic interactions. Genes are essentially selected at two levels: (i) how well their products perform in response to the external environment; and (ii) how well they interact with other gene products. The selection of whole groups of genes together by both internal and external forces is called co-adaptation. Evidence for this process has come from several different sources, including: (i) gametic phase disequilibria, where whole groups of alleles are out of Hardy–Weinberg equilibrium; (ii) supergenes, where several genes with related functions are tightly linked; (iii) the maintenance of balanced frequencies of chromosomal inversion heterozygotes, and (iv) hybrid breakdowns, where hybrids within populations are completely normal, but hybrids between populations are weak or unviable. The genetic cost of maintaining coadapted complexes is high until interacting genes become associated through linkage, inbreeding or apomixis.

Polyploidy and Gene Duplication

4

Introduction

As we discussed in Chapter 1, gene duplications very commonly arise in plant species after a variety of genetic events that affect different numbers of genes. Unequal crossing over and reciprocal translocations result in one or a small number of genes being amplified. Aneuploidy causes a duplication in all the genes of a particular chromosome. Polyploidy results in an amplification of the total genic content. Any type of duplication can have important evolutionary ramifications, but those of polyploidy are often the most dramatic because the whole genome is affected.

One of the most intriguing questions facing plant evolutionists is why there are so many polyploid species. As mentioned earlier, most crop plants have high chromosome numbers and the majority of all angiosperms are polyploid. It has been estimated that 2–4% of all speciation events in flowering plants represent polyploidy (Ramsey and Schemske, 1998; Otto and Whitton, 2000). All 18 of the world's worst weeds are polyploid (Brown and Marshall, 1981; Clegg and Brown, 1983).

As DNA sequence data accumulate, it is becoming clear that many species that were considered to be diploid, based on disomic chromosome behaviour may actually be ancient polyploids whose diploid progenitors have become extinct. Ancient cycles of genome duplication are evident in the cole crops, cotton, soybean and many important cereals (Wendel, 2000). Probably the most thorough evidence of a paleopolyploid has been accumulated for *Zea mays*, as was described in Chapter 3. Even *Arabidopsis*, popular in genetic analysis due to its small genome size, is probably an ancient allopolyploid, based on the degree of gene duplication apparent in the sequence of its whole genome (Blanc *et al.*, 2000). If we use Stebbins' (1950) criteria that n = 12 or greater denotes polyploidy, it turns out that a large number of angiosperm families with very deep phylogenetic roots contain only polyploids (see Chapter 1, Table 1.2). Multiple isozymes have been shown to be expressed in many of these paleopolyploid families (Soltis and Soltis, 1990).

The majority of all plant species are polyploid, even though the probability of a new polyploid species coexisting with its progenitors or replacing them would be extremely small (Fowler and Levin, 1984). First, their initial numbers would be extremely low and subject to elimination due to chance and, secondly, their fertility would be reduced by the meiotic irregularities caused by genomic duplication or the formation of unviable triploid zygotes after the fertilization of 2n eggs by the more abundant haploid pollen. The derived and progenitor species would also be under strong competition, since they would share the same genetic make-up and therefore would have substantial niche overlap. A similar situation would exist for newly emerged gene or chromosomal duplications, except that their fertility would generally be higher.

Factors Enhancing the Establishment of Polyploids

The probability that a new polyploid form or any duplication will be established is increased if it is repeatedly synthesized. There would be more individuals present to face chance elimination, and a higher proportion of the polymorphism present in the progenitor species might be captured for subsequent evolution. Since many diploid species produce measurable amounts of unreduced gametes (Bretagnolle and Thompson, 1995), it must be relatively common for polyploid species to have multiple origins. The most common polyploids are those with balanced numbers of genomes (4x, 6x, etc.), although triploids and other aneuploids do occasionally persist and form measurable amounts of gametes with complete sets of chromosomes (Husband and Schemske, 1998; Ramsey and Schemske, 1998).

Many of our crop species produce measurable quantities of unreduced gametes, including potato, cassava, blueberry, cotton, strawberry and cherry. In fact, nearly all polyploids that have been examined with molecular markers have been found to be polyphyletic with multiple origins (Soltis and Soltis, 1993, 1999). In probably the best studied polyploid group, *Tragopogon mirus* has been estimated to have four to nine lineages, and the estimates for *Tragopogon miscellus* range from two to 21 (Soltis and Soltis, 1995; Soltis, P.S. and Soltis, 2000). Most of the evidence for recurrent formation comes from nuclear genes, but, in some cases, multiple chloroplast DNA haplotypes have also been documented.

Regardless of a new type's initial numbers, it must still compete with its parental species for space and resources. In some cases, the diploids and polyploids may have sufficiently distinct adaptations to ecologically assort across habitats (Husband and Schemske, 1998). The opening of new disturbed sites may in some instances provide them with an opportunity for establishment. These sites might arise due to natural causes or human intervention. Agricultural disturbances are thought to have contributed to the spread of polyploid wheats in the Old World belt of Mediterranean agriculture (Zohary, 1965). Polyploid forms of *Tragopogon* are not present in Europe, where the genus is native, but they are gradually spreading in the Pacific North-West of North America, where the group was introduced and unique habitats probably exist (Ownbey, 1950; Soltis and Soltis, 1989b). Polyploids of numerous species are found in previously glaciated areas that their progenitors have not invaded (Ehrendorfer, 1979; Lewis, 1979; Soltis, 1984).

The bottleneck of reduced fertility and low numbers could be partially 'solved' by the new polyploid being self-fertile and perennial; this would increase the new variant's chances of producing enough viable offspring to avoid chance elimination and evolve higher levels of fertility. In fact, polyploidy is much more common in perennial species than in annuals and most allopolyploid species are highly self-fertile (MacKey, 1970). Müntzing (1936) and Stebbins (1971) describe numerous examples where polyploidy is more prevalent in perennial than in annual species of the same genus. Gustafsson (1948) showed that annuals in general have low percentages of polyploid species. Stebbins (1971) has even suggested that 'polyploidy in annual flowering plants is almost entirely confined to groups which have a high proportion of self-fertilization in both the polyploids and their diploid ancestors'. This statement is much more accurate for allopolyploids than for autopolyploids, as we shall discuss later.

The simple process of polyploidy can by itself result in a partial breakdown of the self-incompatibility system (Levin, 1983). In the gametophytic system of self-incompatibility, the doubling of genes has been shown to disrupt the recognition system in a wide range of species including members of the *Rosaceae*, *Solanaceae*, *Scrophulariaceae* and *Leguminosae* (Lewis, 1943, 1966; Yamane *et al.*, 2001). The basis of this disruption is unknown, but it may be due to competition between pairs of variant S alleles. In new polyploids with low self-fertility, Miller and Venable (2000) have proposed that such a breakdown in self-incompatibility could result in the evolution of separate genders as a means of avoiding inbreeding depression.

While high self-fertility would seem to be an advantage in polyploid establishment, evolutionary biologists have not found a consistent relationship between ploidy level and degree of inbreeding depression (ID). Polyploid complexes with lower seed or fruit set in diploids include couch grass (Dewey, 1966), maize (Alexander, 1960), clover (Townsend and Remmenga, 1968) and *Epilobium angustifolium* (Husband and Schemske, 1996, 1997). Polyploid complexes with less ID in diploids include cocksfoot grass (Kalton *et al.*, 1952) and *Amsinckia* sp. (Johnston and Schoen, 1996). ID was found to be comparable in diploid, tetraploid and hexaploid *Vaccinium corymbosum* (Vander Kloet and Lyrene, 1987), while diploid *Vaccinium myrtilloides* had significantly greater ID than tetraploid *V. corymbosum*, but not tetraploid *Vaccinium angustifolium* (Hokanson and Hancock, 2000). These discrepancies between prediction and results, can probably be attributed to the complex nature of ID and not to a breakdown in theory, as the influence of ploidy on ID is dependent on not only the buffering effect of multiple alleles, but also on relative population sizes, levels of dominance, higher order gene interactions and the amount of time that has been available to purge deleterious alleles and evolve mating systems (Charlesworth and Charlesworth, 1989; Dudash *et al.*, 1997; Cook and Soltis, 1999, 2000).

Evolutionary Advantages of Polyploidy

Polyploids have several characteristics that may contribute to their long term survival and allow them to effectively compete with their parental species. The most commonly implicated advantages are: (i) the effect of nuclear DNA amount on cell size and developmental rate (nucleotypic effects); (ii) the influence of high enzyme levels (dosage effects); and (iii) increased heterozygosity. These factors may play a role in the adaptation of all types of duplication, although the degree of physiological alteration is often dependent on the number of genes involved.

Nucleotypic effects

Nucleus and cell size are positively correlated with DNA content (Ramachandran and Narayan, 1985; Bennett, 1987; Fig. 4.1) and increased cell size frequently translates into larger plants (Grant, 1971). Larger plants sometimes have greater competitive abilities than small ones, but this potential advantage is often balanced by slower developmental rates due to retarded cell division (Bennett, 1972). In most direct comparisons of diploid and polyploid species in the same non-stressed conditions, polyploids are competitively inferior (Levin, 1983).

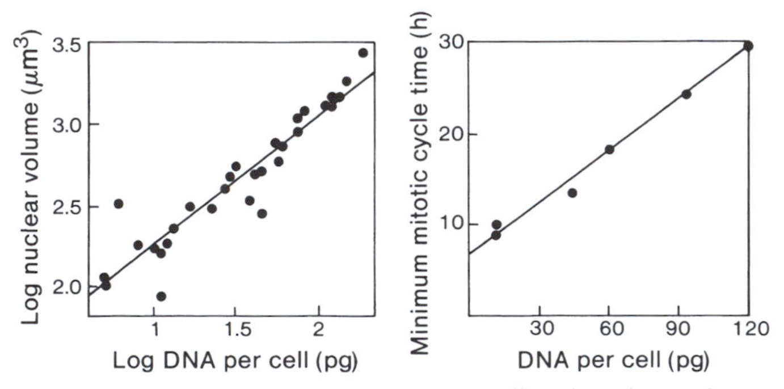

Fig. 4.1. Relationship between DNA content per cell and nuclear volume or mitotic cycle time (redrawn with permission from M.D. Bennett, © 1987, *New Phytologist* 106, 177–200).

The slow developmental rate associated with large, polyploid nuclei may occasionally be adaptive in nutrient- and water-poor environments where resources are easily exhausted (Grime and Hunt, 1975; Levin, 1983). When Stebbins (1972) sowed 2x and 4x seed of *Ehrharta erecta* at several sites in California, the polyploids became most firmly established on steep, shady hillsides, while the diploids predominated on more mesic sites (Stebbins, 1972, 1980). In other studies, tetraploids of *Nicotiana* (Noguti *et al.*, 1940) and *Dianthus* (Rohweder, 1937) were found primarily on dry, calcareous soils where diploids were absent. Such instances of polyploid superiority may be the exception rather than the rule, however, as Stebbins (1971) could find only one study out of nine in western USA where tetraploids were located on more xeric sites than were their diploid relatives. Similar inverse relationships have been observed by others (Johnson and Packer, 1965; Price *et al.*, 1981).

Bennett (1976) discovered a positive correlation between DNA amount and latitude among crop species (Fig. 4.2). This cline may relate in some unknown way to environmental constraints or the differential radiosensitivity of large and small chromosomes (Bennett, 1987). The level of ultraviolet light is higher in tropical than in temperate latitudes (Sanderson and Hulbert, 1955) and increases in DNA amount per chromosome make them a more likely target for ionizing radiation.

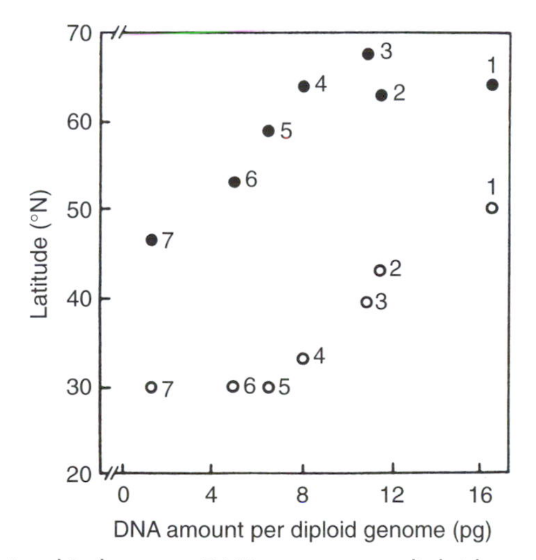

Fig. 4.2. The relationship between DNA amount per diploid genome and the northern limits of cultivation of several cereal grain species. Key to points: ○ for a transect from Hudson Bay to Key West in Florida (approximately 82°W) in winter; ● for a transect from near Murmansk by the Arctic Ocean to Odessa by the Black Sea (approximately 32°E) in summer. 1, *Secale cereale*; 2, *Triticum aestivum*; 3, *Hordeum vulgare*; 4, *Avena sativa*; 5, *Zea mays*; 6, *Sorghum* spp.; 7, *Oryza sativa*. (Used with permission from M.D. Bennett, © 1976, *Environmental and Experimental Botany* 16, 93–108, Pergamon Press, Elmsford, New York.)

Dosage effects

The immediate biochemical effect of gene duplications on structural genes is often increased production of a protein or enzyme. Carlson (1972) was able to determine the chromosomal location of enzyme loci by searching for increased activity in trisomics of *Datura stramonium* (Fig. 4.3; Table 4.1). He looked at 12 enzymes and could assign nine of them to specific chromosomes. Levin *et al.* (1979) discovered 1.5–2.0-fold increases in alcohol dehydrogenase activity in six *de novo* polyploids of *Phlox drummondii*, while Dean and Leech (1982) found regular increases in ribulose bisphosphate carboxylase (Rubisco) levels across 2x, 4x and 6x wheats (764, 1517 and 2242 pg/cell, respectively). Aragoncillo *et al.* (1978) found almost linear increases in protein level as chromosomes were added to nullisomic lines of hexaploid wheat (Fig. 4.4).

Such increases in enzyme activity could have important adaptive consequences if the enzyme plays a critical metabolic role and is rate limiting (Gottleib, 1982; Wilson *et al.*, 1975). However, the biochemical and physiological consequences of gene duplication are unpredictable (Noggle, 1946). Autopolyploids of *Lycopersicon esculentum* (tomato) were shown to have enhanced activity for four enzymes, decreased activity for one and stable activity for two others (Albrigio *et al.*, 1978). Autotetraploid *T. miscellus* had alcohol dehydrogenase (ADH) activities that were intermediate to those of its diploid progenitors (Roose and Gottleib, 1980). Rubisco activity and photosynthetic rate were correlated with ploidy level in polyploids of tall fescue

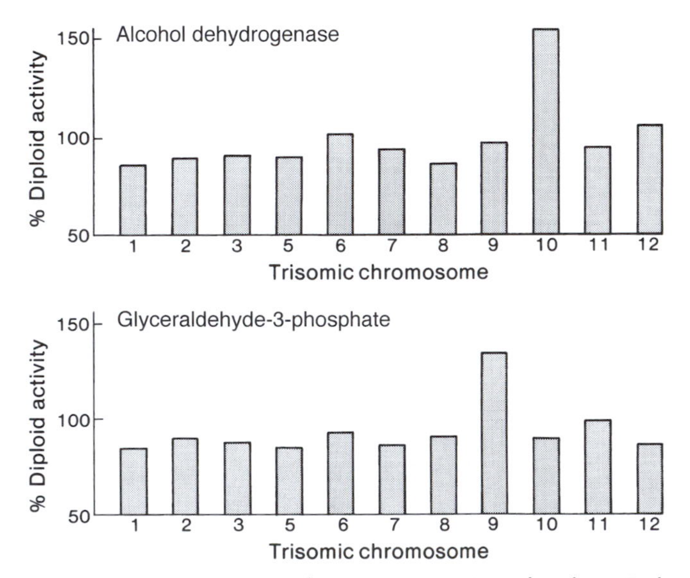

Fig. 4.3. Enzyme activity in trisomics of *Datura stramonium* (data from Carlson, 1972).

Table 4.1. Trisomics of *Datura stramonium* that showed the greatest increase in enzyme activity compared with diploids (from Carlson, 1972).

Enzyme	Diploid activity (%)	Trisomic chromosome
Dehydrogenases		
Alcohol	157	10
Glucose-6-phosphate	141	9
Glutamate	162	11
Glyceraldehyde-3-phosphate	135	9
Isocitrate	147	2
Lactate	138	3
Malate	139	5
6-Phosphogluconate	141	9
Hexokinase	140	2

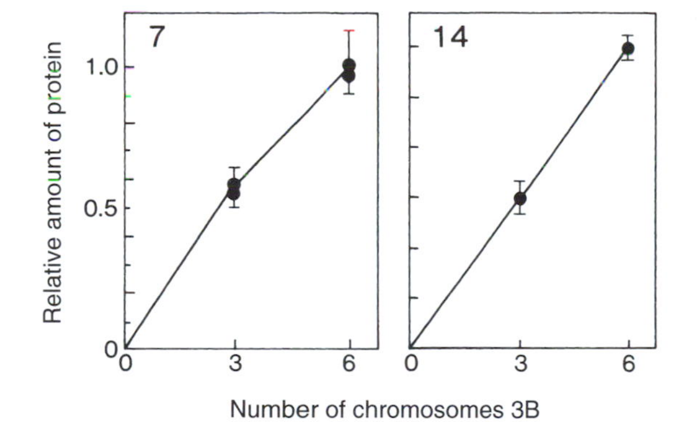

Fig. 4.4. Chromosome-dosage responses for two seed proteins (7 and 14) in allohexaploid wheat (used with permission from C.E. Aragoncillo, © 1978, *Proceedings of the National Academy of Sciences USA* 75, 1446–1450).

(Randall *et al.*, 1977; Joseph *et al.*, 1981), but in castor bean (Timko and Vasconcelos, 1981) and lucerne (Settler *et al.*, 1978) higher ploidies had unchanged or reduced photosynthetic rates even though their Rubisco levels were positively associated with nuclear ploidy. Photosynthetic rates in induced polyploids of *Phlox* varied greatly depending on the diploid progenitor's genotype (Bazzaz *et al.*, 1982).

This unpredictability is probably due to the complexity of the plant genome. As outlined in the chapter on the multifactorial genome, numerous epistatic interactions regulate the relative importance of each gene duplication. Enzyme levels are presumably influenced by factors separate from the structural genes themselves. Regulatory genes exist which affect timing and expression of enzyme genes, and overall changes in surface–volume ratios of cells have cascading influences as membrane sites become limiting and cellular concentrations change. In the case of photosynthesis, dozens of enzymes besides Rubisco play a key role in the fixation of carbon, along with numerous morphological traits, such as cell volume, leaf shape, degree of venation and stomatal density.

Perhaps the most direct examples of the advantage of increased gene product have come when much simpler bacterial and yeast cultures have been challenged with a new substrate. The most common result is an amplification of existing genes which produce an enzyme that is inefficient but has some activity on the novel substrate (Wilson *et al.*, 1975; Jensen, 1976). Tolerance to the herbicide glyphosate has also arisen in plant cell cultures due to a duplication in the gene producing the sensitive protein EPSP synthase (Shah, 1986).

Increased heterozygosity

The presence of duplicated genes can enhance physiological or developmental homoeostasis if two genes with distinct properties are maintained (Barber, 1970; Manwell and Baker, 1970). Multiple enzyme forms might minimize variation in substrate affinity or provide catalytic properties adapted to a broader range of environmental conditions. For example, if one gene produced an enzyme that had high activity under one type of condition and another was most active under another set of conditions, the species might survive under a broader set of environments than would be possible with only one gene.

Polyploids will have higher levels of heterozygosity than their diploid progenitors unless they were produced by somatic doubling. Allopolyploids often carry the divergent alleles of their progenitors on non-pairing homologous chromosomes (Chapter 1). Such ‘fixed heterozygosities’ are maintained indefinitely except when heterogenetic (non-homologous) pairings occur. Fixed heterozygosities appear to be the norm in allopolyploid species for a wide range of electrophoretically detectable enzymes (Gottleib, 1982). In the classic study of Roose and Gottleib (1976), additive patterns of electrophoretic phenotypes were found for 11 loci in *Tragopogon* diploid-tetraploid pairs (Table 4.2). At all these loci, the polyploid species carried the bands of both the progenitors.

Autopolyploids do not carry fixed heterozygosities, but a high percentage of their parents’ heterozygosity is transferred to them via unreduced gametes, and polysomic inheritance maintains a greater amount of heterozygosity than disomic inheritance. As discussed in the first chapter, this occurs because more than two alleles constitute a locus and polysomic inheritance generates fewer homozygotes each generation than disomy. For example, selfing of a heterozygous autotetraploid with the genotype AAaa will produce 94% het-

Table 4.2. Additive electrophoretic phenotypes of diploid and tetraploid *Tragopogon* (Roose and Gottleib, 1976). Enzymes are identified by their migration from the origin. The parental species are represented by arrows.

Gene	Species and ploidy				
	2x *porrifolius*→	4x *mirus*	2x ←*dubius*→	4x *miscellus*	2x ←*pratensis*
EST-2	59	54/59	54	54/57	57
EST-3	53	44/53	44	–	–
EST-4	40	35/40	35,40	–	–
LAP-1	30	30/32	32	32/35	35
LAP-2	29	25/29	25	25/27	27
APH	52	52/55	55	49/55	49
GDH	16	16/21	21	–	–
G6PD	29	29/31	31	–	–
ADH-1	35	30/35,35/40	30,40	30/35	35
ADH-2	–	–	30	18/30	18
ADH-3	–	–	15	5/15	5

erozygous progeny, while a selfed diploid genotype of Aa will produce 50% heterozygous progeny (Table 4.3). Genetic variation has been compared in five native species of diploids and their autopolyploid derivatives, and in all cases higher levels of heterozygosity and mean number of alleles per locus were found in the autopolyploids than in the diploids (Table 4.4).

Table 4.3. Expected segregation groups of balanced heterozygotes in diploids, tetraploids and allopolyploids. The two variant alleles (A,a) are considered fixed on non-homologous chromosomes in the allopolyploid.

Parental genotype	Possible gametes	Ratio	Gametic frequency	Zygotic phenotypes in F_2	Zygotic frequencies	Heterozygote frequency
Diploid						
Aa	A	1	0.50	AA	0.25	
	a	1	0.50	Aa	0.50	0.50
				aa	0.25	
Allotetraploid						
AA aa	Aa	1	1.00	AA aa	1.00	1.00
Autotetraploid						
AAaa	AA	1	0.17	AAAA	0.03	
	Aa	4	0.66	AAAa	0.22	
	aa	1	0.17	AAaa	0.50	0.94
				Aaaa	0.22	
				aaaa	0.03	

Table 4.4. Genetic variation (mean values) in diploid (2n) and autotetraploid (4n) populations (from Soltis and Soltis, 2000).

	P		H		A	
Species	2n	4n	2n	4n	2n	4n
Tolmiea menziesii	0.240	0.408	0.070	0.237	3.00	3.53
Heuchera grossulariifolia	0.238	0.311	0.058	0.159	1.35	1.55
Heuchera micrantha	0.240	0.383	0.074	0.151	1.14	1.64
Dactylis glomerata	0.700	0.800	0.170	0.430	1.51	2.36
Turnera ulmifolia						
var. *elegans*	0.459	0.653	0.11	0.420	2.20	2.56
var. *intermedia*	0.459	0.201	0.11	0.070	2.20	2.00

P, proportion of loci polymorphic; H, observed heterozygosity; A, mean number of alleles per locus.

The initial level of heterozygosity transmitted via unreduced gametes is dependent on the process of 2n-gamete formation. There are a number of events that can result in unreduced gametes (Hermsen, 1984), but the most common are first division restitution (FDR) and second division restitution (SDR). In FDR, homologous chromosomes are not separated during meiosis I, while in SDR sister chromatids remain together in the same gamete due to incomplete meiosis II. Figure 4.5 illustrates the genetic consequences of FDR and SDR with no genetic crossing over. The reduction cell wall (R) is represented by a horizontal line, while the equatorial cell wall (E) is represented by a vertical line. It can be seen in the figure that normal meiosis produces a tetrad with four 1n-gametes, FDR leads to a dyad containing two, genetically identical 2n-gametes and SDR yields a dyad containing two, genetically different 2n-gametes. FDR transmits much more parental heterozygosity to the progeny than SDR because each gamete gets a combination of each parental chromosome rather than just one. Normal crossing over and gene recombination complicate this simplistic picture, but FDR still transmits more heterozygosity than SDR. Assuming one crossover per chromosome, it has been calculated that FDR transmits 80.2% of the parental heterozygosity to the gametes and SDR transmits 39.6% (see Hermsen, 1984, for details). Cytogenetic studies of FDR and SDR have been undertaken in a wide range of species, including potato (Mok and Peloquin, 1975; Mendiburu and Peloquin, 1977; Douches and Quiros, 1988a), lucerne (Vorsa and Bingham, 1979) and maize (Rhodes and Dempsey, 1966).

Gene duplication can also act to relax stabilizing selection such that new enzyme properties might arise (Lewis, 1951; Stephens, 1951; Ohno, 1970). When there is more than one copy of a gene, an organism can survive an accumulation of mutations in one gene ('forbidden mutations') if the other remains functional. Such mutations could result through random events in the 'silencing' of genes or the evolution of unique adaptive properties or novel functions.

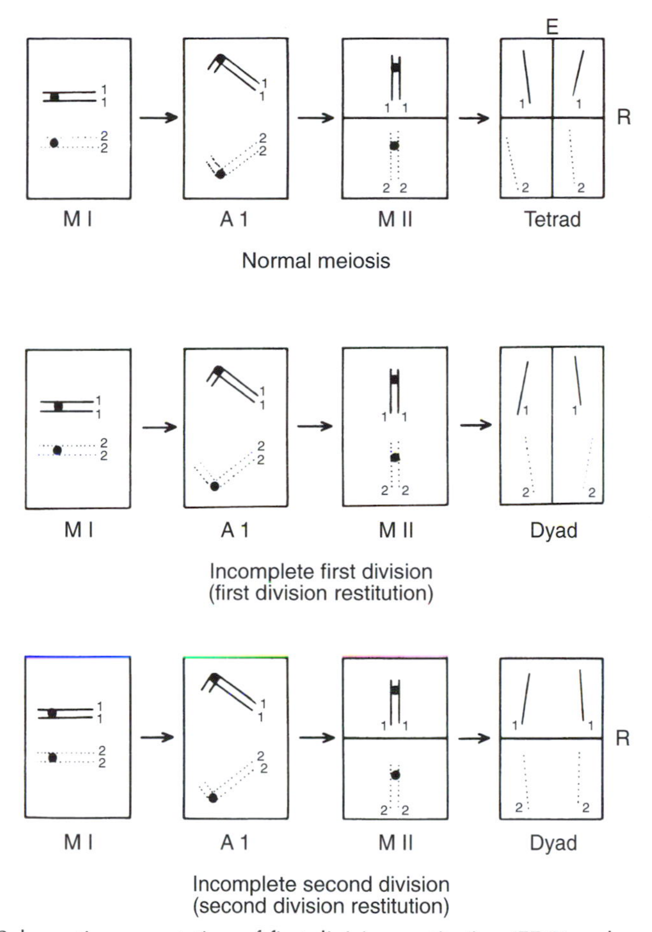

Fig. 4.5. Schematic presentation of first division restitution (FDR) and second division restitution (SDR) assuming one pair of homologous chromosomes and no crossing over (used with permission from J.G. Hermsen, © 1984, *Iowa State Journal of Research* 58, 421–436).

The evidence for beneficial mutations after gene duplication is quite limited. Isozymes generally show little variation in their catalytic properties when compared under laboratory conditions (Roose and Gottleib, 1980; Weeden, 1983), although a few studies have uncovered differences that might have an adaptive benefit. Chickpea alcohol dehydrogenase (ADH) isozymes vary in their heat and acid stability, substrate specificity and inhibitor sensitivity (Gomes *et al.*, 1982). One of the duplicated ADH isozymes of *Gossypium arboreum* is less sensitive to salt than the others (Hoisington and Hancock, 1981). Members of the chalcone synthase (CHS) gene family appear to have diverged in function prior to diversification of the *Asteraceae* (Durbin *et al.*, 1995; Huttley *et al.*, 1997).

Some duplicated genes are differentially expressed in response to developmental or environmental variation (Scandalios, 1969) and the heterogeneous appearance of these isozymes may signal the evolution of specific adaptations to divergent cellular environments. For example, different ADH loci are expressed in the seed, pollen and root tissue of a variety of species (Weeden, 1983). One of the ADH loci is commonly induced by anaerobic conditions associated with flooding while the other loci are not (Freeling, 1973; Roose and Gottleib, 1980). A number of duplicated genes also have distinct subcellular locations even though they all originate from nuclear genes (Table 4.5). These enzymes may have specific adaptations to their subcellular environments (Weeden and Gottleib, 1980; Gottleib, 1982). Gaut and Doebley (1997) have identified a pair of duplicated genes in maize, R and B, that differentially regulate purple pigmentation in variant maize tissues.

Unique intra-allelic interactions might also appear after gene duplication which are of adaptive significance. In the case of 'multimeric' enzymes with more than one subunit, novel hybrid molecules can arise which have unique heterotic properties. Gottleib L.D. (1977a) described such a situation in *Clarkia,* where a duplication in phosphoglucoisomerase (PGI) produced a

Table 4.5. Enzymes of plants whose isozymes are located in different subcellular fractions (from Gottleib, 1982; Newton, 1983).

Enzyme	Location
Amylase	Cytosol and plastid
Carbolic anhydrase	Cytosol and plastid
Enolase	Cytosol and plastid
Fructose-1, 6-diphosphate	Cytosol and plastid
Glucose-6-phosphate dehydrogenase	Cytosol and plastid
Glutamate-oxaloacetate transaminase	Cytosol, mitochondria, plastid and microbody
Glutamine synthetase	Cytosol and plastid
Malate dehydrogenase	Cytosol, mitochondria, plastid and microbody
Malic enzyme	Cytosol and plastid
Phosphofructose isomerase	Cytosol and plastid
Phosphoglucomutase	Cytosol and plastid
Phosphoglucoisomerase	Cytosol and plastid
Phosphoglycerate mutase	Cytosol and plastid
6-Phosphogluconate dehydrogenase	Cytosol and plastid
3-Phosphokinase	Cytosol and plastid
Pyruvate dehydrogenase	Mitochondria and plastid
Pyruvate kinase	Cytosol and plastid
Ribulose-5-phosphate epimerase	Cytosol and plastid
Ribulose-5-phosphate isomerase	Cytosol and plastid
Superoxide dismutatse	Cytosol, mitochondria and plastid
Transaldolase	Cytosol and plastid
Transketolase	Cytosol and plastid
Triophosphate isomerase	Cytosol and plastid

hybrid molecule with a maximum velocity that was higher than the parental forms (Table 4.6). Such a change could potentially have a cascading effect on the whole metabolism.

It seems likely that most of the mutations that arise after gene duplication will be neutral or result in a loss of function (Otto and Whitton, 2000). There may even be a threshold where the energetic cost of replicating and producing unnecessary enzymes becomes a selective disadvantage. The energetic costs of a few duplicated genes might not be substantially deleterious, but the large scale duplications associated with aneuploidy and polyploidy could have major effects. If chromosomal doubling itself does not result in regulating disruptions, 'dosage compensations' may arise as evolution progresses to fit the precise physiological requirements of the species involved (Aragoncillo *et al.*, 1978).

Duplicated genes in plants often remain active after polyploidization for long periods of evolutionary time (Hart, 1988; Gottleib, 1982; Crawford, 1990); however, there are a number of instances where dosage compensations or gene silencing have been documented. Several null or non-active allozymes were found in tetraploid *Chenopodium* (Wilson, 1981; Wilson *et al.*, 1982). An allozyme at a duplicated phosphoglucoisomerase loci was described by Gottleib and Greve (1981) which had a much lower affinity for its substrate. Hoisington and Hancock (1981) found the recently formed tetraploid *Hibiscus radiatus* to have enzyme and protein levels very similar to the sum of its progenitors, while the more ancient species *Hibiscus ace-*

Table 4.6. Biochemical properties of phosphoglucoisomerase (PGI) enzymes from two species of *Clarkia*. PGI is a dimeric enzyme with two subunits. PGI-2B3A is a hybrid molecule composed of one subunit from the ancestral locus PGI-2 and one from the duplicated locus PGI-3, while the other three enzymes are composed of only one type of subunit from a single locus (PGI-1, PGI-2B and PGI-3A). Note that the hybrid molecule PGI-2B3A has a higher maximum velocity (V_{max}) than the parental forms. (From Gottleib, L.D., 1977.)

Enzyme	K_m (mM)	V_{max} (μmol/min)	E_a (kcal/mol)
C. xantiana			
PGI-1	0.29	33.8	9.74
PGI-2B	0.12	59.3	10.58
PGI-2B3A	0.33	83.2	11.19
PGI-3A	1.12	55.5	12.43
C. rubicunda			
PGI-1	0.37	65.8	
PGI-2	0.17	147.0	

K_m, Michaelis constant (substrate affinity); E_a, energy of activation.

tosella had several lower levels. In hexaploid wheats, high-molecular-weight glutenin bands are produced from only two of the three genomes constituting this hexaploid (Feldman *et al.*, 1986). Gottleib and Higgins (1984) discovered that *Clarkia* species with and without a duplication of phosphoglucoisomerase had the same levels of cytosolic activity. Roose and Gottleib (1978) also found that the number of genes coding for electrophoretically detectable enzymes remained constant in seven diploid species of *Clarkia* even though their chromosome numbers ranged from 2n = 6 to 2n = 12.

Many more examples of silenced genes may emerge as more genomes are comprehensively sequenced and their gene expression patterns examined. When Comai *et al.* (2000) generated artificial allotetraploids of *Arabidopsis thaliana* and *Cardaninopsis arenosa* and compared gene expression in the diploids to the derived polyploid, they found that about 0.4% of the genes were silenced in the polyploid. In an examination of the *PgiC2* gene family of *Clarkia mildrediae*, 18 exons were sequenced and nine of these contained insertions or deletions that resulted in frame shifts and truncation (Gottleib and Ford, 1997). A single mutation in one of six duplicate CHS genes was found to block anthocyanin production in the floral limb (Durbin *et al.*, 2000). Genes appear to be silenced after polyploidization via several mechanisms, including methylation, spontaneous mutation and the movement of transposable elements (Wessler, 1998; Comai, 2000; Comai *et al.*, 2000; Wendel, 2000).

Genetic bridge

Repeated cycles of hybridization and polyploidization could also add to the amount of variability found in a polyploid group. Any time a new polyploid individual is formed through the unification of unreduced gametes, there is the potential that new combinations of genes will be injected into the polyploid species. The range of a polyploid species might also be increased directly as it comes in contact with diploid species that produce compatible unreduced gametes. Such introgressions have not been directly documented, but we do know that numerous polyploids have multiple origins, including species of *Tragopogan*, *Tolmiea* and *Triticum*. Much of the extensive variability found in these groups could have been generated by hybridization of polyploids of separate origin and subsequent recombination.

One important evolutionary ramification of genomic duplication is when the polyploids act as a 'genetic bridge' between diploid species (Dewey, 1980). In some cases, two diploids that have limited interfertility at the diploid level may produce an allopolyploid via unreduced gametes. It can then back-cross with the progenitors whenever unreduced gametes from the diploids come in contact with reduced gametes of the allopolyploid. Two newly emerged polyploid types might also hybridize at higher levels than their diploid progenitors (Rieseberg and Warner, 1987). This would afford the allopolyploid more genetic variability than either of the diploids, since it has access to both gene pools. Numerous examples of polyploid bridges have been described in *Gilia* (Grant, 1971), *Bromus* (Stebbins, 1956) and *Clarkia* (Lewis and Lewis, 1955).

Perhaps the classic example of this phenomenon is found in *Aegilops* (Zohary, 1965). Up to 22 wild species of wheat are located in south-western Asia and the Mediterranean basin. Diploids contain at least six genomic groups, which are sexually isolated, and the polyploids group into three clusters of species, each possessing a common 'pivotal genome' (Kihara, 1954; Kihara *et al.*, 1959). Each of the polyploid species clusters also has a common spikelet type. The weedy polyploid *Aegilops* species overlap in their geographical distribution and those with similar pivotal genomes produce partially fertile F_1 hybrids. Zohary and Feldman (1962) and Feldman (1963) examined the linkages between *Aegilops* tetraploids in Israel, Turkey and Greece and found natural hybrids between the seven polyploids carrying the C genome (Fig. 4.6). They are male-sterile, but are repeatedly exposed to parental pollen such that some back-crossed seed is produced. This seed produces vigorous plants that are relatively fertile and, when selfed, they generate an extremely variable progeny. These introgressed genotypes can become genetically fixed due to self-pollination and may act to enlarge the gene pool and ecological amplitude of the species involved.

In many cases it appears that, after the initial polyploidization and diversification, additional episodes of polyploidy occurred to generate what are called polyploid complexes (Grant, 1971; Stebbins, 1971). In the example above, the *Aegilops* complex is composed of the various C, M and S genome species and their overlapping allotetraploid derivatives. The genera *Aegilops* and *Triticum* form an expansive polyploid complex involving dozens of polyploid species at three ploidy levels (Chapter 8). Among the cultivated species, the bread wheat *Triticum aestivum* is an allohexaploid containing three genomes (AABBDD), the durum and emmer wheats *Triticum turgidum* are allotetraploids (AABB) and the einkorn wheat *Triticum monococcum* is a diploid (AA). Similar wide-ranging complexes exist for oats and rice (Chapter 8).

One particularly elegant example of a polyploid complex has been described in Australian species of soybean. *Glycine tomentella* is a perennial whose range is anchored in Australia but which extends into Timor, New Guinea, the Philippines and Taiwan. It is cytogenetically quite complex and is composed of diploids, tetraploids and aneuploids (2n + 38, 40, 78 and 80) (Newell and Hymowitz, 1978). The diploids contain several races (D1–D6) that are reproductively isolated from each other and the other species (Brown *et al.*, 2002). Isozyme and histone H3-D sequences indicate that some diploid races have more than one origin and have undergone lineage recombination. Phylogenetic and network analysis of alleles from diploids and polyploids revealed that polyploid *G. tomentella* is a complex composed of several diploid genomes (T1–T6) and, in many cases, the diploid genomic origins of the vari-

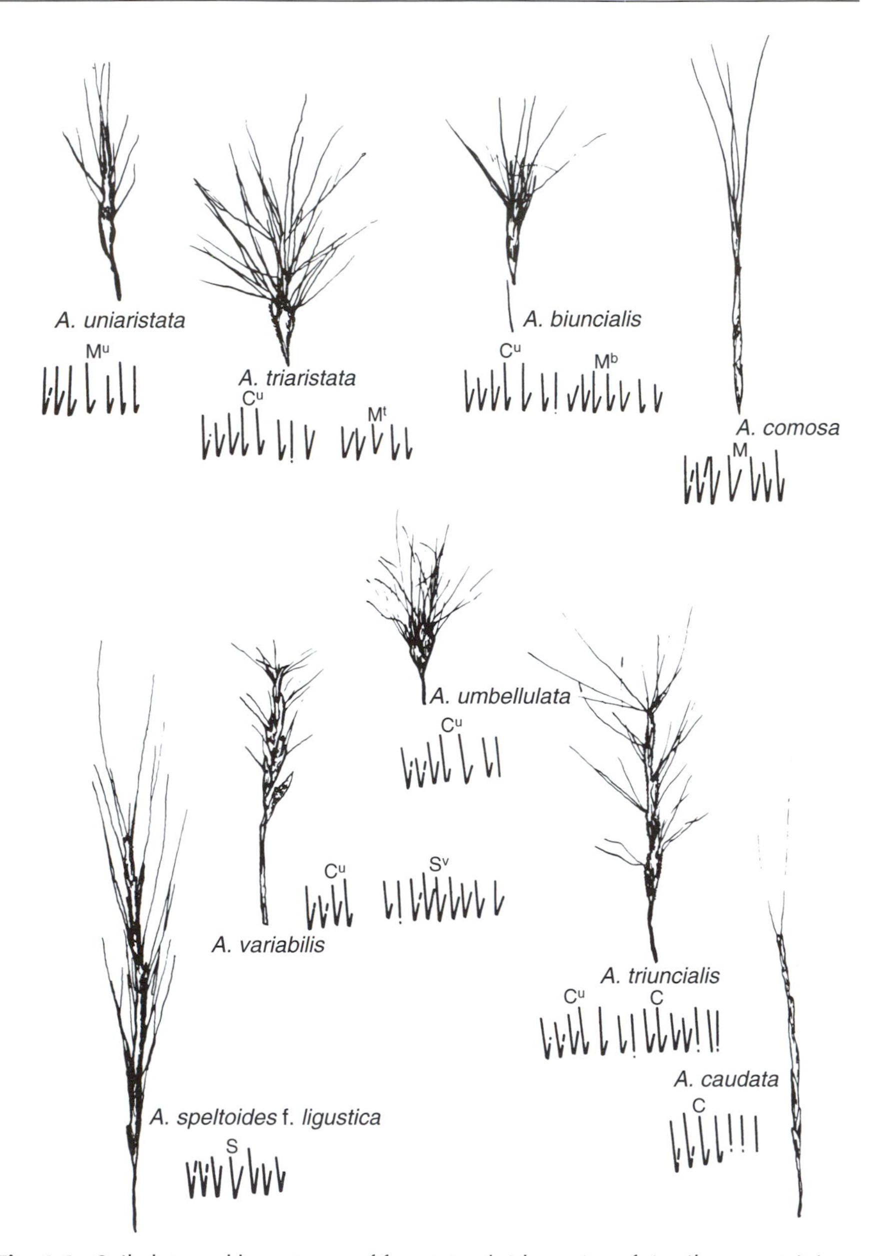

Fig. 4.6. Spikelets and karyotypes of four tetraploid species of *Aegilops* containing the pivotal genome C^u of *A. umbellulata* (centre) and of the four modern diploid species which contain counterparts of the other ancestral genomes: M^u, *A. uniaristata* (top left); M, *A. comosa* (top right); S, *A. speltoides* f. *ligustica* (bottom left); C, *A. caudata* (bottom right). The tetraploids are as follows: C^uM^t, *A. triaristata*; C^uM^b, *A. biuncialis*, C^uS^v, *A. variabilis*; C^uC, *A. triuncialis*. (Used with permission from G.L. Stebbins, © 1971, *Chromosomal Evolution in Higher Plants*, Addison-Wesley Publishing Company, Reading, Massachusetts.)

ous polyploid races could be traced to specific diploid races (Fig. 4.7; Doyle *et al.*, 2002). *Glycine tabacina* has two genomic origins. The polyploid races are thought to have arisen within the last 30,000 years, during the human occupation of Australia and subsequent environmental disruption of the habitat (Doyle *et al.*, 1999).

The origin of agriculture and the subsequent domestication process stimulated the development of many polyploid complexes. The hexaploid wheats are thought to have originated when cultivated tetraploids hybridized with wild populations of diploids. The banana, *Musa acuminata* (A genome), was first cultivated in the Malay region of South-East Asia and as its cultivation spread north it hybridized with another species, *Musa balbisiana* (B genome) forming an array of edible genomic assemblages (AAB, ABB, AAAB, AABB and ABBB) (Chapter 10). Repeated cycles of hybridization and polyploidy have played an important role in the development of sugar cane, potato and sweet potato (Chapter 10).

Genetic Differentiation in Polyploids

In spite of the many potential advantages associated with polyploidy outlined above, polyploidy has long been considered a conservative rather than a progressive factor in evolution (Stebbins, 1950, 1971; Grant, 1971). Polyploids were considered an 'evolutionary dead end' doomed to decline in importance because they had less capacity for genetic differentiation than diploids. It was thought that the presence of multiple alleles reduced the effect of single alleles and that genetic differentiation was further restricted by the lack of segregation due to fixed heterozygosities in allopolyploids and the reduced rate of segregation due to tetrasomic inheritance in autopolyploids. Stebbins (1971) provided a number of examples where polyploid complexes appeared to have gone through a pattern of growth and demise. In these examples, the polyploids were initially much rarer than their progenitors; then they predominated, and eventually their importance diminished to the point where they become relictual.

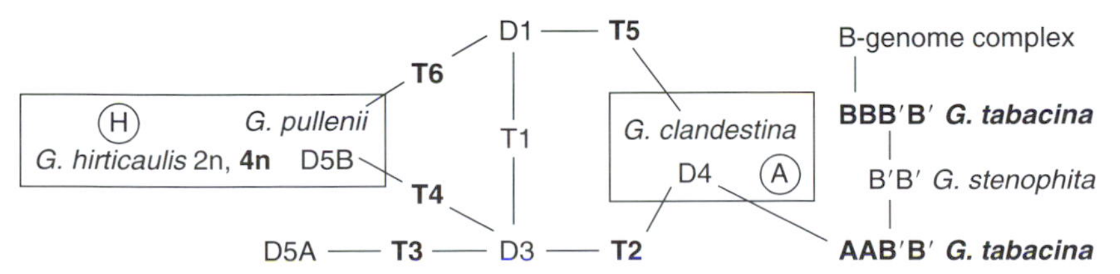

Fig. 4.7. The *Glycine* subgenus *Glycine* polyploid complex. Lines connect tetraploids with diploid progenitors. All taxa lacking names are classified as *G. tomentella*. Boxed taxa marked Ⓐ are members of the A genome group, and Ⓗ are members of the H genome species. (Used with permission from J.J. Doyle *et al.*, © 2002, *Evolution* 56, 1388–1402.)

This conservative opinion about the evolutionary potential of polyploids has dramatically changed in recent years. Polyploids may indeed evolve more slowly than diploids due to the buffering effect of multiple alleles, but they may actually have a broader adaptive potential. There is a greater dose span between additive alleles in polyploids, allowing for a greater range in phenotype (Table 4.7), and the higher levels of genetic variability generally found in polyploids allow for more possible assortments of genes. A number of polyploid species have been shown to have undergone substantial genetic differentiation. The oat populations used as an example of diversifying selection in the second chapter are in fact hexaploid (Hamrick and Allard, 1972). Smith (1946) observed considerable morphological variation in tetraploid *Sedum*, while Adams and Allard (1977) found a large amount of isozyme diversity among hexaploid individuals of *Festuca microstachys* (Fig. 4.8). Among Californian populations of strawberries, octoploids have a greater ecological range than their diploid progenitor *Fragaria vesca* and they show greater divergence in most morphological traits (Table 4.8). Bringhurst and Voth (1984) were able to increase cultivated strawberry yields by 500% over 25 years by artificial selection even though the crop was an octoploid.

Table 4.7. Potential allelic dose span in diploids, tetraploids and hexaploids. It is assumed that the duplicated loci carry two alleles (A, a).

	Number of A alleles						
Ploidy	0	1	2	3	4	5	6
Diploid	AA	Aa	aa				
Tetraploid	AAAA	AAAa	AAaa	Aaaa	aaaa		
Hexaploid	AAAAAA	AAAAAa	AAAAaa	AAAaaa	AAaaaa	Aaaaaa	aaaaaa

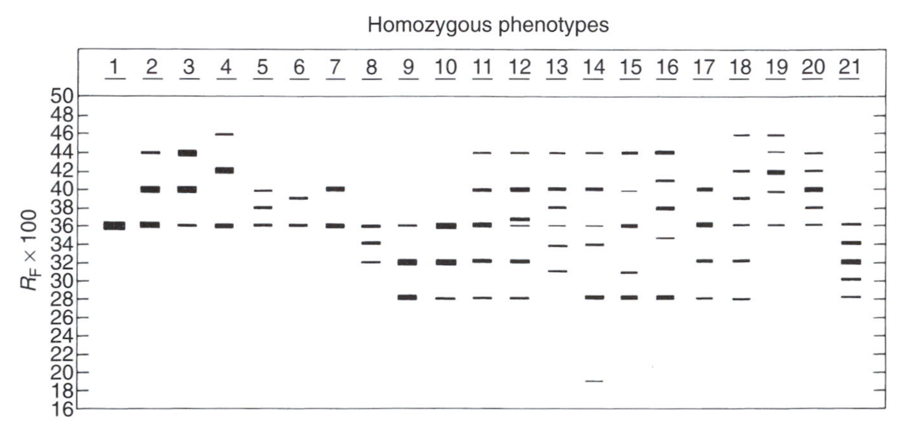

Fig. 4.8. Homozygous PGI phenotypes found in hexaploid *Festuca microstachys* (used with permission from W.T. Adams and R.W. Allard, © 1977, *Proceedings of the National Academy of Sciences USA* 74, 1652–1656).

Similar dramatic improvements have been made in many other polyploid crops by plant breeders.

Even in species propagated primarily through asexual means, each newly emergent type can potentially contain new gene combinations that are of adaptive or horticultural importance. These hybrid derivatives can then be propagated indefinitely through shoot cuttings, ramets or tubers. Since triploid bananas are sterile, varietal development has been very dependent on the repeated formation of unique polyploids. In taro, numerous cultivated races have been selected from the wild even though they rarely flower (Kuruvilla and Singh, 1981). Multiple hybridizations and polyploid formation have played an important role in the development of many other asexually propagated species, including cassava, sweet potato, white potato and yam.

Chromosomal Repatterning

A major source of novelty in polyploids may come through genome rearrangement and gene silencing after polyploidization, as was mentioned before. Grant (1966) was the first to experimentally demonstrate this possibility when he produced a hybrid of tetraploid *Gilia malior* × *Gilia modocensis* that had seed fertility of 0.007%. He selfed the lines for 11

Table 4.8. Morphological and environmental ranges of diploid and octoploid *Fragaria* in California. A value less than 1.0 indicates that the octoploids had a greater range. (From Hancock and Bringhurst, 1981.)

Character	Range 2x/8x	Character	Range 2x/8x
Environmental		Morphological	
January temperature	0.85	Stolon width	0.52
April temperature	0.79	Stolon numbers	1.59
July temperature	0.74	Branch crowns	1.59
Rainfall (mm)	0.95	Petal index	0.82
pH	0.71	Petal area	0.21
Salinity (ppm)	0.52	Peduncle length	1.52
% carbon	0.91	Flower number	0.40
% sand	0.82	Trichome number	0.59
% silt	0.78	Fruit index	1.19
% clay	0.96	First flower	0.86
		Last flower	0.88
		Flowering period	0.65
		Achene weight	0.42
		Fruit weight	0.05
		Leaf area	1.15
		Leaf index	0.55
		Petiole length	1.06
		Sclerophylly	1.93

generations and was able to isolate a highly self-fertile plant that was reproductively isolated from its parents and had a unique combination of parental traits, presumably through chromosomal translocation. More recent work utilizing molecular markers has indicated that DNA sequence elimination may be a major, immediate response to allopolyploidization in at least *Brassica* and *Triticum/Aegilops*, and in many cases duplicated genes may even be silenced through DNA methylation (Eckardt, 2001).

Song *et al.* (1995) produced reciprocal hybrids between the diploids *Brassica rapa* and *Brassica nigra*, and *B. rapa* and *Brassica oleraceae*. The F_1 individuals were colchicine doubled and progenies were generated to the F_5 generation by selfing. They then conducted a restriction fragment length polymorphism (RFLP) analysis of F_2 and F_5 individuals of each line, using 89 nuclear DNA probes, and found substantial genomic alterations in the F_5 generation, including losses of parental fragments and gains of novel fragments (Fig. 4.9). Almost twice as much change was observed in the combinations involving the two most distant relatives, *B. rapa* and *B. nigra* (Table 4.9), and they observed more change in some nuclear/cytoplasmic combinations than others.

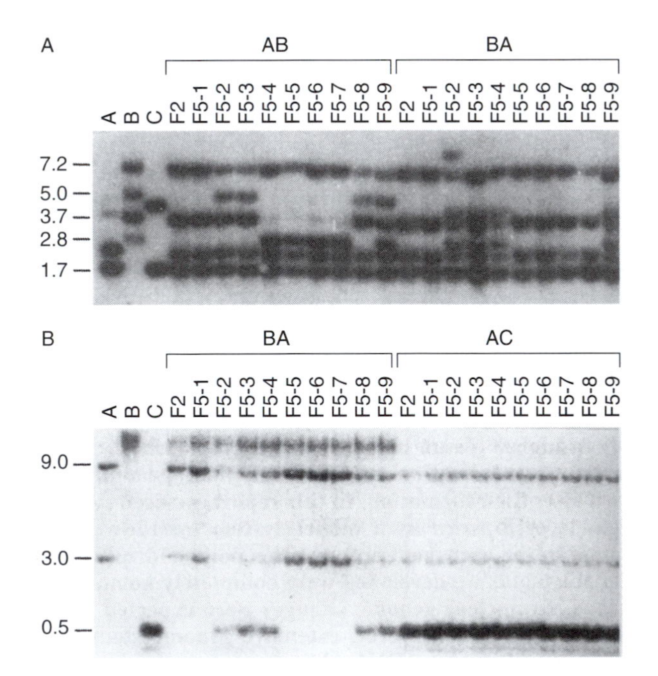

Fig. 4.9. Nuclear RFLP patterns of *Brassica rapa*-A, *Brassica nigra*-B, *Brassica oleraceae*-C, F_2 hybrids between them and F_5 populations. (A) *Hin*dIII-digested DNAs, probed with EZ3, which show a loss of fragments and a gain of fragments in some F_5 plants (5.0 kb and 2.8 kb). (B) *Hpa*II-digested DNAs probed with EC3C8 showing a gain of a 0.5 kb fragment in five BA F_5 plants, which does not exist in either the A or B parental genome, but which is present in the C genome parent and all AC F_5 plants. (Used with permission from K. Song *et al.*, © 1995, *Proceedings of the National Academy of Sciences USA* 92, 7719–7723.)

Table 4.9. Frequencies and types of genomic changes in F_5 progenies of synthetic polyploids of *Brassica* compared with their parents (modified from Song *et al.*, 1995).

	Types of fragment changes				
	Loss/gain of fragments in F_5[b]			Fragments gained in F_2[c]	Fragments found only in F_5[d]
Polyploid line[a]	A	B	C		
AB F_5	9/13	25/12		9	19
BA F_5	8/12	14/0		5	51
AC F_5	7/1		19/4	4	1
CA F_5	15/1		16/5	3	4

[a] A = *B. rapa*, B = *B. nigra*, C = *B. oleraceae*.
[b] Loss = fragments present in diploid parent and the F_2 but not present in F_5 plants; gain = diploid parental fragments absent in F_2 plants but present in F_5; A, B, and C = fragments specific to the various parents.
[c] Fragments found in the F_2 but not in either parent.
[d] Fragments found in F_5 plants but not in the diploids or F_2s.

In the work on wheat, Feldman *et al.* (1997) began by examining RFLP patterns in natural diploid and allopolyploid species. They used 16 probes that were from low-copy, non-coded DNA. Nine of these probes were found in all the diploid species, indicating that they were conserved, but, when they examined aneuploid and nullisomic lines, they found that each sequence was only retained in one of the allopolyploid genomes. In follow-up work, Liu *et al.*, (1998a,b) examined RFLP profiles of both coding and non-coding sequences in synthetic tetraploid, hexaploid and octoploids of *Triticum* and *Aegilops* that had been selfed for three to five generations. They obtained similar results to Feldman *et al.* (1997), observing non-random sequence elimination in all the allopolyploids studied, along with the occasional appearance of unique fragments. They also found that some of the changes were brought about by DNA methylation. By comparing crosses with and without the PH1 gene, which regulates bivalent pairing, they were able to deduce that intergenomic recombination did not play a role in the sequence change, as both types of crosses yielded about the same amount of change.

In two further studies, the Feldman group found that the direction of sequence change in wheat followed a different pattern from that observed by Song's group in *Brassica*, and confirmed that some sequences were silenced by elimination, while others were silenced through methylation. Ozkan *et al.* (2001) analysed diploid parental generations, F_1 progeny and the first three generations (S_1, S_2 and S_3) of synthetic hybrids of several species of *Aegilops* and *Triticum*. When they followed the rate of elimination of eight low-copy DNA sequences, they found in contrast to Song *et al.* (1995) that sequence elimination began earlier in the synthetic allopolyploidal whose species composition most closely represented natural occurring ones, and sequence elimination was not associated with cytoplasm. Shaked *et al.* (2001) used amplified fragment length polymorphism (AFLP) and methylation-sensitive amplification polymorphism (MSAP) fingerprinting to evaluate another set of diploid and tetraploid hybrids within and between genera. They also found considerable sequence elimination after polyploidy and that it occurred most rapidly in allopolyploids of the same species rather than in different species. Further analysis indicated that some of the sequences were eliminated, while others were altered by cytosine methylation.

The Feldman group has suggested that the observed sequence alterations may play a physical role in how chromosomes pair, resulting in the bivalent meiotic behaviour of newly formed allopolyploids. However, sequence losses do not appear to be necessary for bivalent pairing behaviour, as Liu *et al.* (1998) found little evidence of change in 22,000 AFLP loci in artificial hybrids of cotton, even though pairing in cotton tetraploids is strictly bivalent. It is also unknown why the most divergent genomes were altered the most in *Brassica*, while the opposite was true in wheat. Perhaps transposons play a role, with the direction and degree of perturbations being associated with the unification of genomes with or without unique mobile elements.

Genome Amplification and Chance

It is important to realize that, in all these discussions, it is assumed that alterations in gene copy number are maintained because they make significant adaptive differences. However, this may not always be the case. Some polyploids may become established simply due to chance, because unreduced gametes are continually produced and chromosome numbers can go up in a species but not down. Simple doublings of genome size may produce individuals with relatively well balanced gene complexes but going down in size will generally result in genic imbalances as individual chromosomes are lost. In only a few instances have natural polyploids been shown to produce diploid types (deWet, 1968; Ramsey and Schemske, 2003). Since most plant species produce a small frequency of unreduced gametes, a small percentage of polyploids may be continually produced and, by chance alone, some may become established. If the pressure is always towards higher ploidy and not reductions, the higher numbers will eventually accumulate.

Random processes of gene duplication and deletion can also act within genomes to produce 'multigene' families and vestigial sequences. Loomis and Gilpin (1986) have performed computer simulations that generated random duplications and deletions, and found that genome size did not become stable until the amount of 'dispensable sequences had increased to the point that most deletions affected vital genes'. It may be that genome size fluctuates greatly under 'normal' environmental conditions and only becomes important when environmental contingencies demand the amplification or alteration of specific genes.

Orgel and Crick (1980) and Doolittle and Sapienza (1980) propose that much of this amplification involves the proliferation of what they call 'selfish' or 'parasitic' DNA. Sequences like transposable elements may ensure their survival in a cellular environment by maximizing their copy number. This scenario is possible only so long as the sequences have little effect on the fitness of the total organism.

Summary

At least 50% of all angiosperm species are polyploid. The adaptive benefit of polyploidy is largely unknown, although the larger nuclear size, higher levels of heterozygosity and greater enzyme content of polyploids may play a role. Polyploids are thought to undergo less ecological differentiation than diploids because they are highly buffered genetically, but they still carry sufficient genetic variability to evolve unique ecotypes. Polyploids may act as a genetic bridge between diploid species when the polyploid accumulates genes from both progenitors and transfers them via partially fertile hybrids. Chromosomal repatterning may also be common after polyploidization. The high number of polyploids found in nature may be due in part to chance, since low levels of unreduced gametes are continually produced by many species and genomic amplifications are generally less disruptive than chromosome reductions.

Speciation 5

Introduction

Up to this point, we have been discussing primarily populations. We are now ready to move up the evolutionary ladder to species. Plants can be thought of as evolving along two dimensions, anagenesis and cladogenesis. Changes within a specific lineage or species represent anagenesis. Cladogenesis occurs when a lineage splits and the new lines begin to evolve separately. Speciation is the most fundamental cladogenetic process.

Traditionally, most evolutionists felt that lineages evolved at essentially the same rate before and after speciation. This view, called 'phyletic gradualism', assumed that evolution was a slow continuous process. Patterns of change in the fossil record have led some scientists to suggest that species may often undergo rapid changes as they come into existence, but then remain largely unchanged (Mayr, 1963; Eldridge and Gould, 1972; Gould and Eldridge, 1977). Thus, speciation occurs in spurts rather than through gradual change. This view, called 'punctuated equilibria', is one way to explain some of the abrupt discontinuities found in the fossil records. The general consensus is that punctuated patterns can fit into the framework of Darwinian evolution as long as we accept variations in rate due to environmental and genetic perturbations. Darwin himself was primarily a gradualist, but he also 'accepted the influence of both local, episodic speciation and migration' (Rhodes, 1983).

What is a Species?

Before we can begin to discuss speciation, we must first attempt to define a species. Commonly, populations that can be distinguished by prominent morphological differences are considered to be species. This 'taxonomic species concept' is often successful in identifying separately evolving groups, but in some instances can lead to artificial groupings. Quite distinct morphotypes can retain the ability to freely interbreed, making them in reality one genetic entity; and large differences in morphology are sometimes influenced by only a few genes, which do not necessarily reflect the divergence of the whole genome (Chapter 1). These ambiguities have led to frequent revisions of important crop assemblages, such as oats, rice, wheat and sorghum, as different taxonomists have reviewed the available data on natural populations.

Another way to distinguish species is by directly testing their capacity to successfully reproduce. Mayr (1942) defined the 'biological species' as 'groups of actually or potentially interbreeding natural populations, which are reproductively isolated from each other'. If taxa are reproductively isolated, they are evolving separately and therefore must have an identity of their own. This concept has gained widespread approval and is currently the most popular (Howard and Berlocher, 1998; Schemske, 2000).

While the biological species concept allows for the unambiguous delineation of species, it is still not without occasional problems. Strongly divergent groups of plants often maintain some degree of interfertility even though they differ at numerous loci and are effectively evolving on their own. As we shall discuss below, sunflower and violets provide particularly striking examples. Plants also show great ranges in fertility from obligate outcrossing to complete selfing to apomixis (uniparental). In a highly inbred or apomictic group, every individual would be a species according to the biological species concept. As we have already mentioned, many of the grain and legume species are highly inbred, and most of the starchy staples, such as banana, cassava, potato, sugar cane, sweet potato, taro and yam, are only propagated through asexual means.

Harlan and deWet (1971) developed the 'gene pool system' to deal with varying levels of interfertility between related taxa (Fig. 5.1). They recognized three types of genic assemblages:

1. Primary gene pool (GP-1) – hybridization easy, hybrids generally fertile.
2. Secondary gene pool (GP-2) – hybridization possible but difficult, hybrids weak with low fertility.
3. Tertiary gene pool (GP-3) – hybrids lethal or completely sterile.

The primary gene pool is directly equivalent to the biological species. The recognition of GP-2 and GP-3 allows other levels of interfertility to be incorporated into the overall concept of species. These are related taxa which share a considerable amount of genetic homology with GP-1, but are divergent enough to have greatly reduced interfertility. Several agronomically important groups have been described using this system, including legumes (Smartt, 1984), wheat (Fig. 5.2) and most of the other cereals (Table 5.1).

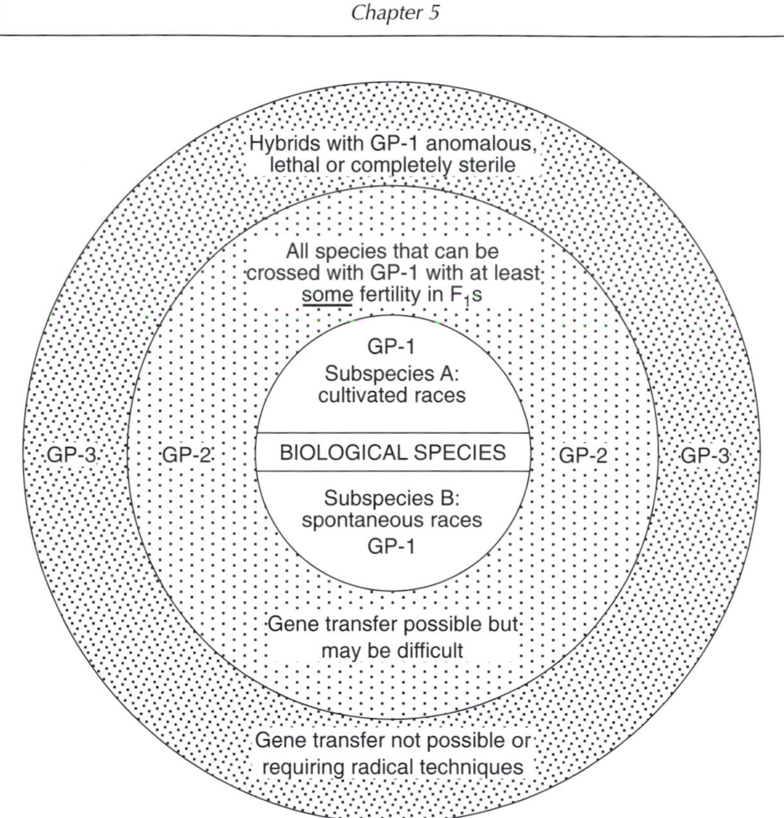

Fig. 5.1. Schematic diagram of primary gene pool (GP-1), secondary gene pool (GP-2), and tertiary gene pool (GP-3) (used with permission from J. Harlan, © 1975, *Plants and Man*, American Society of Agronomy, Madison, Wisconsin).

Simpson (1961) developed the idea of an 'evolutionary species' to minimize the problems associated with uniparental species. He suggested that a species must meet four criteria: (i) is a lineage; (ii) evolved separately from other lineages; (iii) has its own particular niche or habitat; and (iv) has its own evolutionary tendencies. This definition fits uniparental species better than the biological species concept, but we are still left to decide on what constitutes a lineage. Templeton (1989) expanded this theme by using molecular data to construct phylogenies in his 'cohesion species concept'. He defined a species as an 'evolutionary lineage, with the lineage boundaries arising from the forces that create reproductive communities (i.e. cohesive mechanisms)'.

Numerous other concepts have been developed to include ecological with reproductive criteria in defining species (Levin, 2000; Schemske, 2000). Nevertheless, it is clear that no model solves all of the potential problems in trying to define separate evolutionary units. Each has its own strengths and weaknesses. Levin (2000) suggests that 'the choice of a species concept has to do, in part, with the perspective that gives one satisfaction'. Probably the most definitive definition is the concept of biological species since it can be

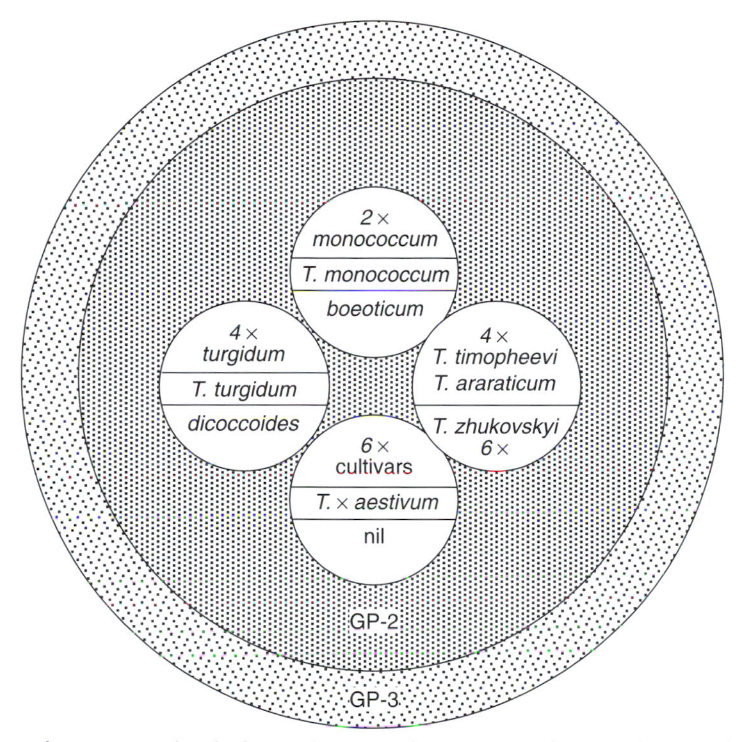

Fig. 5.2. The gene pools of wheat. The secondary gene pool is very large and includes all species of *Aegilops, Secale* and *Haynaldia,* plus at least *Agropyron elongatum, Agropyron intermedium,* and *Agropyron trichophorum.* The tertiary gene pool includes several species of *Agropyron,* several of *Elymus,* and *Hordeum vulgare.* (Used with permission from J. Harlan, © 1975, *Plants and Man,* American Society of Agronomy, Madison, Wisconsin.)

directly tested – two individuals can either successfully reproduce or they cannot. Of course, environmental and genotypic variation can cloud even this approach, but it does minimize the number of subjective judgements. As a rule of thumb, most evolutionists consider species to be those groups that are reproductively isolated, although plant scientists are more willing to accept some degree of hybridization between otherwise distinct species.

Reproductive Isolating Barriers

There are many different ways that plants can be reproductively isolated (Table 5.2). These reproductive isolating barriers (RIBs) are generally broken down into two classes: (i) premating or prezygotic mechanisms that prevent the formation of hybrid zygotes; and (ii) postmating or zygotic isolating mechanisms that reduce the viability or fertility of hybrid zygotes.

Table 5.1. Primary and secondary gene pools of the major cereals (used with permission from J. Harlan, © 1992 *Crops and Man*, American Society of Agronomy, Madison, Wisconsin).

		Primary gene pool (GP-1)			
			Spontaneous subspecies		
Crop	Ploidy level	Cultivated subspecies	Wild races	Weed races	Secondary gene pool (GP-2)
Wheats					
Einkorn	2x	*Triticum monococcum*	*Triticum boeoticum*	*T. boeoticum*	*Triticum*, *Secale*, *Aegilops*
Emmer	4x	*Triticum dicoccum*	*Triticum dicoccoides*	None	*Triticum*, *Secale*, *Aegilops*
Timopheevi	4x	*Triticum timopheevi*	*Triticum araraticum*	*T. timopheevi*	*Triticum*, *Secale*, *Aegilops*
Bread	6x	*Triticum* × *aestivum*	None	None	*Triticum*, *Secale*, *Aegilops*
Rye	2x	*Secale cereale*	*S. cereale*	*S. cereale*	*Triticum*, *Secale*, *Aegilops*
Barley	2x	*Hordeum vulgare*	*Hordeum spontaneum*	*H. spontaneum*	None
Oats					
Sand	2x	*Avena strigosa*	*Avena hirtula; Avena wiestii*	*A. strigosa*	*Avena* spp.
Ethiopian	4x	*Avena abyssinica*			
		Avena vaviloviana	*Avena barbata*	*A. barbata*	*Avena* spp.
Cereal	6x	*Avena sativa*	*Avena sterilis*	*A. sterilis; Avena fatua*	*Avena* spp.
Rices					
Asian	2x	*Oryza sativa*	*Oryza rufipogon*	*O. rufipogon*	*Oryza* spp.
African	2x	*Oryza glaberrima*	*Oryza barthii*	*Oryza stapfii*	*Oryza* spp.
Sorghum	2x	*Sorghum bicolor*	*S. bicolor*	*S. bicolor*	*Sorghum halepense*
Pearl millet	2x	*Pennisetum americanum*	*Pennisetum violaceum*	*P. americanum*	*Pennisetum purpureum*
Maize	2x	*Zea mays*	*Zea mexicana*	*Z. mexicana*	*Tripsacum* spp., *Zea perennis*

Table 5.2. Types of reproductive isolating barriers (RIBs) found in sexually reproducing organisms.

Premating – formation of hybrid zygotes is prevented
Ecogeographical – habitats are distinct and rarely come in contact
Temporal – flowering times have little overlap
Floral – flowers attract different types of pollinators
Gametic incompatibility – foreign pollen grains fail to fertilize ovules
Postmating – viability or fertility of zygotes is reduced
Hybrid unviability – hybrids are weak and have poor survival
Hybrid sterility – hybrids do not produce functional gametes
Hybrid breakdown – F_2 and back-cross hybrids have reduced viability or fertility

In ecogeographical isolation, the habitats of two species are sufficiently different for them to rarely have the opportunity to interbreed. The low-bush blueberry, *Vaccinium angustifolium*, is generally found on dry hillsides and gravelly barrens, while the high-bush blueberry, *Vaccinium corymbosum*, is located in boggy wetlands. The common bean, *Phaseolus vulgaris*, lives in warm, temperate areas in Mexico–Guatemala, in contrast to the runner bean, *Phaseolus coccineus*, which inhabits the cool, humid uplands of Guatemala. *Tradescantia canaliculata* grows on rocky slopes with full sun, while *Tradescantia subaspera* subsp. *typica* prefers rich soil and deep shade (Anderson and Hubricht, 1938). In all these cases, the species remain distinct from each other as long as their habitats are not in close proximity. When they come in contact with each other, they can hybridize and form viable hybrids that survive when an intermediate habitat is present.

Species that are isolated temporally have distinct flowering times that do not have sufficient overlap to allow for hybrid formation. The blueberry and *Tradescantia* species mentioned above are separated not only by habitat but also by bloom date (Fig. 5.3). This further reduces the production of

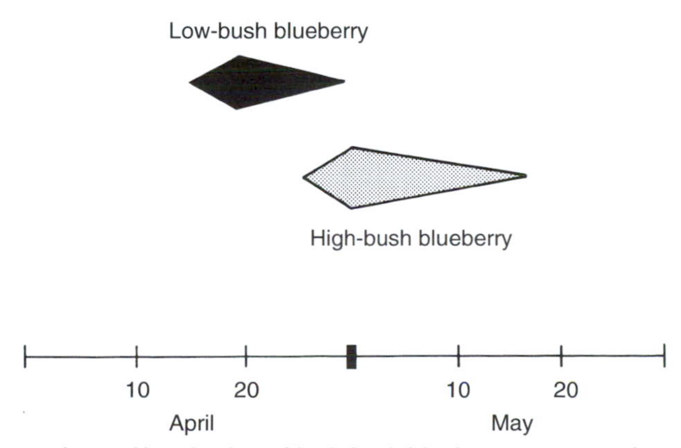

Fig. 5.3. Bloom dates of low-bush and high-bush blueberries at Otis Lake, Michigan, in 1986. The bar widths represent the proportion of individuals in full bloom on each date.

hybrids. The weedy lettuces *Lactuca canadensis* and *Lactuca graminifolia* live together in the south-eastern USA but rarely hybridize because they bloom in different seasons (Whitaker, 1944). *L. graminifolia* flowers in the early spring, while *L. canadensis* blooms in summer.

Floral isolation occurs when species have flowers that attract different types of pollinators. *Penstemon* in California have flowers of different shapes, colour and sizes that are pollinated by four completely different creatures: hummingbirds, wasps and two different sized carpenter bees. Species of columbines in western North America have distinct-looking flowers with nectar at the base of spurs that can only be reached by specific pollinators (Fig. 5.4). *Aquilegia formosa* has short, stout spurs that provide nectar for hummingbirds whose bill is about as long as the spurs. *Aquilegia longissima* and *Aquilegia chrysantha* have much longer, thinner spurs, which can be successfully harvested by hawk moths, which have a much longer proboscis. *Aquilegia pubescens* has an intermediate sized spur that hummingbird bills can barely reach so it hybridizes freely with the normally hawk-moth-pollinated *A. formosa* (Grant, 1952).

Schemske and Bradshaw (1999) directly tested the genetic basis of pollinator discrimination in bee-pollinated *Mimulus lewisii* and hummingbird-pollinated *Mimulis cardinalis.* They developed a quantitative trait locus (QTL)-based map of the key traits associated with pollinator preference and then tracked the activity of bees and hummingbirds in an F_2 population set in the field. They found that bees preferred large flowers with low pigment content, and hummingbirds favoured nectar-rich flowers with high anthocyanin content. Most remarkably, they were able to uncover a single allele that increased petal carotenoids and reduced bee visitations by 80%, and another allele that increased nectar production and doubled hummingbird visits.

One of the strongest RIBs involves gametic incompatibility, where foreign pollen grains cannot germinate on another species' stigmata, or they cannot successfully grow down the style to the ovaries. Species show a broad range of interaction, from no hint of germination to successful pollen growth but failed fertilization. Fruit-trees in the subgenus *Amygdalus* of *Prunus* freely cross, but gene transfer outside this section is generally obstructed by pollen and/or ovule sterility barriers (Chapter 11). Crosses of *Gilia splendens* × *Gilia australis* (Latimer, 1958) and *Iris tenax* × *Iris tenuis* (Smith and Clarkson, 1956) often fail because the foreign pollen tubes grow more slowly than native pollen tubes; native pollen of *Iris* species reach the ovule in 30 h, while foreign pollen takes 50 h. *I. tenuis* pollen tubes sometimes even burst in the style of *I. tenax*. Crosses between white- and purple-flowered species of *Capsicum* display a range of crossing barriers, from a lack of pollen germination to the restriction of pollen tube growth to an inability to penetrate the egg cell (Fig. 5.5).

Hogenboom (1973, 1975) suggested that such prefertilization barriers between divergent taxa often arise due to poor intergenomic coadaptation.

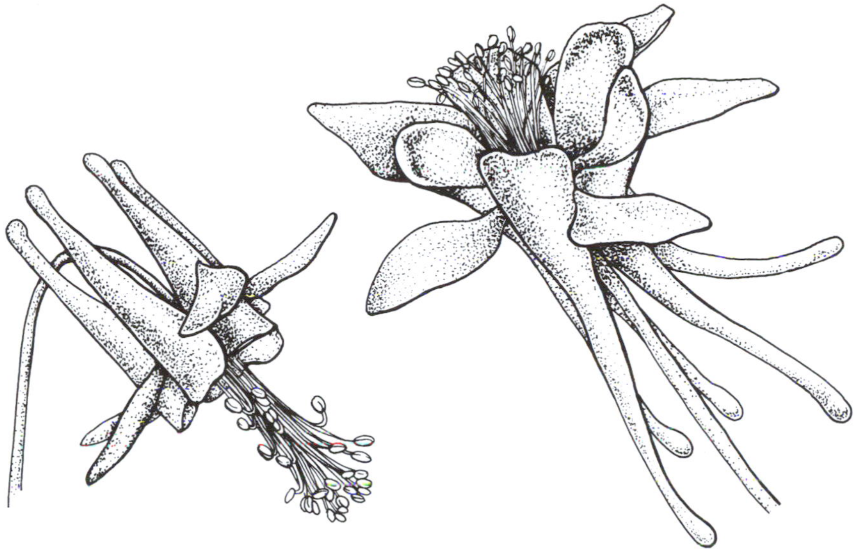

Aquilegia formosa *Aquilegia pubescens*

Species	Spur length (cm)	Primary pollinator	Bill/proboscis length (cm)
A. formosa	1–2	Hummingbirds	1–3
A. pubescens	3–4	*Celerio lineata*	3–4.5
A. chrysantha	4–7	*C. lineata*	3–4.5
		Phlegethontius sexta	8.5–10.0
A. longissima	9–13	*P. sexta*	8.5–10.0
		Phlegethontius quinquemaculatus	10.0–12.0

Fig. 5.4. Flower structure and pollinators of *Aquilegia* species in California (modified with permission from V. Grant, © 1963, *Plant Speciation*, Columbia University Press, New York).

One species may lack the genetic information necessary to properly coordinate the critical functions of the other. This phenomenon called 'incongruity', arises due to disrupted gene regulation, the absence of a gene or poor genomic–cytoplasmic interactions. A number of studies have provided circumstantial evidence of this phenomenon by showing that the vigour of interspecific crosses can be improved through subsequent breeding, presumably through selection for the most coordinated genes (Haghighi and Ascher, 1988).

In some cases, there is genotypic variation within a species such that some genotypes will cross with another species, while others cannot. Diploid species of *Vaccinium* are generally isolated from each other, but combinations can be found which produce some viable seed (Table 5.3; Ballington and Galletta, 1978). Often, the success of an interspecific cross

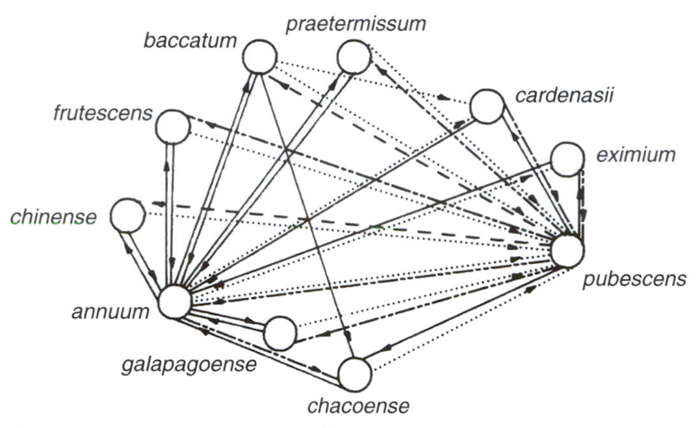

Fig. 5.5. Pollen tube growth in crosses between *Capsicum*. Arrows point in direction of female parent. The connections between species denote: (i) no penetration into the egg cell (—); (ii) barrier in ovary (–––); (iii) barrier in style (----); and (iv) barrier in stigma (.....). (Used with permission from S. Zijlstra, C. Purimahua and P. Lindhout, 1991, *HortScience* 26, 585–587.)

Table 5.3. Cross-compatibility and cross-fertility (poorest and best) among diploid *Vaccinium* species (from Ballington and Galletta, 1978).

Species hybridization		Return/100 pollinations: Fruit	Seed	Germinating seeds	Vigorous seedlings
atrococcum–caeseriense	Poorest	7	7	0	0
	Best	76	798	551	523
atrococcum–darrowi	Poorest	0	0	0	0
	Best	91	807	544	483
atrococcum–tenellum	Poorest	6	8	0	0
	Best	8	92	44	22
caesariense–darrowi	Poorest	14	27	0	0
	Best	55	277	145	109
caesariense–tenellum	Poorest	4	6	0	0
	Best	58	705	336	263
darrowi–tenellum	Poorest	14	72	4	4
	Best	76	274	228	223

is dependent on the genotype used as the maternal parent. For example, interspecific back-crosses in *Phaseolus* are much more successful when the F_1 is used as the maternal parent (Hucl and Scoles, 1985). Such variations have allowed interspecific hybridizations between otherwise strongly isolated species.

Sometimes fertility blocks arise after fertilization in the form of hybrid unviability. Here zygotes either do not develop completely, or are weak and have reduced viability. Crosses between the cultivated rices, *Oryza sativa* and *Oryza glaberrima*, produce mostly sterile hybrids. In *G. australis* × *G. splendens* (Grant and Grant, 1954) and *Papaver dubium* × *Papaver rhoeas* (McNaughton and Harper, 1960), many weak plants are produced which grow very slowly and remain dwarfed. Most of the GP-2 species described by Harlan and deWet fall into this class, along with some of those in GP-3 (Figs 5.1 and 5.2; Table 5.1). In most cases, the embryos abort before they can be successfully grown, but, sometimes, tissue culture techniques can be used to 'rescue' them. These techniques have been successfully employed by breeders in such diverse groups such as *Triticum*, *Gossypium*, *Solanum* and *Prunus* (Briggs and Knowles, 1967).

Hybrid sterilities occur when hybrids do not produce functional gametes. The basis of these sterility barriers can be genic in nature or the result of meiotic irregularities due to chromosomal imbalances. Crosses between the wild chickpea species *Cicer echinospermum* and *Cicer reticulatum* yield very few hybrids and those that are produced are sterile due to poor chromosome pairing. The weak F_1 plants of *Gilia* and *Papaver* mentioned above are generally sterile even though their chromosome pairing is regular. This sterility is probably due to genic discordances affecting normal flower development (Grant and Grant, 1954). Most hybrids of *Gilia ochroleuca* × *Gilia latifolia* are vigorous, but they are largely sterile due to poor pairing of chromosomes between the two genomes (Grant and Grant, 1960).

Interploidy crosses generally result in hybrid sterility as unbalanced sets of chromosomes are distributed to the different gametes. The pentaploid hybrids arising from the natural hybridization of the strawberries *Fragaria vesca* ($2x = 14$) × *Fragaria chiloensis* ($2x = 56$) produce a whole range of gametes containing 19–70 chromosomes with varying levels of fertility (Bringhurst and Senanayaka, 1966). Most banana cultivars are triploid and produce no viable gametes. This trait is critical to edibility, as normally pollinated diploids have flinty, teeth-breaking seeds (Chapter 10).

The final class of isolating mechanisms is hybrid breakdown, where the F_2 or back-cross hybrids have reduced viability or fertility. Low vigour is found in many of the F_2 populations of interspecific bean crosses (Hucl and Scoles, 1985). As we discussed in Chapter 3, *Zauschneria cana* × *Zauschneria septentrionalis* produces a vigorous, semifertile hybrid, but, when Clausen *et al.* (1940) examined 2133 F_2s, all were weak and dwarfish. Crosses of *Layia gaillardiodes* × *Layia hieracioides* (Clausen, 1951) and *Gilia malior* × *Gilia modocensis* (Grant, 1966) also produce high percentages of subvital, sterile F_2 individuals. A preponderance of weak *Gilia* plants is still carried in the F_6 generation. Hogenboom's concept of incongruity probably applies to these cases, since non-harmonious gene combinations are presumably segregating and producing poorly adapted individuals. Disruptive selection would have to be performed to purge the populations of these associations.

Any kind of RIB reduces the amount of gene flow between species, but combinations of mechanisms result in the tightest isolation. Most 'good' species are separated by multiple combinations of RIBs. For example, the leafy-stemmed *Gilia* of central California, *Gilia millefoliata* and *Gilia capitata,* are isolated in five ways according to Grant (1963):

> 1) Ecological isolation: *G. capitata* occurs on sand-dunes and *G. millefoliata* on flats. 2) Floral: *G. capitata* is large-flowered and bee pollinated, while *G. millefoliata* is small flowered and self-pollinating ... 3) Seasonal isolation: *G. millefoliata* blooms earlier than *G. capitata*. 4) Incompatibility: Hybrids are very difficult to produce by artificial crosses in the experimental garden. 5) Hybrid sterility: The F_1s when they can be obtained are chromosomally sterile to a high degree, producing only about 1 percent of good pollen grains and no F_2 seeds.

Another clear example of multiple RIBs is found in crosses between *P. vulgaris* and *P. coccineus*. They rarely hybridize in the first place because they have different habitats (as mentioned earlier) and, when they do cross, there is poor seed set and high seedling mortality and the surviving hybrids have a 'crippled' morphology, represented by dwarfism and abnormal development. In addition, the F_1s produce only a low proportion of viable gametes (Hucl and Scoles, 1985). In cases like these, it is highly unlikely that any successful hybridization will occur at all, even if the species are in proximity.

Modes of Speciation

New species are thought to arise through a number of different pathways (Table 5.4; Fig. 5.6). One way in which the different modes are distinguished is by the degree of separation between the speciating populations. The populations are geographically well isolated in allopatric speciation, there is no separation between populations in sympatric speciation, and the populations touch along one axis in parapatric speciation.

Table 5.4. Factors distinguishing major modes of speciation in sexual plants.

Mode of speciation	Separation	Population size	Differentiation before RIBs
Allopatric			
Geographical (type 1)	Wide	Large	Much
Peripatric (type 2)	Wide	Small	Little
Parapatric	Touching	Large	Much
Sympatric	None	One	Little

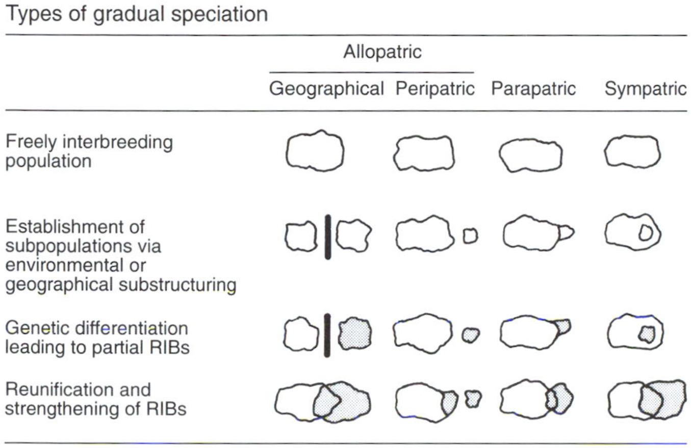

Types of gradual speciation

	Allopatric			
	Geographical	Peripatric	Parapatric	Sympatric
Freely interbreeding population				
Establishment of subpopulations via environmental or geographical substructuring				
Genetic differentiation leading to partial RIBs				
Reunification and strengthening of RIBs				

Fig. 5.6. Cartoons illustrating the different modes of gradual speciation. The drawings found below each of the individual headings (geographical, peripatric, parapatric and sympatric) represent the various stages of species development. See the text and Table 5.4 for more details.

Other important criteria used to describe speciation pathways are: (i) how large the speciating groups are; and (ii) how much genetic differentiation precedes the formation of strong RIBs. Allopatric speciation is broken into two groups – geographical and peripatric. In geographical speciation, large populations are thought to gradually diverge and form RIBs, while, in peripatric speciation, isolating barriers are thought to appear quickly in small populations without much differentiation. Sympatric speciation occurs when a single individual or small group arises that is reproductively isolated from the surrounding population. Parapatric speciation is closely related to sympatric speciation, except that the gradually diverging crowd of genotypes is on one side of the parent population rather than surrounded by it.

Numerous other speciation taxonomies have been proposed (White, 1978; Templeton, 1981), but the five depicted in Table 5.4 are the ones most commonly discussed in the evolutionary literature. We shall describe synonyms and related concepts in the text where appropriate.

Geographical speciation

The most widely recognized type of speciation is geographical speciation. In its first stage, there is a single population found in a large homogeneous environment. The environment then becomes partly diversified due to physical or biotic factors and populations become isolated. These populations begin to diverge genetically and eventually acquire sufficient variation to become reproductively isolated from each other (semispecies). Further changes in the environment allow some of the newly evolved groups to come back in contact, but they do not produce successful hybrids because of past differentiation. Natural selection against the formation of weak or sterile hybrids promotes the reinforcement of RIBs through additional differentiation.

Most of the evidence for this type of speciation is circumstantial in nature due to the slow speed of the process. Few scientists' careers are long enough to follow the whole scenario. However, populations in the various stages of speciation have been identified by individual investigators. Probably the most complete story in plants has been accumulated by Grant and Grant (1960) in their oft cited *Gilia* studies. In this group, they were able to find races and species in all the hypothesized stages of geographical and ecological differentiation (Fig. 5.7), and they were able to show a subsequent build-up of isolating mechanisms as the species became more divergent (Table 5.5).

Peripatric speciation

This type of speciation is very similar to the geographical mode except the speciating population is much smaller. Other terms used for this type of speciation are quantum (Grant, 1971), speciation by catastrophic selection (Lewis, 1962) and founder-induced speciation (Carson, 1971; Carson and

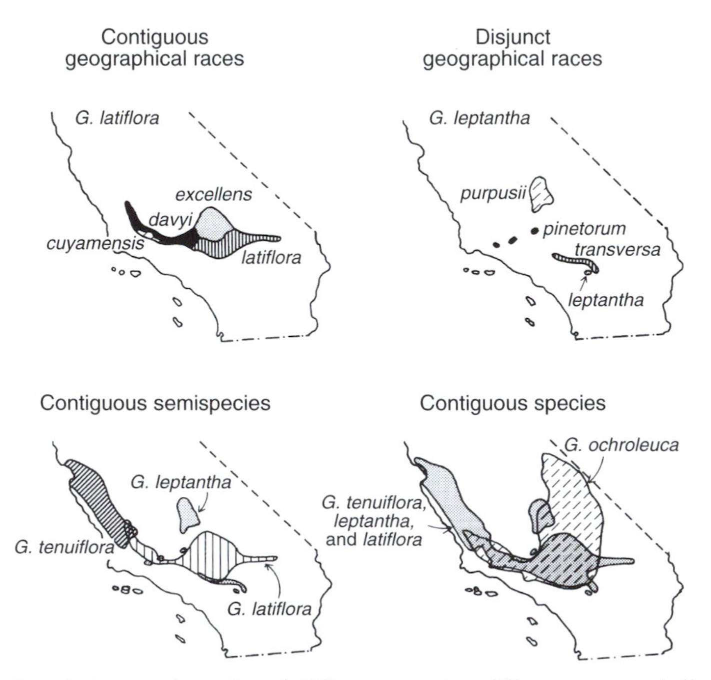

Fig. 5.7. Populations and species of *Gilia* representing different stages of allopatric speciation (redrawn with permission from V. Grant, © 1963, *The Origin of Adaptations*, Columbia University Press, New York).

Table 5.5. Relative ease of crossing diploid cobwebby *Gilia* at different levels of divergence (from Grant and Grant,1960).

Type of cross	Entities crossed	Number of flowers pollinated	Average no. plump seeds per flower	Number of hybrid individuals per ten flowers pollinated
Interindividual	Different individuals belonging to the same population	116	17.8	22
Inter-racial	Different geographical races of the same species	562	15.2	12
Interspecific	Different diploid species of cobwebby *Gilia*	2016	3.7	3
Intersectional	Diploid species of cobwebby *Gilia* with diploid species of leafy-stemmed or woodland *Gilia*	528	0.004	0.032

Templeton, 1984). Because of reduced population size, genetic drift becomes more important and the rate of speciation is accelerated. Peripatric speciation can result in reproductively isolated species that are otherwise quite similar to their progenitors or they may become morphologically quite distinct, depending on how many genes are affected. Most of the evidence for peripatric speciation is also circumstantial, although the process can occur during the lifetime of an individual investigator.

One of the most completely documented cases of peripatric speciation associated with few genetic changes concerns two *Clarkia* species in southern California (Lewis, 1962). *Clarkia biloba* is a relatively widely distributed species, while *Clarkia lingulata* is rare and is found on only two sites at the extreme edge of the *C. biloba* range. The two species are very similar electrophoretically and morphologically except for flower petal shape (Fig. 5.8); however, they are reproductively isolated by a translocation, several paracentric inversions and a chromosomal fusion. Lewis suggested that *C. biloba* arose when an isolated *C. lingulata* population crashed during a drought to only a few individuals, which by chance contained the chromosomal rearrangements. The cultivated rye, *Secale cereale*, may also have arisen via parapatric speciation from the wild species, *Secale montanum*. The two species vary by two reciprocal translocations and are reproductively isolated, but in most other respects are similar. White (1978) uses the term 'stasipatric speciation' to describe the formation of new species due to the fixation of chromosomal rearrangements.

The dramatic reduction of a population due to an environmental catastrophe or the establishment of a 'founder population' of a few individuals can also lead to a morphologically distinct species when the remaining sam-

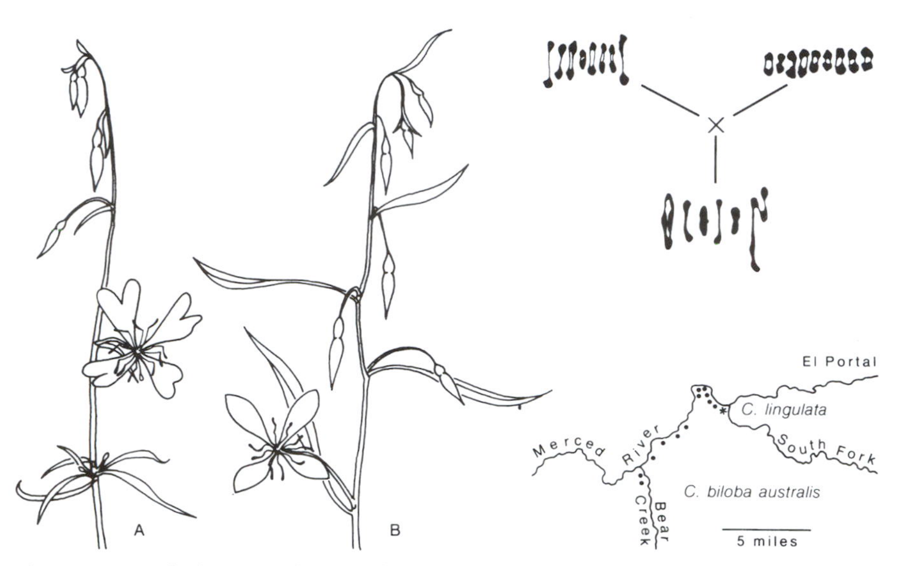

Fig. 5.8. Morphology, cytology and geographical range of *Clarkia biloba* (A) and its peripatric derivative species *Clarkia lingulata* (B) (used with permission from H. Lewis, 1962, *Evolution* 16, 257–271).

ple of the gene pool is unbalanced and 'undergoes a genetic revolution' (Mayr, 1954) or 'genetic transilience' (Templeton, 1981). Genetic drift and changes in selection pressure can result in a shift of many genes into new coadapted complexes. A wide range of distinct plant and animal species in the Hawaiian islands are thought to have arisen in this manner (Carson and Templeton, 1984). The most thoroughly documented cases involve representatives of the *Drosophila*, but the silversword alliance of the *Compositae* includes giant herbs, small trees and ecologically diverse shrubs (Carr and Kyhos, 1981). Small founder populations do not always undergo dramatic alterations, however, as most of our crop species are based on relatively few genotypes and still retain a strong resemblance and interfertility with their progenitors (Chapter 7).

Parapatric speciation

In this mode of speciation, RIBs evolve without geographical separation. The diversifying population is adjacent to the progenitor population (neighbouringly sympatric (Grant, 1985)). The process occurs when a subgroup diverges in response to environmental challenges and isolating barriers begin to form as a by-product of ecological differentiation.

Whether populations can diverge sufficiently without geographical separation has long been a matter of debate. Mayr (1942, 1963) contended that portions of populations are unlikely to differentiate enough to become genetically isolated in the face of strong gene flow. This would be particularly true among sympatric subgroups where the diverging race is completely contained within the parent population and is bombarded on all sides by pollen and seed. There are, however, numerous examples where parapatric populations have undergone substantial differentiation in the face of one-dimensional gene flow. We have already described the mine-tailing experiments of Bradshaw where the differentiated populations varied not only in heavy-metal tolerance but also in flowering date (Chapter 2). McNeilly and Antonovics (1968) have catalogued numerous similar scenarios where ecologically divergent populations have different bloom dates (Table 5.6). Artificial selection experiments have also shown that relatively strong RIBs can arise over a few generations when strong selection is placed on unrelated characteristics. Paterniani (1969) planted white flint and yellow sweet maize in the same field together. Each generation, he selected the purest ears for subsequent planting and, after six generations, fewer than 5% of the seeds resulted from an outcross (Fig. 5.9). These experiments demonstrate that parapatric speciation is at least theoretically possible; it is up to the field biologists to document the complete process in nature.

Sympatric (instantaneous) speciation

Occasionally, new species arise through spontaneous mutation without any ecological or geographical separation. A new isolated type appears in only a few generations without substantial genetic differentiation. Most of these types face a high likelihood of rapidly becoming extinct due to their low numbers, but, even against these odds, many species are known to have had their origin in this manner. Probably the most frequently cited example of sympatric speciation occurred in the apple maggot, *Rhagoletis pomonella* (Bush, 1975). The original host plant of *R. pomonella* in the USA was hawthorn (*Crataegus*), but the fruitfly began to infest introduced populations of apples in the mid-1800s. Apparently, there was a change in a single gene trait affecting host recognition that isolated the two populations with minimal genetic change. The genus *Rhagoletis* has a large number of very similar species that infest fruits of different plant families.

R. pomonella – Rosaceae
R. mendax – Ericaceae
R. carnivora – Cornus
R. zephyria – Caprifoliaceae

Examples of instantaneous speciation abound in plant species. We have already discussed polyploidy at length, where a chromosomal duplication instantly isolates a progeny plant from its parents; at least half of all plant species are polyploid. Even the appearance and fixation of simple chromosomal rearrangements can result in a new species. *Stephanomeria mal-*

Table 5.6. Differences in flowering time of ecotypes compared with the 'normal' type (from McNeilly and Antonovics, 1968).

Flowering species	Ecotype	Time
Gilia capitata	Sand dune	Later
Madia elegans[a]	Subsp. *vernalis*	Spring
	Subsp. *aestivalis*	Summer
	Subsp. *densifolia*	Autumn
Layia platyglossa[a]	Maritime	Later
Hemizonia citrina[a]		April
Hemizonia lutescens[a]		Aug.–Sept.
Hemizonia luzulaefolia[a]		April
Hemizonia rudis[a]		Aug.–Sept.
Lactuca graminifolia[a]		Early spring
Lactuca canadensis[a]		Summer
Ixeris denticulata	Subsp. *typica*	Spring
	Subsp. *sonchifolia*	Autumn
	Subsp. *elegans*	Summer
Pinus attenuata[a]		Later
Pinus radiata		Earlier
Lamium amplexicaule[a]	Vernal race	Earlier
Viola tricolor	Sand dune	Later
Silene cucubalis		Earlier
Silene maritima		Later
Geranium robertianum	Shingle beach	Later
Mimulus guttatus	Coastal	Late
	Mountain	Latest
	Valley and foothills	Early
Geum urbane[a]		Later
Geum rivale[a]		Earlier
Succisa pratensis	Northern race	Earlier
Ranunculus acer	Alpine	Earlier
Solidago virgaurea	Alpine and coastal	Earlier
Rumex acetosa	Alpine	Earlier
Leontodon autumnale	Coastal	Earlier
Clarkia xantiana[a]	Self-compatible race	Earlier
Salvia mellifera		Early spring
Salvia apiana		Late spring

[a]Some evidence given by author that 'ecotypes' are adjacent.

heurensis is a self-compatible species that was probably derived from the self-incompatible *Stephanomeria exigua* subsp. *coronaria* (Gottleib, 1974, 1977b). They are morphologically and electrophoretically quite similar and share a single habitat, but *S. malheurensis* carries a chromosomal translocation that reproductively isolates it from *S. exigua*. The appearance of self-

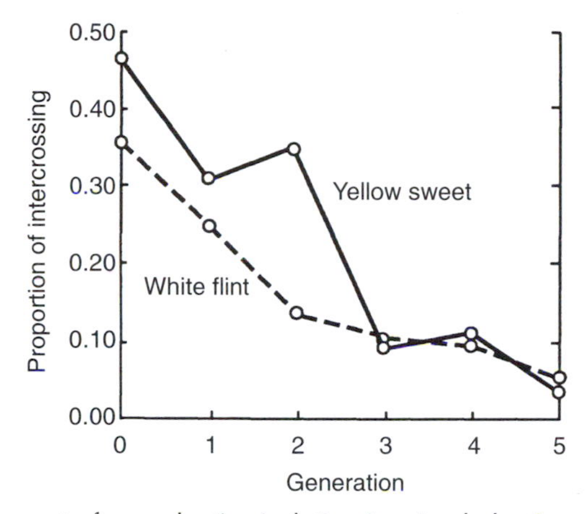

Fig. 5.9. Development of reproductive isolation in mixed plantings of maize when only the purest ears were selected in each generation (used with permission from E. Paterniani, 1969, *Evolution* 23, 534–547).

compatibility in the new type probably allowed the translocation to become homozygous through selfing rather than being lost in an outcrossed flood of parental chromosomes.

Genetic Differentiation during Speciation

During the course of this chapter, it has been repeatedly suggested that RIBs arise when 'sufficient' genetic variability has accumulated. We are now left with the question of just how much genetic variation is necessary to begin the speciation process. We have shown that genetic differentiation is important, but have not yet defined what is 'sufficient'. This question has been most recently posed as whether reproductive isolation occurs at the genome-wide or genic level (Wu, 2001).

Many investigators have been concerned with quantifying the number of genes involved in gradual speciation through differential adaptation. Much of the available information has come from populations that have diverged to the point of ecotypes, but have not speciated. Clausen and Heisey (1958) demonstrated that the races of *Potentilla glandulosa* found at different elevations in the Sierra Nevada mountains of California were genetically distinct by growing them in a common garden. They then made crosses to determine the minimum number of alleles separating the sub-alpine and timberline ecotypes and found dozens of allelic differences for 19 morphological and physiological traits (Table 5.7). Similar high numbers of differences were also discovered between coastal and inland *Viola tricolor* (Clausen, 1922, 1926, 1951) and winter and summer annual races of *Lamium purpureum* (Müntzing, 1932; Bernstrom, 1952).

Table 5.7. Minimum number of alleles governing the inheritance of characters in hybrids of subalpine and foothill races of *Potentilla glandulosa* (from Clausen and Heisey, 1958, as summarized by D. Briggs and S.M. Walters, © 1984, *Plant Variation and Evolution*, Cambridge University Press, Cambridge).

Character and action in genes	Estimated number of pairs of alleles at unlinked gene loci
1. Orientation of petals: 2 erecting, 1 reflexing	3
2. Petal notch: 1 producing notch, 2 inhibiting	3
3. Petal colour: 2 whitening, 2 producing yellow, 1 bleaching	5
4. Petal width: 4 widening, 1 complementary,[a] 1 narrowing	6
5. Petal length: 4 multiples[b]	*c.* 4
6. Sepal length: 3 or 4 multiples for lengthening, 1 for shortening, 1 complementary	*c.* 5
7. Achene weight: 5 multiples for increasing, 1 for decreasing	*c.* 6
8. Achene colour: 4 multiples of equal effect	4
9. Branching, angle of	*c.* 2
10. Inflorescence, density of	*c.* 1
11. Crown height	*c.* 3
12. Anthocyanin: 4 multiples (1 expressed only at timberline), 1 complementary	5
13. Glandular pubescence: 5 multiples, in series of decreasing strength	5
14. Leaf length: transgressive[c] segregation; many patterns of expression in contrasting environments; possibly different sets of multiples activated	*c.* 10–20
15. Leaflet number in bracts	c. 1
16. Stem length: transgressive segregation, 5–6 multiples plus inhibitory and complementary genes; many patterns of expression in contrasting environments	*c.* 10–20
17. Winter dormancy: 3 multiples of equal effect	3
18. Frost susceptibility: slight transgression towards resistance	*c.* 4
19. Earliness of flowering: strongly transgressive; many patterns of altitudinal expression; possibly different sets of genes activated	Many

[a]Complementary effects: factor A or B no effect but combination A + B produces phenotypic difference.
[b]Multiples: factors with comparable effects.
[c]Transgressive effects: in F_2 segregants, values for 'extreme' individuals exceed parental values.

Examinations have also been conducted on closely related species that can still be successfully hybridized. Baur (1924) crossed two species of snapdragon, *Antirrhinum majus* and *Antirrhinum molle* and found considerable phenotypic variability in the F_2 population. Most individuals had combinations of parental characteristics, but a few had unique characteristics that must have come from novel genic assortments. Baur estimated that the two species were separated by more than 100 allelic differences.

Gel electrophoresis has also been used to assess levels of variation during the speciation process. Comparisons have been made between populations at various levels of evolutionary divergence in a wide range of plants and animals (Ayala, 1975). Local populations of plants have average genetic identity values of I = 0.95, while conspecific populations average I = 0.67 (Gottleib, 1981). If we take the natural logarithm of the conspecific value, the genetic distance value of 0.40 is obtained, which represents about 40 changes per 100 loci in an average pair of divergent plant species.

These studies show that extensive amounts of genetic differentiation can be associated with speciation, but it is important to realize that not all of this variability is directly related to the speciation process. For taxonomic species, only those genes which control the characters used to distinguish the species are associated with speciation and, in the case of biological species, RIBs can arise at any time during the process of divergence, regardless of the level of differentiation. There is increasing evidence that relatively strong RIBs can be regulated by relatively few genes. When Bradshaw *et al.* (1998) searched for QTL associated with the 12 differences in floral morphology assumed to reproductively isolate *M. lewisii* and *M. cardinalis*, they found only one to six QTL for each trait, and most traits (9/12) were regulated by a single QTL, which influenced over 25% of the total phenotypic variability. Major QTL have also been shown to control differences in inflorescence architecture in *Zea* (Doebley and Stec, 1993), flowering time in *Brassica* (Camargo and Osborn, 1996) and the growth and flowering of *Arabidopsis* (Mitchell-Olds, 1996). Similar results have been obtained in studies elucidating the genetics of the domestication process (Chapter 7). This body of work implies that a relatively small number of genes with cumulative effects can have rather dramatic effects on the evolution of reproductive isolation and speciation.

Hybridization and Introgression

Thoughts about the relative importance of interspecific hybridization in plant evolution have switched back and forth over time. In the middle of the last century, the role of hybridization was considered to be substantial, based on a number of morphological studies of native and crop species (Anderson, 1949). In the 1970s, enthusiasm waned (Heiser, 1973), but with the advent of molecular markers in the 1980s support grew dramatically (Rieseberg, 1995; Rieseberg *et al.*, 2000). Most plant systematists now believe that hybrid speciation is at least common, with estimates of the percentages of hybrid species in different floras ranging from 25% to 80% (Whitman *et al.*, 1991; Abbott, 1992; Masterson, 1994). Ellstrand *et al.* (1996) found 16–34% of the families in five biosystematic floras to have at least one pair of species that hybridized locally, and 6–16% of the genera.

Hybridizations between native and introduced species have often led to the development of new taxa and have even been implicated in the evolution of a number of new invasive species. Abbott (1992) estimated that 45% of the British flora was alien, and 7% of those introduced species were involved in the production of hybrids now prominent in the native flora. One of the most widespread examples is *Senecio vulgaris* var. *hibernicus*, a hybrid of native *S. vulgaris* var. *vulgaris* and introduced *Senecio squalidus*, which escaped from the Oxford Botanical Garden in 1794 (Abbott, 1992). Highly invasive thistles from Europe have widely hybridized in Australia (O'Hanlon *et al.*, 1999). Ellstrand and Schierenbeck (2000) found 28 examples 'where invasiveness was preceded by hybridization' and at least half of these hybrid lineages were the product of native × non-native hybridizations.

As we have discussed previously in this chapter, many plant species retain the ability to hybridize with their relatives, even when they are quite distinct and have relatively strong RIBs. Numerous hybrid populations or 'hybrid swarms' have been identified where closely related species come into contact. These hybrid zones are often narrow and stable when hybrid fitness is low or the species are adapted to very distinct habitats (Levin and Schmidt, 1985; Campbell and Waser, 2001); however, interspecies hybridization can act as the nucleus for evolution if hybrids are at least partially viable and suitable habitats exist to support them (Buerkle *et al.*, 2000).

Hybridization can stimulate evolutionary change in two ways: (i) the adaptive potential of one or both parents might be increased through back-crossing, or 'introgressive hybridization' (Fig. 5.10; Anderson,

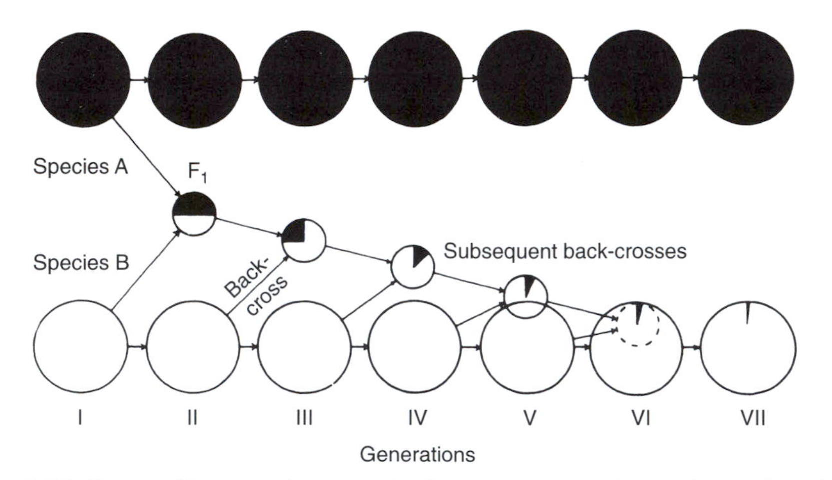

Fig. 5.10. Cartoon illustrating introgression between two species. Back-crossing of the F_1 hybrid to species B ultimately results in the absorption of some genes from species A into at least some individuals of species B. (Used with permission from D. Briggs and S.M. Walters, © 1984, *Plant Variation and Evolution*, Cambridge University Press, New York.)

1949); this trickle of genes might expand the adaptive potential of the recipient population; or (ii) the hybrid population itself may evolve unique adaptations through genetic differentiation and genomic reorganization, or 'recombinational speciation' (Stebbins, 1957; Grant, 1958). The latter process is generally thought to occur over a number of generations, although Rieseberg and Ellstrand (1993) have identified many examples where F_1 hybrids had unique phenotypes that could be a nucleus for speciation. Most hybrids are less fit than their progenitors, but several cases have been found where hybrids had higher fitness and unique adaptations (Emms and Arnold, 1997; Burke and Arnold, 2001; Johnston *et al.*, 2001).

Most hybrid species are polyploid, but speciation via homoploid hybridization has also been documented in several instances (Gallez and Gottleib, 1982; Crawford and Ornduff, 1989; Wolf and Elisens, 1993). In one of the earliest studies of hybridization, Riley (1938) described natural hybridization between *Iris fulva* and *Iris hexagona* in Louisiana, USA, using a 'hybrid index'. He compiled a list of differences between the two species and one was arbitrarily selected at the low end (Fig. 5.11). Plants in the natural environment were then scored for each character and the

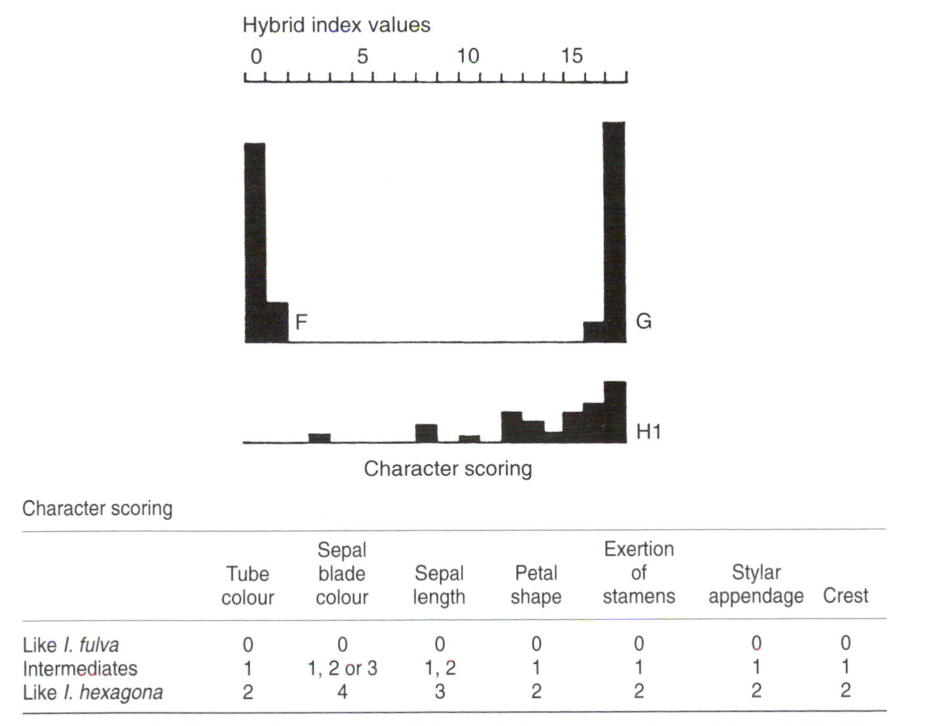

Character scoring

	Tube colour	Sepal blade colour	Sepal length	Petal shape	Exertion of stamens	Stylar appendage	Crest
Like *I. fulva*	0	0	0	0	0	0	0
Intermediates	1	1, 2 or 3	1, 2	1	1	1	1
Like *I. hexagona*	2	4	3	2	2	2	2

Fig. 5.11. Hybridization between *Iris fulva* and *Iris hexagona* in Louisiana, USA. Two relatively pure (F and G) and one hybrid population are shown. (From Riley, 1938.)

grand total was calculated for each individual. Many populations contained individuals with intermediate scores, suggesting that they were indeed hybrids. Randolph (1966) later identified a population of *Iris* called 'Abbeville Reds' that had a bright red-purple colour distinct from that of any other *Iris* species, different shaped capsules and a unique habitat of deep shade and high water. He gave it a new species name, *Iris nelsonii*, and speculated that it was derived from the hybridization of *I. fulva* and other local species.

Arnold (1993) used a broad array of nuclear and cytoplasmic markers to confirm that the Louisiana irises were indeed hybridizing. He found individuals with many different combinations of the cytoplasmic DNA (cpDNA) and nuclear genes of three species, *I. fulva, I. hexagona* and *Iris brevicaulis* (Fig. 5.12). He also found that *I. nelsonii* carried markers from all three species, although every individual had the *I. fulva* cytotype and only a few individuals carried species-specific randomly amplified polymorphic DNAs (RAPDs) from the other two species. This suggested that the taxon was the product of repeated back-crossing of the original F_1 hybrid to *I. fulva*, a likely scenario, as the appearance of hybrids in nature is rare and the species have several prezygotic and postzygotic RIBs (Carney *et al.*, 1994; Carney and Arnold, 1997; Emms and Arnold, 2000).

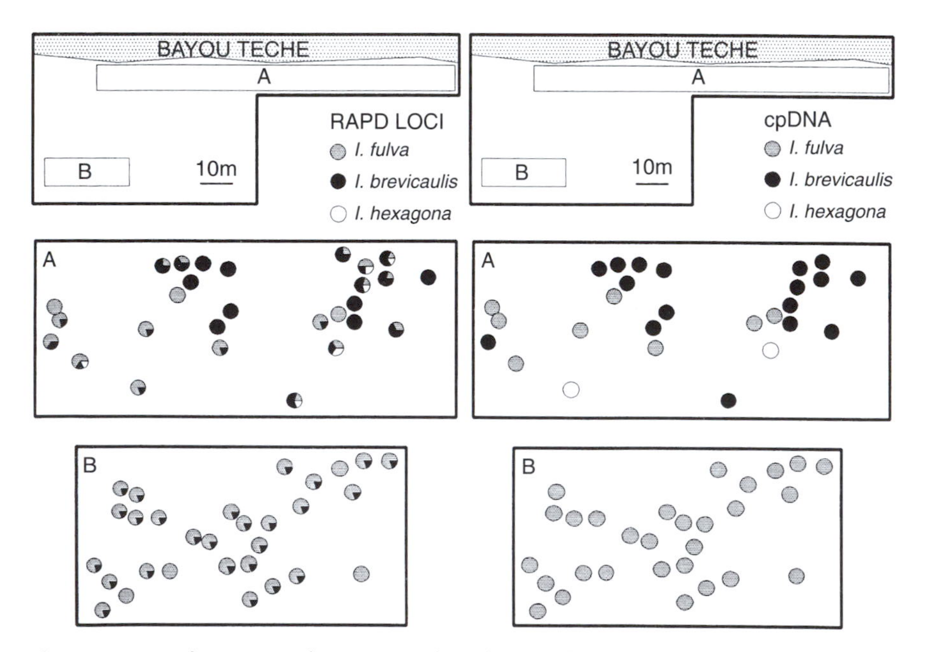

Fig. 5.12. Combinations of cpDNA and nuclear markers (RAPDs) in a hybrid swarm of *Iris* at Bayou Teche, Louisiana. The relative proportion of *Iris fulva, Iris hexagona* and *Iris brevicaulis* markers in each plant is represented by the pie charts. (Used with permission from Arnold, 1993, *American Journal of Botany* 80, 577–583.)

Using morphological and cytological data, Heiser (1947, 1949, 1951, 1958) documented widespread hybridization and introgression among *Helianthus* taxa in the south-eastern USA. He proposed two instances of introgression, *Helianthus annuus* subsp. *texanus* arising due to hybridization with *Helianthus debilis* subsp. *cucumerifolius* and a weedy race of *Helianthus bolanderi* being the product of introgression with *H. annuus*. He also suggested that three species, *Helianthus paradoxus*, *Helianthus anamolus* and *Helianthus neglectus*, were the stabilized products of hybridization between *H. annuus* and *Helianthus petiolaris*. Later molecular work by Rieseberg and associates (1988, 1990a,b, 1991) verified Heiser's prediction of a hybrid origin for *H. paradoxus* and *H. anamolus*, and the proposed introgression between *H. annuus* and *H. debilis*. They also found evidence of the introgression of cpDNA from *H. annuus* into southern California populations of *H. petiolaris* (Dorado *et al*., 1992). However, the hybrid origin of *H. neglectus* was not supported, nor was any evidence found of gene transfer between *H. annuus* and *H. bolanderi*.

Rieseberg *et al.* (1995) went on to show how the genomes of *H. annuus* and *H. petiolaris* were rearranged to produce *H. anamolus*. They developed a molecular linkage map of the two progenitor species and then compared it with the proposed derivative. They discovered six linkage groups that were conserved across all three species (collinear), and another 11 chromosomal regions whose gene order differed across taxa. Of these 11, *H. anamolus* shared four of them with one or the other parent, while the other seven were distinct from either parent, suggesting that substantial reorganization had occurred within these linkages (Fig. 5.13).

To determine if particular gene assemblages were maintained by selection during the formation of *H. anamolus*, Rieseberg *et al.* (1996) produced three hybrid lineages through artificial crosses and compared them with the ancient hybrid species, using a map-based approach. The genomic composition of the ancient and synthesized genomes were highly concordant, indicating that the particular gene assemblages maintained in the hybrids were probably under selection. Some parts of the genome were less subject to introgression than others, suggesting that there was coadaptation between blocks of parental species genes. Jiang *et al.* (2000) also found introgression to be confined to specific chromosomal regions in interspecific populations of polyploid *Gossypium*.

To further confirm the role of coadapted complexes in hybrid species formation, Rieseberg *et al.* (1999) examined 88 markers across 17 chromosomes in three natural hybrid zones of *H. petiolaris* and *H. annuus*. They found patterns of introgression to be quite similar across all three hybrid zones, with 26 chromosomal locations having significantly lower levels of introgression than according to neutral expectations. They were even able to link pollen sterility with 16 of these segments.

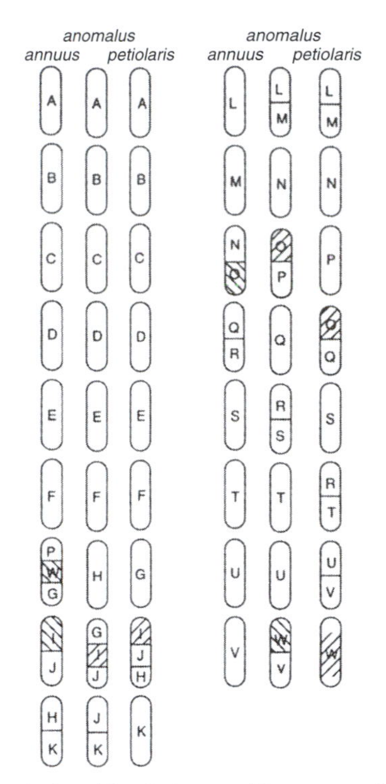

Fig. 5.13. Major linkage relationships between *H. annuus, H. petiolaris* and its hybrid derivative *H. anomolus.* Note that six linkage groups are collinear between the two species and 11 are structurally different. Lines of shading indicate inversions. (Used with permission from Rieseberg *et al.*, 1995, *Nature* 375, 313–316.)

Hybridization and Extinction

In some instances, hybridization between native species can result in the extinction of one of the species (Levin *et al.*, 1995; Buerkle *et al.*, 2000). There can be 'genetic assimilation', where the hybrids are fertile and they replace pure conspecifics of either or both hybridizing taxa. This is thought to be happening on the Santa Catalina Island, where *Cerocarpus betaloides* appears to be assimilating *Cerocarpus traskiae*, which is represented by a single population (Rieseberg and Gerber, 1995). *Clarkia speciosa polyantha* is replacing *Clarkia nitens* in regions of overlap in the Sierra Nevada foothills of central California (Bloom, 1976). Substantial introgression has occurred between *Gossypium darwinii* on the Galapagos Islands and the crop, *Gossypium hirsutum* (Wendel and Percy, 1990). The cultivated radish, *Raphanus sativus*, and the jointed charlock, *Raphanus raphanistrum*, have completely merged in California (N.C. Ellstrand, personal communication). There can also be 'demographic swamping', where population growth in a numerically inferior taxon is retarded by the formation of hybrid seed, and population growth falls below the replacement level. Hybridization with cultivated rice is thought to have led to the near extinction of the endemic Taiwanese taxon, *Oryza rufipogon* subsp. *formosana* (Kiang *et al.*, 1979). In fact, most populations of native Asian subspecies of *O. rufipogon* may be endangered through hybridization with the crop (Chang, 1995; Ellstrand *et al.*, 1999).

A number of factors influence the risk of extinction after hybridization, including the strength of the RIBs separating the taxa, habitat differentiation, hybrid viability and fertility, population sizes and growth rates and environmental stochasticity (Wolf *et al.*, 2001). Interspecific hybridization is a particular threat to rare species, where extinction can occur whether the hybrid seeds are viable or abort. Small populations can be so swamped by alien pollen that they produce too few non-hybrid individuals to replace themselves.

Crop–Weed Hybridizations

As human beings carried domesticants far from their origins, the crop species would have occasionally come into contact with previously isolated relatives. When RIBs were not complete, the domesticated and native species could then hybridize and produce novel progeny. The adaptations of the native congeners could be altered through subsequent introgression or back-crosses to the crop, and unique types might be selected by humans. The natural disturbances caused by farming would have greatly facilitated the hybridization process, by providing novel habitats for the hybrids to gain a foothold (Anderson, 1949).

Hybridization still occurs between many domesticated and wild species (Anderson, 1961; Harlan, 1965; Zhukovskii, 1970). In many cases, the RIBs are so minimal that the wild and domesticated taxa are considered subspecies. In a recent survey of hybridization between crops and congeners, Ellstrand *et al.* (2000) found that hybridization has played an important role in the development of most of the major crop species (Table 5.8). In fact, introgression between crop and wild species has been documented for 34 species, including barley, beets, cabbages, canola, carrots, cassava, chilli, cocoa, common bean, cotton, cowpea, finger millet, foxtail millet, hemp, hops, lettuce, lucerne, maize, oats, pearl millet, pigeon-pea, potato, quinoa, radish, raspberry, rice, rye, sorghum, soybean, squash, sunflower, tomato, water melon and wheat (Ellstrand *et al.*, 1999; Jarvis and Hodgkin, 1999).

Most F_1 hybrids of cultivated and wild congeners have reduced fitness in nature and the genes of domesticated plants rarely travel far from narrow hybrid zones. However, some wild plants have acquired crop genes that make them more effective agronomic weeds. In some instances, genes have introgressed into wild plants that allow them to 'mimic' the habit of domesticated ones, and thus escape removal in agronomic sites by farmers. The weed may look identical to the crop until seed dispersal, or the weed seeds

Table 5.8. Role of hybridization in the evolution of major crop species and their relatives (modified from Ellstrand *et al.*, 2000).

		Role of hybridization	
Crop	Origin	Crop development	Weed evolution
Wheat	Allopolyploid	Great (intra[a] and inter)	Great (intra and inter)
Rice	Complex diploid hybrid	Great (intra and inter)	Great (intra and inter)
Maize	Single progenitor	Little	Little
Soybean	Single progenitor	Little	Moderate (intra and inter)
Barley	Single progenitor	Great (intra)	Great (intra)
Cotton	Allopolyploid	Little	Little (inter)[b]
Sorghum	Single progenitor	Great (intra and inter)	Great (intra and inter)
Finger millet	Single progenitor	Moderate (intra)	Great (intra)
Pearl millet	Single progenitor	Moderate (intra)	Great (intra and inter)
Dry beans	Single progenitor	Great (inter)	Great (inter)
Rapeseed	Allopolyploid	Great (intra and inter)	Moderate (intra and inter)
Groundnut	Allopolyploid	Little?	Little
Sunflower	Single progenitor	Great (intra)	Great (intra and inter)
Potato	Complex autopolyploid	Great (intra and inter)	Great (intra and inter)
Sugar cane	Complex autopolyploid	Great (inter)	Great (inter)

[a]Intra-species and inter-species.
[b]Regionally important in the Galapagos (Wendel and Percy, 1990).

may be impossible to separate from the agronomic source. Crop mimicry has been particularly prevalent among the grains (Harlan *et al.*, 1973; Harlan, 1992). A classic example of crop mimicry can be found in sugar beet fields in Europe, where weed introgressants bolt and scatter seeds before crop harvest (Viard *et al.*, 2002). There are weedy bolters that probably resulted from the contamination of seed producers by pollen from wild individuals, and bolters that emerge from the seed bank containing wild/crop introgressants. The bolters carry the dominant B allele, which cancels any cold requirement. A number of additional examples are described in Chapter 7.

Wild/crop hybridizations may also occasionally have resulted in alterations of the crop itself, when farmers noticed new, useful genetic combinations. Farmer selection of crop/weed introgressants may have played a particularly important role in the early development of crops, as agriculture spread out of the centres of origin. Local adaptations would have been greatly enhanced by hybridization with native populations. It is difficult to document the historical introgression of native genes into crops, but Jarvis and Hodgkin (1999) have identified nine examples where today's farmers are selecting crop/wild introgressants. Such farmer-based selection is probably widespread even today, as a large percentage of the world's agriculture is still conducted by subsistence farmers who grow traditional crop varieties in the native range of their antecedents.

Risk of Transgene Escape into the Environment

A recent concern is that genes will escape from genetically engineered crops into their wild progenitors and produce more noxious weeds (Ellstrand and Hoffman, 1990; Ellstrand, 2001). The possibility of transgene flow from engineered crops to their wild relatives has been one of the primary concerns associated with the unrestricted release of genetically modified crops (Colwell *et al.*, 1985; Ellstrand, 1988; Dale, 1992). It has been pointed out that such gene flow could result in the evolution of increased weediness in wild relatives and the evolution of pests that are resistant to the newly developed control strategies (Rissler and Mellon, 1996; Snow and Palma, 1997; Hails, 2000). Numerous studies have now measured the likelihood of hybridization between crops and their wild relatives (Ellstrand, 2001). While the early consensus was that such hybridizations occurred infrequently, research in the last decade has shown that they are relatively common (Ellstrand *et al.*, 1999) and the crop alleles can persist for long periods of time in natural populations (Klinger and Ellstrand, 1994; Arriola and Ellstrand, 1997; Linder *et al.*, 1998). Factors such as breeding system, flowering time, hybrid viability and isolation distance can alter the rate of gene escape (Hancock *et al.*, 1996; Hokanson *et al.*, 1997a), but, if compatible relatives are within the cloud of crop pollen, genes will escape.

The impact of transgene escape into wild populations will be strongly associated with the phenotypic effect of the gene itself and the invasiveness of its wild progenitors (Hancock and Hokanson, 2001; Hancock, 2003). Transgenes that have a neutral effect on fitness, such as the marker genes used to recognize transformants, might spread in natural populations through genetic drift, but would have no subsequent impact on native fitness. Genes with detrimental effects on growth and development, such as male sterility or reduced lignin biosynthesis, would most probably be selected against in the natural environment and would not spread beyond a narrow hybrid zone. The transgenes associated with pest resistance would have variable effects, depending on the invasiveness of the recipient species and the level of natural control. If a wild species is an agronomic weed, the escape of a herbicide resistance gene could make it a more noxious pest. Virus, fungal and pest resistance genes could increase the fitness of wild populations and make them better competitors if the pest is currently controlling natural populations. Transgenes with direct positive effects on fitness, such as those associated with environmental tolerances, could result in dramatic adaptive shifts and have a major impact on the fitness of native populations.

Summary

Many different classes of RIBs exist in plant species, including both premating (ecogeographical, temporal, floral and gametic incompatibility) and postmating (hybrid unviability, sterility and breakdown) mechanisms. Numerous different modes of speciation have been proposed, based on the size of the speciating populations, how far they are separated and how much they differentiate before RIB formation. The most widely recognized type of speciation is called geographical, where the speciating populations are well separated geographically and undergo substantial differentiation before they become reproductively isolated. There are two types of geographical speciation: (i) allopatric, where the speciating population is large; and (ii) peripatric, where the speciating population is small. Other forms of speciation are called parapatric and sympatric. In parapatric speciation adjacent populations gradually differentiate enough to become reproductively isolated, while in sympatric speciation reproductively isolated individuals arise within the borders of a population. Sympatric speciation can occur instantaneously when mutations arise that immediately isolate an individual from its progenitors. High levels of gene flow prohibit gradual sympatric speciation. In some instances, interspecific hybridization has increased the adaptive potential of both parents through back-crossing or introgression, and in other cases the hybrid population itself has evolved into a new species. This phenomenon is most commonly associated with polyploidy, but has also occurred among homoploids. Crop–weed hybridizations have contributed to the adaptations of crops as they were dispersed by humans and has led to the evolution of weeds that mimic the crop in such a way that their removal becomes difficult. A recent concern is that transgenes will escape from engineered crops into their wild progenitors and alter their adaptive characteristics. Before the deployment of genetically modified crops, the invasive biology of a crop species and the nature of the transgene need to be carefully evaluated.

The Origins of Agriculture 6

Introduction

Up to this point, we have concerned ourselves primarily with evolutionary mechanisms and have placed little emphasis on the emergence of humans and their crops. We are now ready to expand our discussion to the development of land plants and people. Today, our landscape is dotted with farms, from small garden plots to huge corporate giants. Virtually everyone on earth relies on farms or farmers for their daily sustenance, and in only a few remote corners of Africa, Australia and North America do humans still rely on the ancient hunter–gatherer strategy. Even these societies are greatly endangered, tainted by their use of industrial technologies, such as guns and snowmobiles and backed into an ever-diminishing corner due to deforestation. Remarkably, humans did not start farming until about 12,000 years ago, and domesticated plants and animals did not become the major source of food until the last few thousand years. In this chapter, we shall outline the series of changes that led to contemporary plants and humans, and speculate on the circumstances associated with their appearance.

Rise of our Food Crops

The angiosperms provide most of our food crops. They first appeared in the early Mesozoic or late Paleozoic era about 200–250 million years ago, but fossil evidence of them is extremely limited until they began to dominate during the Cretaceous (136–190 million years ago). The angiosperms were the first plants to have double fertilization and the enclosure of seeds in fruit. Double fertilization provided zygotes with copious resources to help them get established and fruits attracted animals for dispersal. It was the appearance of the angiosperms that set the stage for the development of our mammal ancestors.

One of the most intriguing mysteries left by an incomplete fossil record is the sudden widespread appearance of the flowering plants or angiosperms in the Cretaceous period. There is evidence of earlier origins (Sun *et al.*, 1998), but most of the record of angiosperm diversification is among Cretaceous fossils (Sporne, 1971). Dramatic variations arose in flower shape, symmetry, arrangement, part number, location and aggregation (Stebbins, 1974; Dilcher, 2000). We are forced to conclude that the evolution of flowering plants was extraordinarily rapid or that, for some unknown reason, most pre-Cretaceous fossils of angiosperms have disappeared.

Dicotyledons were the first angiosperms to appear, followed by monocotyledons. Both groups are thought by most experts to be monophyletic in nature, with the monocotyledons being derived from primitive dicotyledons 135–75 million years ago. Both dicotyledons and monocotyledons have undergone considerable genetic differentiation: there are currently 200,000 living species of dicotyledons and 50,000 monocotyledons (Simmonds, 1979).

A wide range of hypotheses implicating selection have been presented for the rapid emergence and diversification of the angiosperms (Beck, 1976). The most popular hypothesis is that the concomitant rise of pollinating insects led to powerful divergent selection as foragers and hosts developed complex relationships (Takhtajan, 1969; Proctor and Yeo, 1973; Faegri and van der Pijl, 1979; Armstrong *et al.*, 1982). Whitehouse (1950) and de Nettancourt (1977) proposed that the appearance of closed carpels and self-incompatibility restricted self-pollination and enhanced cross-pollination, which increased genetic diversity and the potential for divergence. Mulcahy (1979) took this hypothesis a step further and suggested that the pollen tube growth associated with reaching a closed carpel 'enhanced the ability of natural selection to act on the gametophytic phase of the life cycle'. Poorly balanced genomes would be selected in the style according to their 'metabolic vigour'. Other hypotheses involve selection stress, the overall competitive ability of plants containing closed carpels as a defence against predation, the value of an endosperm in seedling establishment, the improved water conducting ability of vessel elements, the defensive nature of higher-plant alkaloids and the wide dispersion of fruit eaten by animals (Mulcahy, 1979). Drift may also have played an important role in the refinement of angiosperms via the shifting balance process. As with the emergence of land plants, the population crashes and genetic reorganizations associated with invading new environments must have catalysed the emergence and establishment of these totally unique types.

It was the angiosperms that ultimately provided us with most of our crops and their emergence predated the appearance of our species, *Homo sapiens*. In fact, most of our food families or their close relatives were in existence long before we began farming. The only completely new crop type to appear after the advent of agriculture was maize, *Zea mays*, which has an ear and tassel arrangement not found in its progenitors (Chapter 8). In most cases, human beings did not influence the overall structure of crop species, only the size of their edible organs and their ease of harvest. This topic will be discussed more fully in the next chapter.

Human beings now consume a diverse array of plant structures (Table 6.1), and at least 64 families of angiosperms and 180 genera are utilized as crops (Simmonds, 1979). This is a broad systematic group, but represents only a small fraction of the total number of angiosperm families (300) and genera (3000). The dicotyledons provide the highest number of crop plants (Table 6.2); however, the bulk of the world is fed by a few monocotyledonous grains (maize, rice and wheat).

Emergence of *Homo*

The emergence of *Homo sapiens* came at the end of a long, slow process of change from a quadrupedal, arboreal existence to a bipedal, grassland lifestyle (Haviland, 1996). These changes occurred gradually over a period of 40–50 million years as a chain of species emerged, evolved and became

Table 6.1. Diversity of plant structures eaten by *Homo sapiens*.

Plant part	Example
Root	Beet, radish and carrot
Above-ground stem	Sugar cane
Underground stems	
Tuber	Potato, yam and cassava
Corm	Taro
Bulb	Onion
Leaf	Cabbage, lettuce and tea
Inflorescence	Cauliflower and broccoli
Fruit	
Multiple	Pineapple, fig and breadfruit
Aggregate	Raspberry and strawberry
Pome	Apple and pear
Drupe	Peach, olive, coconut and mango
Hesperidium	Orange and lemon
Pepo	Cucumber, water melon and cantaloup
Nut	Walnut and hazelnut
Grain	Wheat, rice, maize and barley
Achene	Sunflower and safflower
Legume	Bean, pea, groundnut and soybean

Table 6.2. Selected food families.

Family	Crop
Dicotyledoneae	
Anacardiaceae	*Mangifera* – mango
Camelliaceae	*Camellia* – tea
Caricaceae	*Carica* – pawpaw
Chenopodiaceae	*Beta* – sugar beet
Compositae	*Carthamus* – safflower
	Helianthus – sunflower, Jerusalem artichoke
	Lactuca – lettuce
Convolvulaceae	*Ipomoea* – sweet potato
Cruciferae	*Brassica* – kale, cabbage, turnip, rape
Cucurbitaceae	*Cucumis* – cucumber, melons
	Cucurbita – squash, gourds
Euphorbiaceae	*Manihot* – cassava
Lauraceae	*Persea* – avocado
Leguminosae	*Arachis* – groundnut
	Cajanus – pigeon-pea
	Cicer – chickpea
	Glycine – soybean
	Lens – lentil
	Phaseolus – beans
	Pisum – peas
	Vicia – field bean
	Vigna – cowpeas
Moraceae	*Artocarpus* – breadfruit
	Ficus – fig
Oleaceae	*Olea* – olive
	Phoenix – date palm
Pedaliaceae	*Sesamum* – sesame
Rosaceae	*Fragaria* – strawberry
	Malus – apples
	Prunus – cherry, peach, almond
	Rubus – raspberries, blackberries
Rubiaceae	*Coffee* – coffee
Rutaceae	*Citrus* – orange, lime
Solanaceae	*Capsicum* – peppers
	Lycopersicon – tomato
	Solanum – aubergine, potato
Vitaceae	*Vitus*, *Muscadinia* – grapes
Monocotyledoneae	
Araceae	*Colocacia* – taro
Bromeliaceae	*Ananas* – pineapple
Dioscoreaceae	*Dioscorea* – yams

Continued

Table 6.2. *Continued.*

Family	Crop
Monocotyledoneae *(Continued)*	
Gramineae	*Avena* – oats
	Eleusine – finger millet
	Hordeum – barley
	Oryza – rice
	Pennisetum – pearl millet
	Saccharum – sugar cane
	Secale – rye
	Sorghum – sorghum
	Triticum – wheat
	Zea – maize
Liliaceae	*Allium* – onion
	Musa – bananas
Palmae	*Cocos* – coconut
	Elaeis – oil-palm

extinct. Our ancestry can be traced back to the earliest tree-dwelling primates that inhabited the forests over 50 million years ago. Monkeys and apes diverged into separate lineages 20–30 million years ago, and the first hominids evolved from apes 15 million years ago (Fig. 6.1).

Several traits changed dramatically during the evolution of *H. sapiens* to make us what we are: (i) faces became flatter and craniums increased in size; (ii) teeth became more generalized, as incisors and canines became reduced and molars became more massive and flat; (iii) cheek teeth became less separated and the jaw more interlocking; (iv) individuals gained the ability to swing in limbs or brachiate, as forearms became shorter than hind limbs and the size of the thumb became reduced so that the hand could function as a hook; and (v) upright or bipedal motion evolved, as a lumbar curve developed in the vertebral column, the pelvis became larger and the head and hip became repositioned. These changes were initially associated with increased diversification in a forest environment, but ultimately led to the effective exploitation of the surrounding grasslands.

The initial emergence of hominids was probably stimulated by the climatic changes associated with the Miocene 15 million years ago. The climate became much drier and tropical forests began to break up into mosaics of forest surrounded by open bush country. The ape *Sivapithecus* evolved at the forest margins with the ability to occasionally stand upright, carry food, scan the countryside for predators, perhaps throw things and scamper back to the safety of trees. Doubtless, this creature was in stiff competition with other mammals for food, and had to keep a sharp eye out for predators.

Our first fully upright ancestor, *Aripithecus ramidus*, appeared between 4.4 and 3.9 million years ago, but is represented by only a few fragmented

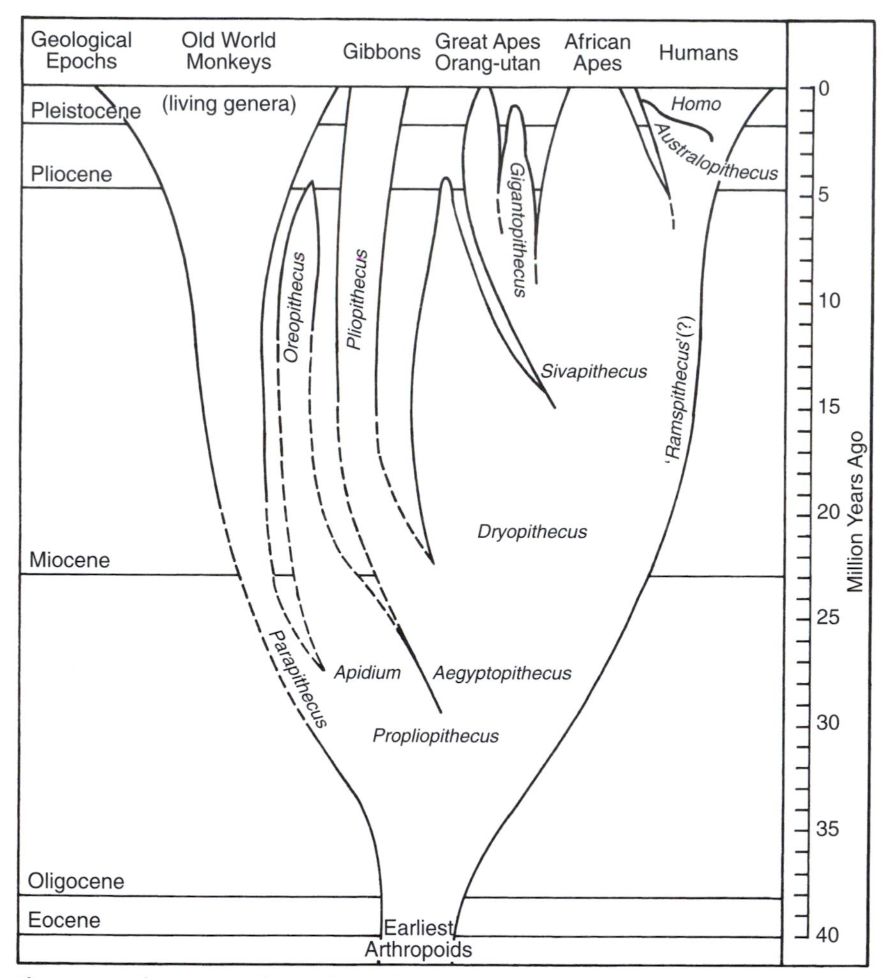

Fig. 6.1. Evolutionary relationships of apes and humans (used with permission from W.A. Haviland, © 1996, *Human Evolution and Prehistory*, Reinhart and Winston, New York).

skeletons from Ethiopia and Kenya (Tattersall, 1998). The first widespread bipedal species to appear belonged to the genus *Australopithecus*, which emerged about 4 million years ago. Convincing fossil evidence of this genus has been found at several locations in Africa, including a perfect set of footprints found by Mary Leakey in Tanzania (Johanson and White, 1979). These protohumans were omnivores, with small canines and heavily enamelled molars that could process abrasive seeds. It is not known whether they just scavenged or hunted for food, but among their remains are high numbers of limb bones from hoofed animals, suggesting meat was brought back from butchering sites. It may have taken primitive cutting tools to do this, and the bones themselves may have been those tools.

The most famous member of this genus is 'Lucy', discovered by D.C. Johanson in northern Ethiopia (Johanson and Eday, 1981). Almost 40% of the skeleton of this young women remains, along with fossils of dozens of other members of her species, *Australopithecus afarensis*. No one really knows why the remains of so many individuals ended up together, but it has been speculated that a group of them may have been killed together by a flash flood or a particularly virulent disease. They could also have been the pile of bones left behind by a predator.

While the story is still unfolding, numerous species of *Australopithecus* appear to have roamed the earth at one time or the other, some even as contemporaries (Johanson and White, 1979; Lasker and Tyzzer, 1982; Leakey and Lewin, 1992). There were small and slightly built 'gracile' forms that weighed as little as 60 pounds (*A. afarensis* and *Australopithecus africanus*), and more 'robust' forms (*Australopithecus robustus* and *Australopithecus boisei*) that approached 150 pounds. The relationships between species of these two groups and later *Homo* are very much in dispute, but most scientists agree that Lucy's species, *A. afarensis*, is in the direct ancestry of our genus, *Homo* (Leakey and Lewin, 1992; Fig. 6.2). The robust forms represent derivatives that evolved on one or more separate branches from ours. One won-

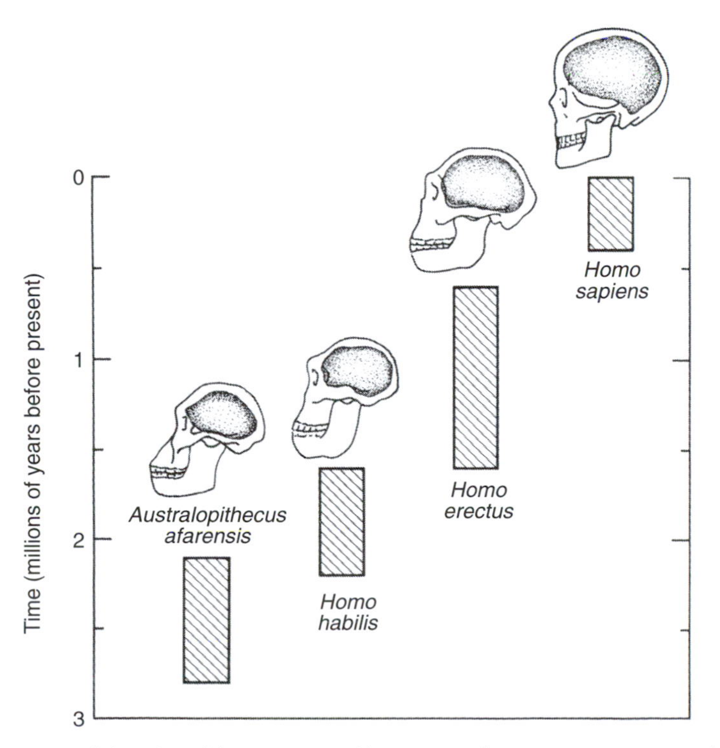

Fig. 6.2. One of the plausible continua of human evolution. Many proposed intermediate forms are not included (see text).

ders how much contact these various species had and if competition between them actually stimulated our own emergence.

Even though *Australopithecus* species underwent considerable morphological change during their 3 million years on earth, their cultural development remained stagnant. In many ways, these 'southern apes' were no better equipped to cope with the vagaries of nature than other contemporary organisms. The invention of agriculture was far from their capabilities. They existed for millions of years, so it is unfair to call them an evolutionary failure, but real cultural advancement awaited the appearance of their brighter cousins, *Homo*, about 2–2.5 million years ago. These early *Homo* showed a dramatic increase in cranial capacity from the 380–500 cm^3 of *Australopithecus* to 600–775 cm^3, and, over the next $1\frac{1}{2}$ million years of *Homo*'s evolution, average brain size increased another 900 cm^3 (Haviland, 1996).

While it has long been popular to talk about the 'missing links' in hominid evolution, an almost continual record of gradual morphological change is now available, which represents changes not only in brain size, but also in skull shape, dental traits, forearm length and pelvis size. Richard Leakey and R. Lewin suggest in their book *Origins Reconsidere*d that we have, in Africa alone, 'fossilized fragments of about a thousand human individuals from the early part of our evolution'. Fossils have been found with almost every conceivable combination of human and ape characteristics. It is becoming increasingly difficult to determine which fossil remains constitute real species change and which represent the simple extremes of species variability.

The earliest fossils that are commonly attributed to our genus have been found in East Africa, eastern Ethiopia and South Africa, and represent a diverse group (Leakey, 1971; Leakey and Lewin, 1992; Haviland, 1996). Louis Leakey named one of these creatures *Homo habilis*, or 'handy person' (Fig. 6.2). Other early collections of bones have been referred to as *Homo ergaster* and *Homo rudolfensis*. Archaic *Homo* were the earliest hominid toolmakers, producing what are called 'Oldowan' tools by battering rocks together to remove a few flakes (Washburn and Moore, 1980; Fig. 6.3). These primitive tools were used to perform a wide array of daily tasks such as cutting, chopping, scraping and perhaps defence.

The early *Homo* were hunters of small and medium-sized animals including rodents, snakes, antelopes and pigs. They probably supplemented their diets by scavenging the left-overs of larger animals and gathering plant material, but there is still no evidence that they practised agriculture. They lived together in base camps of temporary shelters, evidenced by semicircular concentrations of tools and marrow-bearing bones and circles of stones that may have been used to support branches. These early people are unlikely to have had well developed speech, but they had clearly begun down the human path, as they made tools, shared food and returned to a base camp.

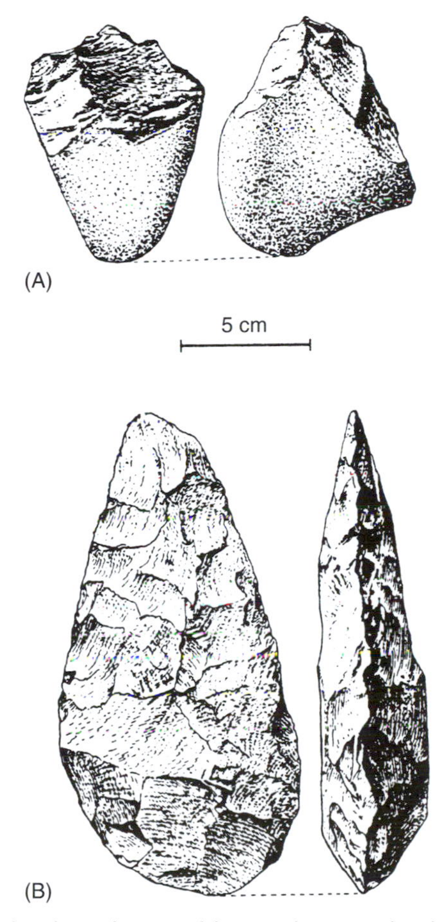

Fig. 6.3. Examples of early tools: (A) Oldowan chopper of archaic *Homo*; (B) Acheulian hand axe of *H. erectus* (used with permission from V. Grant, © 1977, *Organismic Evolution*, W.H. Freeman and Company, San Francisco).

Evolution of *Homo*

The first widely dispersed taxon of *Homo* to appear was *Homo erectus*, which arose as early as 1,500,000 years ago (White and Brown, 1973; Pfeiffer, 1978; Brace *et al.*, 1979; Fig. 6.3). Their brain was larger than the earlier *Homo* and they had larger teeth and jaws. In many respects, they looked like a short, rugged version of modern people. The tools of *H. erectus* consisted of a much more efficient hand axe, which was modified into cleavers, scrapers and other tools that could serve a multitude of jobs from skinning and butchering wild animals to digging up wild roots (Fig. 6.3). These tools were superficially similar to the Oldowan choppers, but were much more pointed and sharp, and have been given another name, 'Acheulian', after the place in France where they were originally discovered. They were produced from flint by chipping off flakes with a hammer stone (Marshack, 1976).

H. erectus was clever enough to begin using fire for warmth and cooking, and had developed systematic means of herding and slaughtering local large animals, such as elephants, rhinoceros, bears, horses, camels and deer. It seems likely that *H. erectus* knew well the habits of every animal they hunted and the seasonal cycles of the plants they gathered, but there is no evidence that they chose to domesticate them. It is not known whether *H. erectus* could speak, but their level of cooperative activity, particularly in hunting large game, suggests they must have had good communication skills.

H. erectus was the first hominid to migrate out of Africa, and by 1 million years before the present (BP) had dispersed throughout Europe, India and South-East Asia (Fig. 6.4). Their increased body size and shifted emphasis towards meat eating may have contributed to this dramatic expansion in range. Regional game scarcities and seasonal animal migrations surely must have stimulated movement, and their success as hunters and gatherers could have resulted in large enough population densities to encourage dispersal. This movement into novel environments may have led to considerable population differentiation as our ancestors adapted to their new challenges, and may even have accelerated the development of human beings.

The population structure of *H. erectus* and all the later hominid species would have encouraged rapid evolution and diversification. There were numerous small groups of hominids scattered all over the world that were isolated from each other, except for occasional contact. Mutations would have periodically arisen in some of these hominid populations, and their small population size would have been conducive to the rapid establishment of novel adaptations.

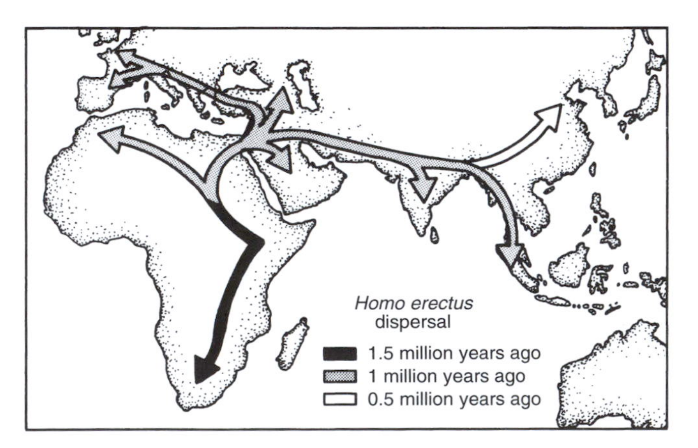

Fig. 6.4. Spread of *Homo erectus* from Africa (used with permission from Sally Black, © 1988, Discover Publications, New York).

The earliest evidence of *H. erectus* evolving towards modern people appears about 300,000 years ago, in a surprisingly broad area across Europe, Africa and Asia. Evidence is generally scanty, but at several sites hominoid fossils have been uncovered that appear to be transitional between *H. erectus* and *H. sapiens* (Tattersall, 1998). The variability between sites is often great, and there is some argument as to whether *H. sapiens* arose directly from *H. erectus* or through other transitional species, variously named *Homo antecessor* or *Homo heidelbergensis*.

The first clearly recognizable members of our species, *H. sapiens*, appear about 100,000 years ago (Brace *et al.*, 1979; Lasker and Tyzzer, 1982). The so-called 'Neanderthal man' was first found in the Neander valley of Germany, but has subsequently been located at many points across the range of *H. erectus* (Constable, 1973). Its first discovery in the mid-1800s caused quite a stir, as it was the first evidence of a prehistoric being that resembled us but was definitely different. The Neanderthal had much larger jaws and eye sockets and particularly strong brow ridges. It still had a low forehead, like its ape ancestors, but its cranium had reached proportions that were slightly larger than those of humans today.

Neanderthals' increased cranial capacity may have freed them for the first time from evolutionary change directed solely by external biotic and abiotic factors. They could now modify their environment and transfer that knowledge from generation to generation. Neanderthal toolmaking progressed to the point where stones were not only trimmed but flaked in what is called the levallois technique (Bordes, 1968). They moved periodically as whole groups to favoured caves and camp-sites, where they made temporary shelters. Neanderthals left only limited evidence of art or symbolling (Appenzeller, 1998), but there is evidence that they took care of their crippled and sick relatives and they buried their dead. It is not known how well Neanderthals' language skills had developed, but they had the clear beginnings of culture (Holden, 1998). They were the first hominid to adapt to cold climes, living in northern Europe during the ice age.

Appearance of Modern Humans

The first evidence of modern *H. sapiens* or 'Cro-Magnon' people (Fig. 6.2) appears in 50,000-year-old deposits in the Middle East and, by 30,000 BP, fossil evidence has shown that these people were found all across Europe and Asia (Campbell, 1982). There has been much discussion over how Cro-Magnon could have appeared so quickly in so many places (Haviland, 1996; Balter, 2001). One theory suggests that our direct ancestors evolved simultaneously at several locations. This seems unlikely as one would not expect such widely dispersed populations to follow such similar patterns of evolution. Another theory is that Cro-Magnon originally emerged in the Middle East and, after a period of local adaptation, dispersed rapidly across the world. This seems more likely than the multiple origin idea, as it identifies a more plausible single origin and there is ample evidence that early humans could spread vast distances. As we shall discuss later, the entire New World was inhabited within 2000 years of the first arrival of *H. sapiens* in Alaska.

There has also been considerable debate about whether Cro-Magnon evolved from Neanderthals or diverged from a common ancestor (Leakey and Lewin, 1992; Cavalli-Sforza and Cavalli-Sforza, 1995; Tattersall, 1998). There are numerous intermediate forms that could have been a progenitor of both, including what has been referred to as *H. heidelbergensis*. It is possible that Neanderthal and Cro-Magnon evolved from a common ancestor in Africa, with Neanderthal diverging into a cold-adapted European subspecies and Cro-Magnon becoming a warm-adapted Middle Eastern subspecies. In Western Europe, Neanderthal appears as the 'classic' form, with massive brow ridges and a compact, squat body, but, in the Middle East, the most prevalent types have more 'generalized' skull features and in some cases appear to be intermediate to modern forms. The evidence for an African origin is not strong, but some very old fossils in Java and Africa show features common to *H. erectus*, Neanderthal and modern people.

Regardless of exactly how they evolved, Cro-Magnon and Neanderthal were very similar beings, and they coexisted for thousands of years (Gibbons, 2001b). A remaining question about these two groups is why Neanderthal became extinct while Cro-Magnon succeeded. Some have suggested that the two groups merged through interbreeding, but there is no fossil evidence of intermediate forms in the late periods of Neanderthal existence, and the cultural and physical differences between the two subspecies probably made sexual contact infrequent. It has also been suggested that Cro-Magnon displaced Neanderthal by conquest and slaughter, but, again, there is no evidence of large-scale slaughters of Neanderthal groups or their imprisonment. Probably the simplest explanation for Neanderthal's disappearance may be that the two groups began to compete for the same resources and Neanderthals were the losers. Neanderthal would not have been the first hominid line to go extinct.

The early Cro-Magnon strongly resembled modern human beings and had greatly developed toolmaking and art, but still no agriculture. They lived in large groups of cooperating families and the sharing of culture was very important to them (Prideaux, 1973). As more and more was learned, it was stored in the 'human data bank' and each new generation was able to build on the previously collected knowledge, rather than relearning from the beginning. Cro-Magnon were proficient big-game hunters who lived in small communities. They probably had semi-permanent camps where most people remained and satellite camps where small foray groups would operate. They learned to make a diverse array of multi-piece tools of stone, wood and bone (Clark, 1967). Antlers were gathered and used for a multiplicity of purposes. Among their food-gathering implements were nets and snares, fish-hooks made of bone, harpoons, spear-throwers and bows and arrows.

They were aesthetic and spiritual, as they buried their dead with ceremony and made jewellery, musical instruments and statues and produced magnificent cave paintings. They also had sewn clothing, which ultimately allowed them to move into regions that were even colder than Neanderthal could tolerate.

So what led to this apparent leap forward in culture? It was not simple brain size, as Neanderthals actually had slightly larger brains than Cro-Magnon. The voice-box may have developed to the point that modern language became possible, and this allowed the further development of culture. It is also possible that there were changes in brain organization that improved cognitive abilities, but left no fossil record. Since Neanderthal had already developed the rudiments of culture, it would not have taken much biological change to greatly accelerate the evolution of human society. In fact, at Châtelperron, France, there is evidence that Neanderthal was using a number of tools very similar to those of Cro-Magnon and may have been making ornaments from teeth and ivory beads. The debate rages as to whether Neanderthals made these advancements on their own or were copying Cro-Magnon. Regardless, it is clear that Neanderthals' culture was at the threshold of modern human culture.

Spread of *H. sapiens*

Once they were established in Europe and the Middle East, *H. sapiens sapiens* rapidly spread all over the rest of the world (Edwards and Cavalli-Sforza, 1964; Fig. 6.5). Japan was settled by 20,000 BP and Australia more than 30,000 years ago. Humans arrived in North America by around 20,000 BP

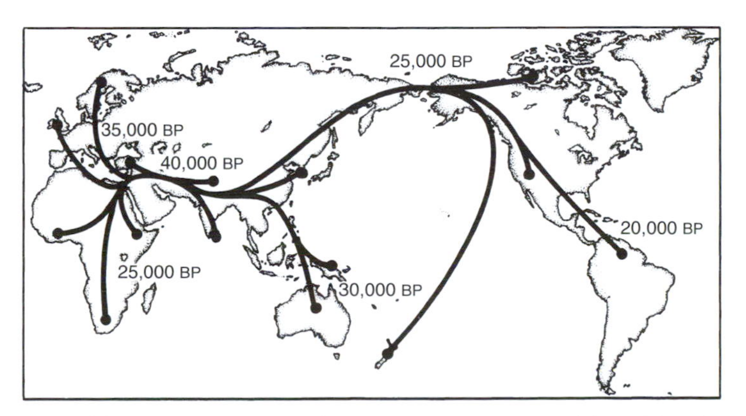

Fig. 6.5. Possible migration routes of *Homo sapiens* estimated from blood group frequencies of existing races of people (used with permission from L.L. Cavalli-Sforza and W.F. Bodmer, © 1971, *The Genetics of Human Populations,* W.H. Freeman and Company, San Francisco).

and by 12,000 BP had migrated to the tip of South America. When agriculture appeared 12,000 years ago, modern peoples were scattered over most of the world. Major mesolithic cultures had appeared in Europe, Australia, Siberia, Greece, northern Africa, the Near East, India, China, South-East Asia, South America, Mesoamerica and eastern and central North America.

The diffusion across the contiguous land mass stretching from Africa to Europe and Asia was probably relatively straightforward as people spread out in response to crowding and in search of food. Similarly, movement on to what are now the Indonesian islands of Borneo, Sumatra, Java and Bali was unimpeded, as the seas between them were dry when the early migrations occurred during the ice age (Diamond, 1998). The further migrations across open water to the other Indonesian islands would have taken the invention of water-craft, but were probably encouraged by the distant sight of land. The very long range migrations across open ocean to places like Australia and beyond took much more faith, courage and perhaps an extreme desire to escape crowding or warfare. The long oceanic migrations into the Pacific were begun by seafaring people from the Bismark Archipelago north of New Guinea about 3000 years ago and continued until 1000 BP with the settling of New Zealand and the central Pacific Islands. These seafarers had their origins in South-East Asia (Gibbons, 2001a; Kayser *et al.*, 2001).

North America was probably first reached by people from Siberia who crossed the Bering Strait on the land bridge exposed during the last great ice age (Marshall, 2001). The first unquestioned remains of humans in Alaska date to about 14,000 BP, but a much earlier arrival of humans is likely. The movement may have been enticed by a much milder climate than we know today. Once people arrived in the New World, they spread out rapidly, leaving fossil evidence all over North America and reaching the tip of South America by 12,000 BP, a migration rate of almost 16 km a year.

It is also possible that some people arrived in North and South America by routes other than the Bering Strait (Dillehay, 2000; Marshall, 2001). Ancient human skulls have been found in Brazil, Middle America and the Pacific North-West that resemble those of South Asians, Polynesians and even African bushmen. If this is true, some early Americans must have arrived by canoe from Asia and/or Europe, hugging the frozen coastlines of North America to the beyond.

As people spread across the world, they learned to use the resources available in each area and became specialized hunters and gatherers. In some parts of Europe, people were primarily big-game hunters who followed a diverse array of herd species, including deer, pig, cattle, horses, elk and foxes. In other parts of Europe, people were more stationary, relying on fishing for the bulk of their calories. Along the coasts of both the New World and the Old, people collected shellfish. Across broad expanses of western parts of California and northern Mexico, people became specialized gatherers of plant food. Hunter–gatherers in the Zagros Mountains of

Iraq relied on a combination of wild goats, sheep and wild cereal grasses for their sustenance. The stage was now set for the emergence of a diverse agriculture, one that was specifically tuned to the natural resources available in each region.

Agricultural Origins

Agriculture arose independently at several locations across the world, beginning about 12,000 years ago. Vavilov (1926, 1949–1950) originally identified eight centres of domestication, based primarily on patterns of crop diversity. These were modified by Harlan (1967) who used a combination of archaeological evidence and the native ranges of crop progenitors (Fig. 6.6). He identified three relatively small geographical regions, which he called centres, and another three rather diffuse regions, which he called non-centres. He felt that the centres were probably independent of each other, but the non-centres may have had some contact with adjacent ones. The three centres he envisioned were the Near East (parts of Jordan, Syria, Turkey, Iraq and Iran), Mesoamerica (Mexico and Central America) and north China. He considered Africa, South-East Asia and South America to be non-centres. Recently, Bruce Smith (1989, 1998) added eastern North America as a centre of crop origin.

Fig. 6.6. Earliest places where agriculture began: A1, Near East centre; A2, African non-centre; B1, north Chinese centre; B2, South-East Asian and South Pacific non-centre; C1, Mesoamerican centre; C2, South American non-centre; D, eastern North American centre (figure modified from Harlan, 1967).

Most evidence indicates that human beings first began farming in the hills above the Tigris River on the western edge of what is now Iran. So many ancient agricultural sites have been located in this region that Harlan and Zohary (1966) have suggested that the Near East be thought of as a 'nuclear area' where events at one location influenced others. Among the first crops domesticated in this region were emmer and einkorn wheat, barley, pea, lentil, chickpea and flax. Goats and sheep were the first animal domesticants.

Agriculture also began at a very early period in the forest margins of West Africa, Ethiopia and along a transcontinental band through the savannah. Ethiopia has provided the earliest evidence of African farming, dating from the 6th millennium BP (Phillipson, 1977), but the other origins are probably just as old (Harlan, 1992). People in the Sahel domesticated sorghum, pearl millet, African rice (*Oryza glaberrima*) and guinea-fowl. West Africans were responsible for African yams, cowpea and the oil-palm. Ethiopians domesticated coffee, finger millet and teff.

Data on China and Asia are still emerging, but they were clearly early areas of plant and animal domestication. South-East Asia is the greatest puzzle since its hot humid conditions result in the rapid degradation of food materials (Li, 1970); however, there is evidence that cultivation was practised by 8000 to 10,000 BP in Thailand (Gorman, 1969) and New Guinea (Golson, 1984). Two areas in China also show evidence of agriculture by about 8000 BP, the Yellow and Wei River basins in the north and the Yangtze Valley in the south (Smith, 1998). Banana and sugar cane were first domesticated in South-East Asia. Millets, rice, soybean and pigs were among the early Chinese domesticants.

Farming probably began independently in the New World, 1000–2000 years later than in the Old World. There was a relatively compact Mesoamerican centre extending from Mexico City to Honduras, while South American crops emerged in a broad area covering most of coastal and central South America. Coastal Peru is often described as a focal point of early South American agriculture, but the data are somewhat biased, since most archaeological work has been performed on dry coastal sites where plant material is readily preserved. Three ecological/geographical regions of domestication have been proposed: (i) an Andean high-elevation complex; (ii) an Andean mid-elevation complex, and (iii) a lowland complex (Pearsall, 1992). The centre in eastern North America spanned the area between the Appalachian Mountains and the western borders of Missouri and Arkansas (Smith, 1989).

A diverse group of crops and animals were originally developed in the Americas (McClung de Tapia, 1992; Pearsall, 1992). Quinoa, potatoes and several other tuber crops were domesticated in the Andean high-elevation complex. Amaranth, groundnut, coca, common bean and lima bean were part of the Andean mid-elevation complex. Avocados, chilli peppers, cotton, manioc, pawpaw, pineapple, squash, sweet potato and yams were domesticated in the tropical lowlands of Central and South America. Recently, fruit phytoliths (silica deposits) have been recovered from domesticated *Cucurbita* in early Holocene archaeological sites in south-western Ecuador (12,000–10,000 BP) (Piperno and Stothert, 2003), and starch grains of manioc, yam and arrowroot have been found on ancient plant milling stones from the Panamanian tropical forest (7000–5000 BP), supporting the independent emergence of plant domestication in the lowland neotropical forest (Piperno *et al.*, 2000). Species of amaranth, avocado, beans, chenopod, cotton, chilli peppers, pumpkin and squash were domesticated independently in Mesoamerica, along with maize. Turkeys were the first animals domesticated in Mexico, while llamas and guinea-pigs were the first animals tamed in South America. North Americans were responsible for sunflower, sumpweed (*Iva annua*), goosefoot (*Chenopodium berlandieri*) and probably squash (Smith, 1998).

Early Crop Dispersals

Near East grain culture spread rapidly to Europe, West Africa and the Nile Valley (Ammerman and Cavalli-Sforza, 1984; Zohary, 1986). By the 8th millennium BP, Near East crops appeared in Greece, Egypt, along the Caspian Sea and in Pakistan. Central Europe was heavily farmed less than 1000 years later, and by 5000 BP farming communities spanned the area from coastal Spain to England to Scandinavia. Wheat and barley reached China about 4000 BP (Ho, 1969). The Chinese literature of 3000 to 2000 BP mentions the 'five grains': millet, glutenous millet, soybean, wheat and rice (Whittwer *et al.*, 1987).

Most of the Near East founder crops (emmer wheat, einkorn wheat, barley, lentil, pea and flax) travelled across Europe as a group (Fig. 6.7), picking up other crops along the way. Oats and flax began as weeds moving with the Near East assemblage, but were eventually exploited and became secondary crops (Harlan, 1992). Many of the vegetables that appeared in Europe and the Mediterranean regions were probably developed through this route (Zohary, 1986).

The spread of agriculture across the Middle East and Europe could have been caused by cultural diffusion, where the new techniques were transmitted by simple learning, or by migration, where the transfer was associated with population expansion and intermating (Ammerman and Cavalli-Sforza, 1984). Sokal *et al.* (1991) tested these two hypotheses by examining 26 polymorphic blood proteins of extant people from 3373 locations across the Near East and Europe. They found a clinal trend in the allelic frequencies at six loci that was significantly correlated with the local dates of agricultural settlement. This lends support to the migration hypothesis, by which the original stock of Near Eastern agricultural peoples was slowly diluted as their descendants moved west and mated with local peoples along the way.

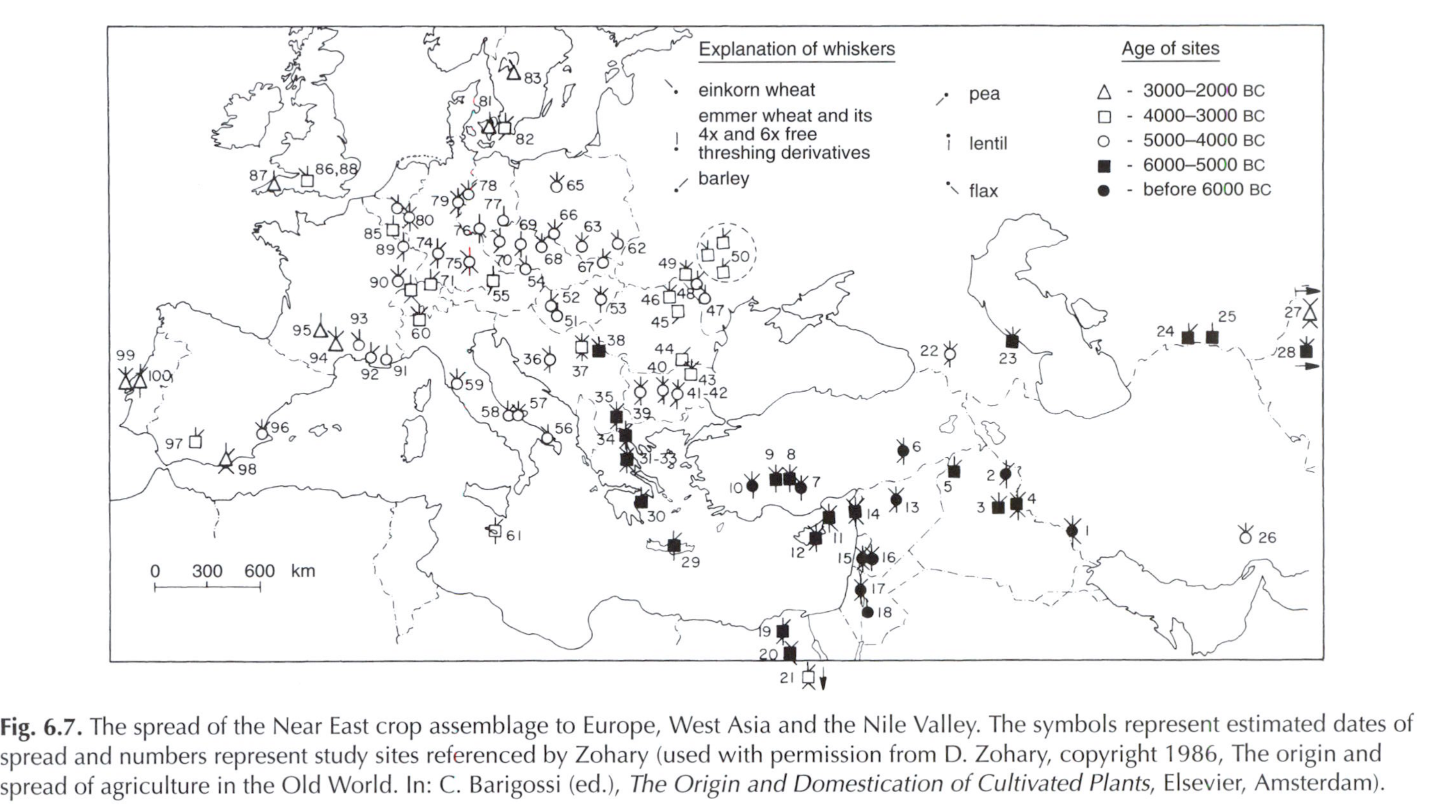

Fig. 6.7. The spread of the Near East crop assemblage to Europe, West Asia and the Nile Valley. The symbols represent estimated dates of spread and numbers represent study sites referenced by Zohary (used with permission from D. Zohary, copyright 1986, The origin and spread of agriculture in the Old World. In: C. Barigossi (ed.), *The Origin and Domestication of Cultivated Plants*, Elsevier, Amsterdam).

Indigenous African agriculture was considerably more diffuse than the Near East agricultural complex, and the African crops lacked cohesion. However, the early agriculture of Africa was associated primarily with the savannah and it probably spread out from the Sahel and Guinea zones southward towards East Africa (Vavilov, 1949–1950; Harlan, 1992). Sorghum, pearl millet (*Pennisetum americanum*), finger millet (*Eleusine corocana*) and cowpea reached India about 4000 years ago from Africa, along with cotton, sesame and pigeon-pea. While these crops became very important across all of Asia, native Asian crops were of little significance in Africa except for Asian rice (*Oryza sativa*), which was utilized where African rice (*O. glaberrima*) was already established. Most of the dispersals out of South-East Asia were seaward towards the Malay Archipelago (2000 BP) and then to the far-off South Pacific Islands. Agriculture reached remote Hawaii and New Zealand from South-East Asia about 1000 years ago (Emory and Sinoto, 1964).

Agriculture gradually spread across China over a period of several thousand years (Wittwer *et al.*, 1987). Rice was probably introduced into South-East Asia from South China 5000 to 4000 years ago (Crawford, 1992; Smith, 1998). Soybean remained close to home until recent history. The millets, *Setaria indica* and *Panicum miliaceum*, were found in neolithic villages in Europe, and may have been introduced from China. However, the possibility of independent domestications cannot be excluded, as the Chinese centres were so remote and no other crops show similar distributions.

Several new crops reached the Near East from Asia and Africa approximately 3000 to 2000 BP. Sorghum, sesame and the Old World cottons came from Africa, while common rice entered from Asia. These initiated summer crop agriculture as an integral part of food production (Zohary, 1986). Fruit-trees also arrived from the east about 2000 years ago, including apricot, peach and citron (Bailey and Hough, 1975; Hesse, 1975).

From the Mesoamerican centre, a maize–bean–squash assemblage gradually moved northward, picking up sunflower and numerous other native species on the way, to eastern North America, where it was well established by 4500 BP (Chomkos and Crawford, 1978; Berry, 1985). Upon its arrival, it displaced the indigenous crops of sumpweed (*I. annua*) and chenopod. There is an ongoing debate as to whether these Mesoamerican crops diffused across the West Gulf Coast Plain or through the American south-east on the march west (Story, 1985). Movement south from Mesoamerica is difficult to trace for most crops, but at least maize had arrived in Central America (Piperno *et al.*, 2000) and the Amazonian basin by 5000 to 4000 BP (Pickersgill, 1969; Bush *et al.*, 1989). The South American domesticants, potato, groundnut and lima bean, had reached north to Mexico by 3000 to 2000 BP, travelling through either the Caribbean Islands from Venezuela or via Central America. American crops were not known in the Old World until the ocean explorations of Columbus (Harlan, 1992).

Transcontinental Crop Distributions

Until the last 500 years virtually all crop dispersal was within continents and not between them. There was spread into Oceania from South-East Asia, but little movement occurred between the two hemispheres. The complete homogenization of world crops began with Columbus in 1492 with his discovery of the New World and his subsequent attempts to settle Hispaniola. Columbus introduced banana, cabbage, carrot, chickpeas, citrus, cocoa, cucumber, grape, melon, olives, onions, radishes, rice, sugar cane and wheat to the New World and took back the sweet potato and maize.

The Spanish and Portuguese sailors completed the bulk of world crop homogenization, through their routes of exploration and trading in the 1500s and 1600s. Ultimately, the settlers tried to transfer the entire European agricultural system to the New World, with varying levels of success, depending on the climate of the settled region (Butzer, 1995; Simmonds, 1995c). In Mesoamerica, Central America and the Andean region, the Spanish introduced barley, chickpea, cucumber, fig and wheat, which were originally domesticated in the Near East, citrus, pear and peach from China, melons from Africa and cabbage, lettuce, grapes and onions from the Mediterranean. They brought back to Europe cotton, potato, chilli pepper and maize from Mesoamerica, and tomato, beans, maize and groundnuts from South America. The Spanish also carried maize, sweet potato and groundnut to the Philippines, and from there they ultimately found their way to China. The Portuguese introduced a wide array of crops into Brazil, including chickpea, faba bean, fig and wheat from the Near East, sugar cane and banana from South-East Asia, peach and citrus from China, sorghum from Africa and grapes from the Mediterranean. They moved cassava, common bean, cotton, lima bean, groundnut, maize and sweet potato from the New World to Africa. In the late 1600s, the English and French explorers/colonists also introduced the full array of European crops to North America and brought back sunflower and the wild strawberry, *Fragaria virginiana*.

Today, the major crops grown in many parts of the world are far from their origins and, in fact, many regions are now dependent on alien crop species (Table 6.3). Sorghum, millet and yam were originally dominant in Africa, but now the most widely grown crops are maize from Mesoamerica, cassava and sweet potato from South America and banana from South-East Asia. Banana was an ancient introduction from South-East Asia, but the other three were post-Columbian. Europe and North America are now almost totally dependent on crops from elsewhere, including wheat and barley from the Near East, maize from Mesoamerica, potatoes from South America and soybean from China. Rice and soybean have remained important in China, but maize, sweet potato and potato are now almost as important. At almost all locations in the world, people are now dependent on crops originally domesticated at distant locations.

Table 6.3. Dependence of various regions of the world on outside crops. Dependence is based on percentage of total production (source – Kloppenburg, 1988).

Region	Dependence	Indigenous crops	Major imports	Origin of imports	Period of introduction
Africa	87%	Millet	Banana	South-East Asia	Ancient
		Sorhum	Cassava	South America	Post-Columbian
		Yam	Maize	Mesoamerica	Post-Columbian
			Sweet potato	South America	Post-Columbian
China	60%	Millets	Maize	Mesoamerica	Post-Columbian
		Rice	Peanut	South America	Post-Columbian
		Soybean	White potato	South America	Post-Columbian
			Sweet potato	South America	Post-Columbian
Europe	90%	Oats	Barley	Near East	Ancient
			Maize	Mesoamerica	Post-Columbian
			White potato	South America	Post-Columbian
			Wheat	Near East	Ancient
South America	56%	Cassava	Maize	Mesoamerica	Ancient
		Sweet potato	Wheat	Near East	Post-Columbian
		White potato			
		Yam			
North America	80%	Beans	Barley	Near East	Post-Columbian
		Maize	Wheat	Near East	Post-Columbian
		Squash	White potato	South America	Post-Columbian
			Soybean	China	Post-Columbian

Summary

Land plants arose from the sea as improved methods of utilizing and conserving water evolved. Some of the earliest adaptations that appeared were cuticles, roots and resistant spores. Later adaptations were conducting systems, stomata, leaves and land-based reproductive systems. The first angiosperms with double fertilization and fruit appeared hundreds of millions of years ago. The history of human evolution was intimately associated with that of plants. The appearance of our primate ancestors was dependent on the emergence of angiosperms, and most of our crop genera had evolved long before we began agriculture. All our ancestors had to do was learn how to effectively exploit the available plant foods. Human evolution went through several stages involving face and teeth structure, mode of locomotion and cranial capacity. The earliest primates lived in trees and had flexible hands and feet, but could not swing. Gradually, quadrupedal monkeys and apes appeared with the ability to brachiate and to chew hard foods. *Australopithecus* evolved from apes at the edge of the forest, with the ability to occasionally stand upright and use tools. Finally our early hominid ancestors arose, with a continuous upright posture and the ability to make tools. Modern humans, *H. sapiens sapiens*, emerged about 50,000 years ago and coexisted for thousands of years with an even earlier form of humans, *H. sapiens neandertalensis*. Agriculture first began in the Near East about 10,000–12,000 years ago, but at least five other areas soon followed, including China, South-East Asia, Africa, South America and Mesoamerica. A separate origin is also postulated for the eastern portion of North America. Crops were gradually moved across continents in antiquity, but it was not until the last 500 years that intercontinental movement began.

The Dynamics of Plant Domestication

7

The sudden, scattered appearance of agriculture all across the globe suggests that farming was an important step in the evolving culture of human beings. A question that has intrigued anthropologists and ethnobotanists alike is why it took so long for farming to emerge (Pringle, 1998). It seems likely that people had the wherewithal to farm long before they actually began doing it. Our ancestors surely gained considerable knowledge about plants and animals through the very acts of hunting and gathering. They had observed seasonal patterns of plant development and animal migrations, and noticed seeds germinating and growing on their dump heaps. Our antecedents burned fields to drive game, and must have noticed the subsequent plant regenerations. They had developed an intimate knowledge of how countless plant species could be used for food and medicine, and knew how to detoxify otherwise poisonous food sources.

Probably the oldest formal idea about why humans began cultivation is Childe's 'oasis theory' (1952). He suggested that, after glaciation, North Africa and south-west Asia became drier and humans began to aggregate in areas where there was water. People first learned how to domesticate the animals that congregated around them and then, as human populations grew, they learned how to raise crops to avoid starvation.

While this theory is an appealing explanation for agriculture at xeric sites, it is now known that mesic areas in South-East Asia and tropical South America also spawned agriculture. In addition, the climate may not have been as harsh as Childe imagined. Evidence suggests that even the dry Zagros Mountains of the Near East may have been shifting from a cool steppe to a warmer and perhaps moister savannah when agriculture was beginning in that area (Wright, 1968).

Sauer (1952) suggested that farming first arose among fishermen in South-East Asia. They had a dependable food source, were sedentary and therefore had the time and strength to experiment with new food production systems. Again, this theory works well in areas where fish and crustaceans were readily available, but it does not explain the origin of agriculture in dry places without seafood, such as Mesoamerica and Central Africa. In addition, there is evidence that not all fishing people took up agriculture readily, as the Natufion fishing people of the Near East were among the last to take up agriculture in the Fertile Crescent, even though they actively gathered wild plant material (Harlan, 1992).

Many anthropologists have related population growth with the rise of agriculture. It is thought that, as populations grew, food requirements rose to the point where alternative sources were needed to supply adequate resources (Cohen, 1977). Cities are thought to have been possible because of agriculture, and the subsequent growth of populations led to specialization in jobs and the need for a farming class. While this theory holds great appeal, it is difficult to eliminate the possibility that population growth was often stimulated by the advent of agriculture rather than the reverse. There is also evidence that quite large cities could arise with only the most rudimentarily developed agriculture (Balter, 1998).

Binford (1968) and Flannery (1968) combined the population pressure and sedentary hypotheses into what is called the 'marginal zone hypothesis'. They envisioned communities of fishermen who were initially sedentary, but as populations grew they moved out to more marginal regions. They were knowledgeable botanists who were used to gathering food in a restricted area, and developed agriculture as a means of feeding themselves. They became farmers not only because of population pressure, but because they were sedentary and in competition with the original gathering people. Again, this theory is plausible for many locations, but not all agricultural people have fishing ancestors.

There are suggestions that agriculture arose as a by-product of religious ceremony (Hahn, 1909; Anderson, 1954; Heiser, 1990). Plants providing ritualistic drugs were gathered and perhaps grown. Seeds may have been scattered on burial mounds. Animals could have been domesticated for sacrifice. While religion would have been a strong impetus for neolithic peoples to apply what they knew about the life cycles of plants and animals, we are still left with our original question of why it took so long for people to begin the farming process. As was outlined earlier, there is considerable evidence that people were spiritual long before they began domesticating plants and animals.

The simple answer to why it took us so long to begin farming is probably that hunting and gathering was a very comfortable way of life, and humans had to have a very good reason to give it up. Juliet Clutton-Brock (1999) states that 'with the abundance of food and excellent raw materials of wood, bone, flint and antler it is difficult to see what the mesolithic people of Europe lacked'. Food gathering did not need to be an intense daily activity. Richard Lee (1968) in an examination of !Kung Bushmen of the African Kalahari Desert found that they spent an average of 2.3 days a week in food gathering. They had sufficient food to consume 2140 calories a day, which is above the US Department of Agriculture (USDA) recommended daily allowance for small vigorous people (1975 calories). Harlan (1992) has shown that enough wild wheat can be collected in a few hours in the Near East to provide adequate nourishment for over a week. Even casual gardeners know that farming is hard work, and early crop production must have been just as subject to the vagaries of nature as gathering, particularly before the advent of irrigation.

Palaeolithic people were complex, intelligent creatures who could readily adapt to the situations at hand. Sauer (1952) suggested, 'We need not think of ancestral man as living in vagrant bands, endlessly and unhappily drifting about. Rather, they were as sedentary as they could be and set up housekeeping at one spot for as long as they might.' They liked hunting and gathering, and were pushed towards farming only by a variety of regionally specific forces, including population growth, climatic change, overhunting, religion or a simple desire for more of something in short supply, be it food, spice, oil, ceremonial colour or fibre. Food production is only one of the possible reasons for bringing plants under cultivation. Harlan (1992) has referred to this possibility as the 'no-model model', where the reasons for farming are as diverse as the people and environments found at the focal points of agriculture. Farming began gradually in a number of diffuse areas as something desirable became scarce, and only began in earnest when natural plant and animal populations were not sufficient to feed the growing population of people. Already sedentary fishing peoples may have made this transition more easily than nomads, but there is no reason to assume that all human populations could not make the transition when they had good reason to do so. We waited so long to farm, because we could.

Evolution of Farming

The shift from the hunter–gatherer strategy to farming probably occurred in stages (Fig. 7.1). For millions of years, our ancestors subsisted on the bounty provided by our natural environment. Our earliest upright ancestors may not have had a particularly orderly approach to finding food, but, by the time of *Homo erectus*, hominids were surely collectors, who planned the use of resources whose location was known and monitored. By the time Cro-Magnon appeared, *Homo sapiens* must have had considerable knowledge about how plants and animals developed, and were returning to the same areas year after year to harvest and hunt dependable sources. This probably led to significant changes in the plant populations, a topic we shall return to later.

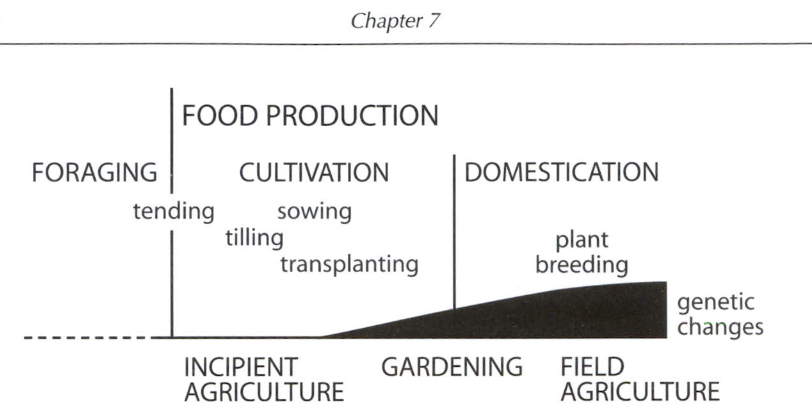

Fig. 7.1. Development of food production methods over time (used with permission from R.I. Ford, © 1985, *Prehistoric Food Production in North America*, Museum of Anthropology, University of Michigan, Ann Arbor).

Once Cro-Magnon was returning regularly to the same spot, it may not have taken them long to become cultivators, who enhanced the productivity of native fields by weeding, pruning and burning. They probably began tilling with a digging stick or hoe to reduce competition and encourage germination. They may also have discovered at an early stage that crops did better the following year if the soil was turned after harvest. This may have been particularly apparent in sunflower, which produces alleochemicals that inhibit germination, but readily detoxify if exposed to air (Wilson and Rice, 1968).

Eventually, Cro-Magnon became producers, who transplanted small numbers of plants and held a few animals captive. These early gardens were probably very small and in close proximity to residences, and remained small until humans decided to make a major commitment to agriculture. Edgar Anderson (1954) has suggested that the original idea for planting may have come from waste dumps, where seeds were observed to germinate and grow. Larger farms may have first appeared when a specific farmer class emerged. During the early stages of the domestication process, *H. sapiens* may not have been consciously selecting superior plant types, but it would not have taken us long to become domesticators who saved seeds and clonal material of superior types for replanting.

Early Stages of Plant Domestication

Evidence for plant domestication comes from a variety of sources including: (i) carbonized remains formed by high temperature 'baking' under low oxygen; (ii) impressions pressed into pottery and bricks; (iii) parched plant remains produced under extreme dry conditions; (iv) plant material sunk in peat bogs or mud under anaerobic conditions; (v) impregnations of metal oxides; (vi) mineralization where cell cavities are replaced by minerals such as silica (phytoliths); and (vii) faecal remains (Smith, 1998). In these deposits, domestication is thought to be signified by substantial increases in the size of seeds (Fig. 7.2), dramatic reductions in seed- or fruit-coat thickness (Fig. 7.3) and the apparent loss of dispersal organs. This evidence is often supplemented by human artefacts that give clues about diet, such as sickles and grinding wheels.

Soil from ancient settlements is sometimes passed through screens to obtain small objects, but this technique often allows valuable material such as seeds to be lost (Smith, 1998). Flotation techniques are now widely employed, where the archaeological soils are poured into water so that organic materials will float to the top and can be recovered. Cloth filters are utilized to catch the materials as the water is poured off.

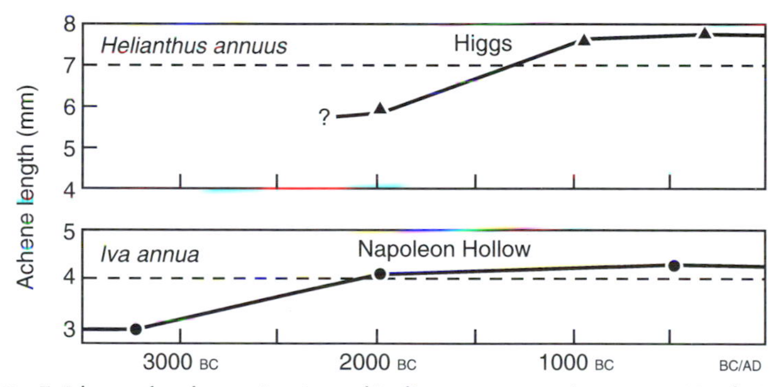

Fig. 7.2. Evidence for domestication of indigenous crops in eastern North America based on increases in achene size. The dashed line represents the baseline used for domestication. (Used with permission from B.D. Smith, 1989, Origins of agriculture in eastern North America, *Science* 246, 1566–1571).

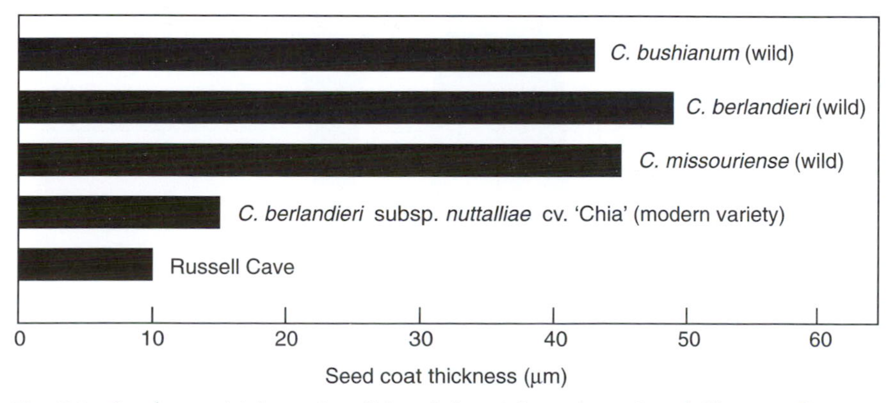

Fig. 7.3. Seed-coat thickness in wild and domesticated species of *Chenopodium*. The seed-coats of domesticated *C. berlandieri* subsp. *nuttalliae* from the Russell Cave were much thinner than the wild forms. (Data from B. Smith, 1998.)

Dating of remains is done by analysing the ratio of the isotopes ^{12}C and ^{14}C. After incorporation into organic material, ^{12}C remains stable while the radioactive isotope ^{14}C gradually disintegrates, with a half-life of 5568 years. Most estimates are based on the assumption that ancient atmospheres were similar to those of today, but corrections have been made by evaluating carbon isotope ratios in the annual rings of trees. Ratios are available for the last 8000 years from bristlecone pines in California and for the last 9000 years from oaks in Ireland.

At most of the early agricultural sites, the transition from hunter–gatherer to farmer was a gradual one that took thousands of years. A very early record of this slow transition is found in the excavations of Richard MacNeisch *et al.* (1967) in the Tehuacán Valley of Mexico (Fig. 7.4). He excavated 12 sites and uncovered 12,000 years of agricultural history in the area. Initially, the people lived on wild plant food and small animals, such as jack rabbits, deer, peccary and lizards. They collected plant foods on a scheduled round of annual activities. About 9000 years ago, game became more scarce and the people began to shift more of their energy into the collection of wild plants, including squashes, chilli peppers and avocados. They scattered in small foraging groups during the dry season, and came together

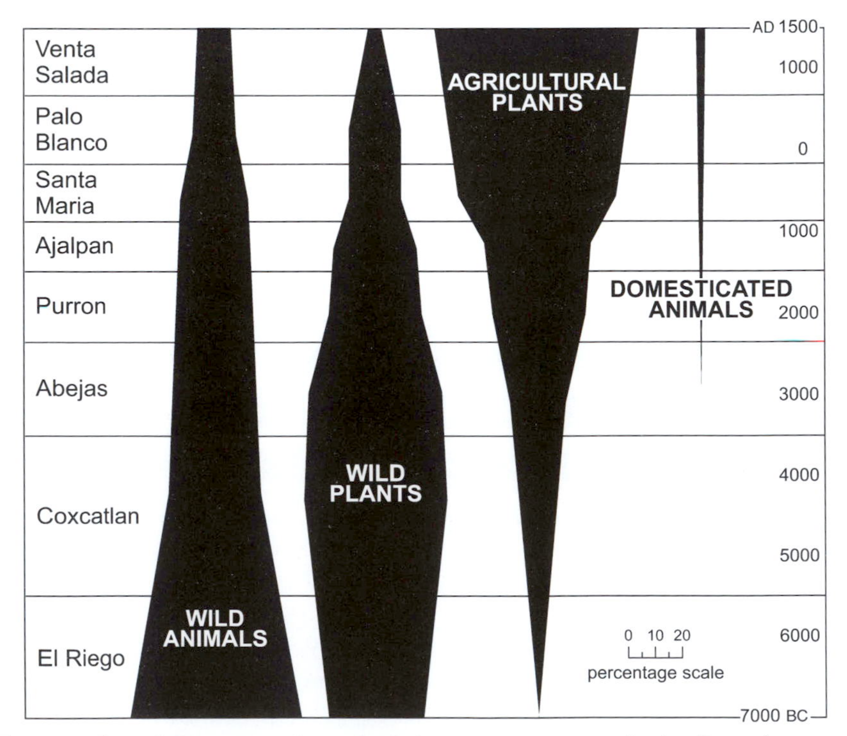

Fig. 7.4. The relative proportions of subsistence components in the diets of Tehuacán Valley inhabitants over time (from MacNeisch *et al.* 1967).

during seasons of plenty. They may have begun the sporadic cultivation of wild plants during this period, but the effort was minimal.

Over the next 5000 years, the people of the Tehuacán Valley gradually increased their use of domesticated plants, such that, by 7000 before present (BP), about 10% of their diet came from cultivated plants. They were outside the original areas of domestication, but by this time were growing a large group of presumably introduced crops including maize, amaranth, beans, squashes and chillies. The maize ears were only about the size of a pencil eraser, but the plant now existed in its modern form. The dog appeared about 5000 years ago. As time went on, people continued to devote more and more effort to farming and, by 3000 BP, the majority of their food came from domesticated sources, with maize being grown, along with avocados, amaranth, squash and cotton. Turkeys were domesticated about 2000 years ago.

Similar evidence of transitions from hunters to farmers can be found at numerous locations across the Near East. One such site is Jericho in the Jordan Valley, where a continuous record of 9000 years of habitation was left as people built new mud huts on top of others as they deteriorated over time. In the earliest period, the settlement consisted of Natufians, who were primarily hunters of gazelles and foxes and who tended a few cereals but had no domestic animals (Hopf, 1969). About 9000 years ago, they began to raise cereals in earnest, and there is the first evidence that sheep and goats were being domesticated. A similar long-term record of successive settlement is recorded at Catalhöyük, Turkey, where people were initially foragers who raised a few cereals on the side, but by 10,000 BP had domesticated cattle and were large-scale farmers (Balter, 1998).

Origins of Crops

Almost all of our major crops had been domesticated by 5000 BP, with the earliest grains and legumes appearing in the Near East over 10,000 years ago (Table 7.1). The first crops were as diverse as the people and places where agriculture began. Different native plant assemblages were located in each of the early farming areas and, as a result, variant plant and animal species were domesticated in each of them. In the Middle East, there were huge natural stands of wheat and barley, and as a result, the early farmers in this region exploited these as their staple crops. In South-East Asia, wheat and barley were absent, but large-grained rice was plentiful and, as a result, rice became one of the crops of choice. Wheat, barley and rice were not present in Mesoamerica and Africa, so people exploited the locally abundant monocotyledons: sorghum in Africa and maize in Mesoamerica. No large-grained species of any kind existed in South America and, as a result, the early farmers there domesticated the tuberous species potato, sweet potato and cassava and the pseudograins chenopod and amaranth.

Table 7.1. When and where our major crops were probably first domesticated.

	Years before present			
Place	12,000 to 8000	8000 to 4000	4000 to recent	Recent
Early centres				
Africa (except Egypt)	Yam	Cowpea Finger millet Musk melons Pearl millet Sorghum Water melon	Coffee Oil palm	
China	Broom-corn millet Rice	Foxtail millet Hemp Peach Pear Soybean Tea	Onion	
Mesoamerica	*Cucurbita* squash *Lagenaria* gourds	Avocado Chilli peppers Common beans Cotton (*Gossypium hirsutum*) Grain amaranth Maize	Agave Cocoa Tobacco Tomato	
Near East	Barley Chickpea Garden pea Lentil Wheats	Date-palm Faba bean Fig Flax Olive Grape		
North America		*Cucurbita* gourds	Goosefoot Sumpweed Sunflower	Blueberry
South America	Chilli peppers Common bean Sweet potato	Cassava Cotton (*Gossypium barbadense*) *Cucurbita* squash Groundnut Guava *Lagenaria* gourds Potato Quinoa Sweet potato	Pineapple Quinine Strawberry	
South-East Asia	Yam	Coconut Sugar cane Taro	Banana Citrus	Rubber
Later centres				
Central Asia		Apple	Black mustard Pistachio	
Europe		Rye Rape	Oats	Strawberry

Continued

Table 7.1. *Continued.*

	Years before present			
Place	12,000 to 8000	8000 to 4000	4000 to recent	Recent
Later centres *(Continued)*				
Indus Valley		Cotton (*Gossypium arboreum*)	Aubergine	
		Cotton (*Gossypium herbaceum*)	Black pepper	
		Cucumber	Pigeon-pea	
		Sesame		
Mediterranean (including Egypt)		Almond	Cabbage	
		Lettuce	Celery	
		Onion	Safflower	
		Radish	Sugar beets	
			Turnip	

Starchy staples were among the first domesticants at all the centres of crop origins, but they were always complemented with a high-protein vegetable and fibre crop. Vegetables in the legume family were domesticated in all the major regions, including cowpea (Africa), soybean (China), groundnut (South America), lentil and chickpea (Near East). Amaranth and chenopod were also very important sources of vegetable protein in the New World. Fibre was provided by different cotton species in Africa and South America, flax in the Near East and hemp in China.

To the core group of crops in each region, additional leafy vegetables, spices, oil crops and fruits were gradually added. Among the last group of plants to be domesticated were the fruits. While the grape and fig are very ancient and may have been cultivated for 10,000 years, most of our other woody fruit crops were among the last additions to farming. This may be due, in part, to tree fruits taking so long to mature, since after planting the farmer must wait 5–10 years for a harvest. In addition, the fruit crops are outcrossed species, whose seedlings would frequently be inferior to the mother plant due to cross-pollination. Complex pruning and grafting techniques had to be developed to fully exploit their potential.

Characteristics of Early Domesticants

Among the earliest crops, three were grains in the *Gramineae* (barley, rice and wheat), two were fleshy tubers in *Convolvulaceae* (sweet potato) and *Dioscoreaceae* (yams), four were seeds in the *Leguminoseae* (chickpea, garden pea, lentil and common bean) and two were fruits in the *Solanaceae* (chilli peppers) and *Cucurbitaceae* (squashes and gourds). Some species were domesticated more than once (dry beans and chilli peppers), while others had relatively restricted origins (maize). Some were domesticated for more than one plant part (beets, mustard and squash).

This early group of domesticants represents a broad array of families, but most of them are herbaceous annuals, capable of selfing. It is not hard to speculate as to why so many of these early crop species are annual and self-pollinating. Annuals allocate a high proportion of their energy to reproduction and, as a result, produce unusually large seeds, fruits or tubers that are readily harvested. As Sauer suggests in his book *Historical Geography of Crop Plants* (1993):

> the annual habit offered farmers a syndrome of advantages – quick results after planting, heavy seed production in a sudden burst for a single harvest, escape from unfavourable cold or dry seasons by storage of seed indoors allowing spread to diverse climatic regions, and periods of fallow or rotation preventing buildup of pests and parasites.

Self-pollinating species breed true to type and any seed collected will be similar to the parent. Also, they do not need a pollinator for successful reproduction, increasing the dependability of their reproduction and ensuring seed production even if only a few genotypes were selected by early farmers. They would also be at least partially isolated from their wild progenitors, which would aid in the maintenance of their integrity. In most cases, the annual, selfing habit of individual crops was derived directly from native progenitors; however, the cultivated annual cottons evolved from wild perennial shrubs and the self-pollinating soybean was converted from a wild outcrosser to a cultivated inbreeder.

The only major crop species that are not annual selfers are: (i) maize, pearl millet and quinoa, which are annual but outcrossed; (ii) cassava and sugar cane, which are both outcrossed and perennial; and (iii) a number of woody fruit crops that are also both outcrossed and perennial. Maize, pearl millet and quinoa were domesticated in areas where there were few large-grained grasses available, and so primitive farmers had no other option but to work with outcrossed grains. Pearl millet is not large seeded, but does produce a large inflorescence. Cassava and sugar cane, while being perennial outcrossers, are relatively easy to propagate by cuttings and as a result were easily domesticated. Woody perennials were very difficult to propagate, but their fleshy fruits offered great rewards to the people who figured out how to enable them to proliferate.

Both Harlan (1992) and Diamond (1998) have made the observation that most of the large-seeded grain and tuber crops were domesticated in the savannah or Mediterranean climates of the Middle East, China, Mesoamerica and South America. These climates have long annual periods of drought, which would have favoured plants with large annual seeds or tubers. Seeds can survive long periods of drought and germinate when rains come, and likewise tubers can remain quiescent until moisture levels encourage sprouting. The few grains that were domesticated in eastern North America were not drought adapted, but large seeds are advantageous in areas with periods of extreme cold. Here seeds remain dormant during the winter, and then germinate when temperatures warm up in the spring.

Changes During the Domestication Process

Once plants were domesticated, they were dramatically altered by humans through both conscious and unconscious selection. Simmonds has made the point in his book on *Principles of Crop Improvement* (1979) that 'probably the total genetic change achieved by farmers over the last 9,000 years was far greater than that of scientific breeders in the last 100 years'.

At early stages in the domestication process, a number of changes began to appear in the genetic and physiological make-up of many crop species (Table 7.2). Some of these changes were due to conscious selection such as increases in palatability and colour, but many of the others were the unconscious by-product of planting and harvesting. Harlan *et al.* (1973) recognized a whole syndrome of traits associated with inadvertent selection due to the broadcasting and harvesting of grain crops (Table 7.3). Domestications of other broadcast seed crops such as legumes produced similar patterns (Zohary, 1989). Harvesting resulted in the selection of the non-shattering trait, more determinate growth, more uniform ripening and increased seed production. All of these characteristics would have increased the likelihood that the seed of a genotype would be collected and subsequently planted. Other characteristics associated with enhanced harvests would be selection towards erect types with synchronous tillering, and an increase in the size of inflorescences, number of fertile florets per inflorescence and the number of inflorescences.

Seedling competition caused by planting in close proximity probably increased seedling vigour and rate of germination, as individuals with these characteristics would have been most likely to win the race to reproductive age. Under natural conditions, plants would have evolved a delay in germination from seed maturation until favourable conditions the next growing season. Also, prolonged seed dormancy would guarantee that at least a few seeds would be available to germinate after prolonged stretches of subopti-

Table 7.2. Traits commonly associated with the domestication process.

Increased reproductive effort
Larger seeds and fruit
More even and rapid germination
More uniform ripening period
Non-dehiscent fruits and seeds
Self-pollination
Trend to annuality
Increased palatability
Colour changes
Loss of defensive structures
Increased local adaptations

Table 7.3. Adaptation syndromes resulting from automatic selection due to planting and harvesting seed of cereals (adapted from J.R. Harlan, J.M.J. deWet and E.G. Price, 1973, Comparative evolution of cereals, *Evolution* 27, 311–325).

Selection pressure	Response	Adaptation
Harvesting	Increase in per cent seed recovered	Non-shattering
		More determinate growth
	Increase in seed production	Increase in seed set
		Reduced or sterile flowers become fertile
		Increase in inflorescence size
		Increase in inflorescence number
Seedling competition	Increased seedling vigour	Increase in seed size
		Reduction in protein content of seeds and an increase in carbohydrate
	More rapid germination	Loss of or reduction in inhibitors of germination
		Reduction in glumes and other appendages

mal weather. Once seed were harvested and stored away from the natural environment, these adaptations were no longer necessary. Greater seed size probably contributed to seedling vigour, and a loss of germination inhibitors would have allowed faster germination. Thin seed-coats also evolved under domestication, as a reduced seed-coat is more permeable to water and results in more rapid germination.

Of course, not all crop domestications followed the same evolutionary path as the broadcast grains and legumes. In those crops harvested for some plant part other than seeds, increases in seed size would have been much more modest than in the legumes and grains, as humans selected genotypes partitioning more energy into larger leaves or tubers. In the seeded crops like lettuce and radish, non-dehiscence and high fertility would have remained important, but, among the vegetatively propagated crops, the ability to sexually reproduce would have been far less critical. In fact, many vegetatively propagated lines of taro, banana and potato cannot sexually reproduce at all. These genotypes presumably devote more energy into their harvested parts than sexually reproducing ones and, in the case of banana, there may even have been human selection against hard, flinty seeds.

Sometimes the primary reason for domestication changed as humans began to consciously improve a crop (Anderson, 1954). Squashes and pumpkins started out with small fruits and bitter flesh. They may have first been used as rattles in ceremonies and dances, as dishes and as storage vessels. Only much later were they used as food, first for their seeds and then for their flesh. Other crops that came to have multiple uses include flax (oil and fibre), hemp (oil, fibre and stimulation) and chenopod (seeds and leafy vegetables). One of the most dramatic examples of a species used in multiple ways is *Brassica oleraceae*, whose flowers came to be used as cauliflower and broccoli, while its leaves became kale, cabbage and Brussels sprouts and its fleshy corms became kohlrabi (Chapter 11).

As humans began to save grain seeds and plant the same field every year, local adaptations would have gradually increased over time as the highest proportion of seeds were gathered from the most vigorous individuals. An experimental documentation of such change was briefly described in Chapter 2, where Clegg *et al.* (1972) studied evolutionary change in a synthetic population of 28 worldwide barley varieties. The mixture of seeds was initially sown in a large plot in 1929 and was allowed to reproduce by natural crossing without any artificial selection, and a random sample of seeds was collected and sown annually. Over the ensuing decades, they documented dramatic changes in gene frequency, resulting in higher and more stable grain yields with more compact, heavier spikes with larger numbers of seeds (Allard, 1988) .

Genetic Regulation of Domestication Syndromes

Many of the traits associated with plant domestication are regulated by only a few genes, making rapid evolutionary change possible (Table 7.4). For example, seed non-shattering is controlled by only one or two genes in a broad range of grains and legumes. A single recessive mutant transforms two-rowed into six-rowed barley. Determinate vs. indeterminate growth in maize and bean is regulated by one or two genes, as is branching vs. not branching in sunflower and sesame.

Table 7.4. Genetics and number of loci governing seed non-shattering in cultivated crops (reprinted with permission from G. Ladizinsky, © 1985, *Economic Botany* 39, 191–199, New York Botanical Garden).

Crop	Number of loci	Genetics of domesticated form
Rice – *Oryza sativa*	1	Recessive
Oat – *Avena sativa*		
Spikelet non-shedding	1	Dominant
Floral non-disjunction	1	Recessive
Barley – *Hordeum vulgare*	2	Recessive at both loci
Sorghum – *Sorghum bicolor*	2	Recessive at both loci
Pearl millet – *Pennisetum glaucum*	2	Recessive at both loci
Lentil – *Lens culinaris*	1	Recessive
Wheat – *Triticum monococcum*	2	Recessive complementary

Even in cases where traits are thought to be regulated by a large number of genes or quantitative trait loci (QTL), it is not unusual to find that a few major genes influence a large amount of the genetic variability (Knight, 1948; Hilu, 1983; Gottleib, 1984). Koinange *et al.* (1996) found major genes associated with mode of seed dispersal, seed dormancy, growth habit, gigantism, earliness, photoperiod sensitivity and harvest index in dry beans (Table 7.5). John Doebley and his colleagues in his laboratory (1990) have isolated three key QTL that regulate glume toughness, sex expression and the number and length of internodes in both lateral branches and inflorescences. One of these, *teosinte glume architecture 1,* probably played a particularly important role in the appearance of maize 5000 years ago, as it disrupts reproductive development in such a way that the kernels are naked, rather than encased in tough glumes (Dorweiler *et al.*, 1993; Doebley *et al.*, 1995; Lukens and Doebley, 1999).

Major QTL for domestication traits have also been described in pearl millet (Poncet *et al.*, 1998), sorghum (Paterson *et al.*, 1998; Table 7.6), tomato (Grandillo and Tanksley, 1996; Grandillo *et al.*, 1999) and rice (Xiong *et al.*, 1999; Bres-Patry *et al.*, 2001). The location of many QTL for domestication-related traits have been found to be conserved in aubergine, tomato, pepper and potato (Doganlar *et al.*, 2002). Xiao *et al.* (1998) located two *Oryza rufipogan* alleles on two chromosomes that were associated with almost 20% increases each on grain yield per plant. Two QTL for domestication traits have even been cloned: *fw2.2*, which has dramatic effects on fruit weight in tomato (Frary *et al.*, 2000), and *Hd1*, which regulates photosensitivity in rice (Yano *et al.*, 2000) .

Not only are there commonly major genes for the individual traits associated with plant domestication, but in many cases these genes have pleiotropic effects, where they affect a number of traits simultaneously. As a result, their selection would make change across the whole domestication syndrome much more rapid than if the traits were evolving separately. In maize, Doebley has discovered two QTL that coordinately regulate plant and inflorescence architecture and have much stronger effects together than separately. Allard (1988) identified a number of marker loci for quantitative traits that had significant additive effects on several to many quantitative traits in barley. Koinange *et al.* (1996) found several cases where the large effect of individual genes in bean was further magnified by their having pleiotropic effects. For example, the gene *fin* influenced determinacy, node number, pod number, days to flowering and days to maturity (Table 7.5). Grandillo *et al.* (1999) also discovered several regions of the tomato genome that had effects on more than one trait.

Many of the QTL associated with the domestication syndromes are clustered close together on the same chromosome. Such close associations of genes would reduce the amount of segregation between these adaptively important genes and in a sense 'fix' the crop type, again allowing for rapid

Table 7.5. Genetic factors influencing the domestication syndrome in common bean (modified from E.M.K. Koinange, S.P. Singh and P. Gepts, 1996, *Crop Science* 36, 1037–1045).

General attribute	Trait	Gene or marker	Magnitude of effect on trait
Seed dispersal	Pod suture fibres	*St*	Single gene
	Pod wall fibres	*St* (?)[a]	Single gene
Seed dormancy	Germination	*PvPR*-2–1	18%
		D1132	52%
		D1009–2	19%
		D1066	12%
Growth habit	Determinacy	*fin*	Single gene
	Twinning	*Tor* or *fin*	Single gene
	Number of nodes	*fin*	53%
		D1492–3	20%
		D1468–3	16%
	Number of pods	*fin*	32%
		D1468–3	21%
		D1009–2	14%
	Internode length	D1032	19%
Gigantism	Pod length	D1520	23%
		PvPR-2–1	20%
		*LegH*16	16 %
	100-seed weight	D1492–3	18%
		Phs	27%
		Uri-2	16%
		D0252	15%
Earliness	Days to flowering	*fin*	38%
		Ppd (?)[a]	19%
		D1468–3	12%
	Days to maturity	*fin*	30%
		Ppd (?)[a]	18%
		D1468–3	14%
Photoperiod sensitivity	Delay under 16 h days	*Ppd*	44%
		D1479	44%
Harvest index	Seed yield/biomass	*Ppd* (?)[a]	28%
		D1468–3	28%
Seed pigmentation	Presence vs. absence	*P*	Single gene

[a]Pleiotropy and tight linkage could not be clearly distinguished for genes with question marks.

change. In many crops like wheat and oats, self-pollination would help maintain these linkages, once the genes had reached homozygosity. Koinange *et al.* (1996) found the distribution of domestication syndrome genes to be concentrated in three genomic regions, one of which greatly affected growth habit and phenology, another seed dispersal and dormancy

Table 7.6. Action of QTL for traits related to domestication in *Sorghum propinquum* (used with permission from A.H. Paterson, K.F. Schertz, Y. Lin and Z. Li, 1998, *Case History in Plant Domestication: Sorghum, an Example of Cereal Evolution*, CRC Press, London).

Trait	Number of QTL	Mode of gene action			
		Dominant	Additive	Recessive	Overdominance
Shattering	1	1	0	0	0
Height	6	4	1	0	1
Flowering date	3	2	0	1	0
Seed size	9	2	1	6	0
Tiller number	4	1	2	1	0
Rhizomatousness	8	2	4	1	1
Overall	31	12	8	9	2

and a third the size of fruit and seed (Fig. 7.5). Doebley *et al.* (1990) found five of the QTL that distinguish maize and teosinte in a tight cluster on chromosome 8. These genes regulated: (i) the tendency of the ear to shatter; (ii) the percentage of male spikelets in the primary inflorescence; (iii) the average length of internodes on the primary lateral branch; (iv) the percentage of cupules lacking the pedicellate spikelet; and (v) the number of cupules in a single rank. Many of the genetic factors controlling domestication-related traits were also concentrated on a few chromosomal blocks in pearl millet (Poncet *et al.*, 1998), tomato (Grandillo and Tanksley, 1996) and rice (Xiong *et al.*, 1999).

Evolution of Weeds

In many cases, species that started out as weeds of crops were eventually domesticated themselves. Sauer (1952) states:

> The ancestors of most New World seed plants appear to have been attractive weeds. They were not tenacious intruders that the cultivator had difficulty getting rid of, nor are they such as grow on trodden ground. They were gentle, well-behaved weeds that liked the sunshine, loose earth and plant food of tilled species, and had no great root systems. Such volunteers were first tolerated, then protected, and finally planted.

He mentioned that the cherry tomato in Mexico and Central America is not planted, but protected. *Madia sativa* and *Bromus mango* were initially weeds of root crops in Chile, but are now minor seed crops. In the northern periphery of root crops, amaranth, chenopod, squashes and beans may have started out as weeds, but:

> where climatic advantage shifted from the root plant to seed plant, the attention of the cultivator shifted from the former to the latter. Instead of selecting root variants to meet the local situation, he began to select the attractive weeds.

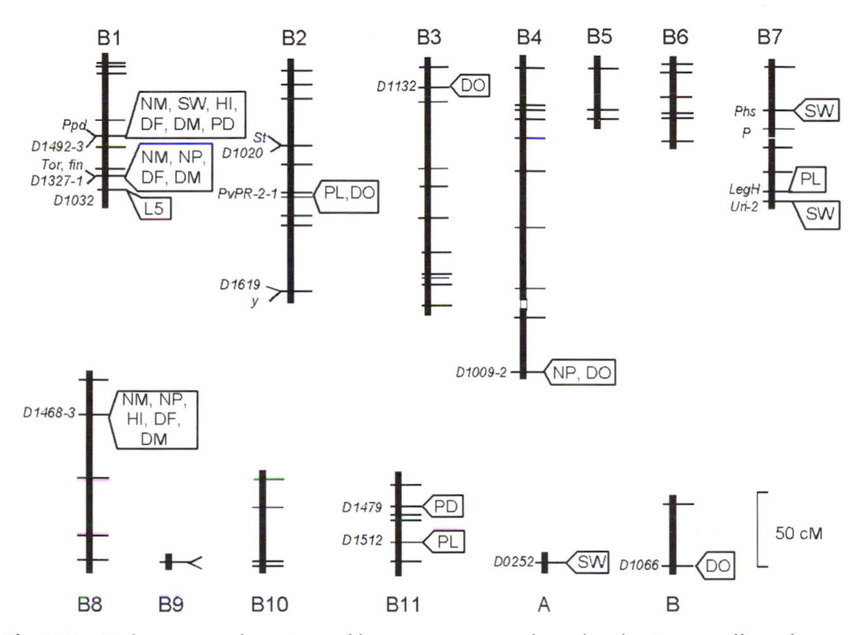

Fig. 7.5. Linkage map location of known genes and marker loci controlling the domestication syndrome in common bean. Symbols for the genes: *fin*, determinacy; *P*, anthocyanin pigmentation; *Ppd*, photoperiod-induced delay in flowering; *St*, pod string; *y*, yellow pod colour. Symbols for the marker loci: DF, days to flowering; DM, days to maturity; DO, seed dormancy; HI, harvest index; L5, length of the fifth internode, NM, number of nodes on the main stem; PL, pod length; NP, number of pods per plant; PD, photoperiod induced delay in flowering; SW, seed weight. (Used with permission from E.M.K. Koinange, S.P. Singh and P. Gepts, 1996, *Crop Science* 36, 1037–1045.)

Rye and oats are thought to have begun as weeds of the Near Eastern assemblage, which were domesticated in more northern climates where their potential was more apparent. Hexaploid wheat may also have begun its existence as a weed in cultivated tetraploid fields before its full benefit was noted.

As farming progressed, humans also began to inadvertently select for weedy crop mimics through tilling, weeding and harvesting (Harlan *et al.* 1973). As we discussed in Chapter 5, many of these weedy races arose after introgression with the crop type, but many non-crop species have also developed weedy races. Two major forms of crop mimics are generally recognized: vegetative, where the weed is similar looking to crop seedlings and their vegetative stage; and seed, where weed seeds have similar density and appearance to those of crops, making it difficult to separate them before planting.

Numerous examples of crop mimicry have been reported. Several species of wild rice and barnyard grass have evolved developmental and growth patterns that make them very difficult to distinguish from cultivated types (Barrett, 1983). In the case of barnyard grass, the crop mimic *Echinochloa crus-galli* var. *oryzicola* is actually more similar to rice in many attributes than to its own progenitors (Barrett, 1983). Weedy races of grain chenopods, teosinte, amaranths, pearl millet and sorghum invade agricultural fields and look almost identical to their related crop species until their inflorescences shatter just before harvest (Sauer, 1967; Harlan *et al.*, 1973; Wilson and Heiser, 1979). Considerable differentiation is often observed in weedy races depending on the types of cultivars grown in a region and the natural diversity present. This is particularly apparent in weedy rice, where there are numerous different indica- and japonica-mimicking races, both where wild species are present and where they are absent (LingHwa and Morishima, 1997; Xiong *et al.*, 1999). One of the most unusual adaptations has been described in maize fields in Mazatlan, Mexico, where fields are cropped one year and then fallowed the next. Here teosinte populations have arisen with an inhibitor that prevents germination for 1 year and therefore protects the plants from being grazed in the fallow years (Wilkes, 1977).

Numerous examples of seed mimicry have also been described. One of the earliest cases involved races of *Camelina sativa*, whose seeds were so similar to those of flax that they could not be separated by winnowing, where chaff and lighter seeds are removed by wind from the crop's heavy seeds (Stebbins, 1950). Seeds of *Vicia sativa* are normally a different shape and size from lentils, but in Central Europe the seeds of the two species are very similar and as a result *V. sativa* can be a very serious weed (Rowlands, 1959). In many cases, vegetative and crop mimicry are combined in weeds, making it almost impossible to identify the invaders. The *Camelina* species have the same growth habit, branching pattern, flowering time and fruit characteristics as flax.

With the advent of herbicides, a third class of mimics has arisen: herbicide mimicry. Resistance to S-triazine herbicides, atrizine and simizine, has been found in several different weed species, including *Brassica campestris* and *Chenopodium alba* (Souza Machado *et al.*, 1977; Warwick and Black, 1980). The distribution is highly localized in most cases, but in *Senecio vulgaris* a wide range of susceptibility was found to simazine among the fruit farms in England. A population's susceptibility was correlated with the number of years of continuous herbicide use, strongly implicating selection (Holliday and Putwain, 1980); however, continuous selection with herbicides has not always been necessary for resistance to emerge. Friesen *et al.* (2000) found native populations of *Avena fatua* in Manitoba, Canada, that were resistant to imazamethabenz, even though this herbicide had not previously been applied. Numerous factors influence the rate that herbicide resistance evolves, including rates of genetic mutation, initial frequency of resistance genes, type of inheritance, mating system and gene flow (Jasieniuk and Maxwell, 1996; Mortensen *et al.*, 2000).

Genetic Diversity and Domestication

An important ramification of domestication was a reduction in levels of genetic variability in both plants and animals. Virtually all domesticants are substantially less diverse than their progenitors due to the bottleneck associated with selecting a few élite types and directional selection. However, several factors had an important influence on the amount of diversity captured during domestication including geographical isolation, sexual structure and levels of diversity in the original species population.

The way crops were planted and their mode of reproduction had a substantial influence on the amount of diversity maintained in them. Those crops that were outcrossed and planted individually would have remained much more diverse than the broadcast, selfed seed crops. Ancient races of maize and chilli peppers show astonishing levels of variability across South America (Fig. 7.6), in large part because they were outcrossed and planted

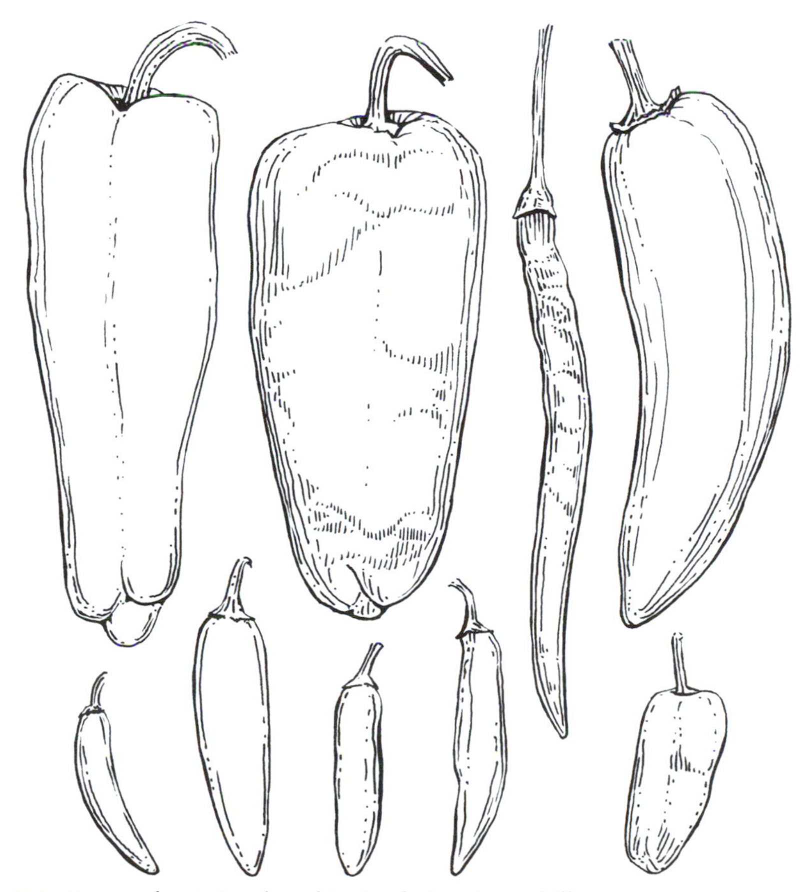

Fig. 7.6. Range of variation found in South American chilli peppers.

individually in hills, so that individual diversity could be recognized and exploited. In the asexually propagated crops like potatoes, similar high levels of diversity have also been maintained, since specific genotypes can be selected for propagation without the necessity of gathering segregating populations of seeds.

Hybridizations between sexually propagated crops and their wild progenitors would have occasionally increased their variability and improved local adaptations, particularly as the crops diffused from their point of origin. As we discussed in Chapter 5, crop–wild progenitor hybridizations played an important early role in the evolution of numerous crops. Some of the most dramatic examples have been maize and kidney beans at a local level, and sorghum and apples across large geographical areas (see later chapters on these crops). In some cases, hybrids may even have been directly accepted as new cultigens, possible examples being the grain amaranths, mangos and the allopolyploid bread wheat. In the selfed grains and legumes, crop–wild hybridizations would have been restricted, but even the strongest selfers occasionally outcross (Allard and Kahler, 1971).

Domestication and Native Diversity Patterns

In his landmark work on cultivated plants, N.I. Vavilov (1926) used the centres of diversity of native crop species to predict where they were initially domesticated. This system works well for a high percentage of crops, but a number, such as wheat, sorghum, pearl millet and beans, do not have a true centre of diversity and others, such as barley and rice, were domesticated far from their centres of diversity (Harlan, 1992). To adjust for this problem, Vavilov developed the concept of secondary centres to describe those cases where centres of diversity and origin were not the same. This approach helps with several crops, including wheat, barley and rice, but there are still a number of crop species that appear to have been domesticated completely outside their native ranges. These transdomestications might have resulted from long-range oceanic drift (bottle gourd outside Africa, sweet potato in Polynesia), dispersal by migratory birds (perhaps tomato in Mexico, arabica coffee in Arabia), human trade of wild material (cotton from Africa to India, perhaps pepo gourds in eastern North America) or original movement as a weed (rye and oats).

Harlan (1975, 1976) has classified domestication patterns into five classes: (i) endemic – the domesticant occupies a well defined, small geographical region (guinea millet); (ii) semi-endemic – the domesticant occupies a small range with some dispersal out of it (African rice and teff); (iii) monocentric – the domesticant has a wide distribution with a discernible centre of origin (the later plantation crops, such as coffee, rubber, cacao and oil-palm); (iv) oligocentric – domesticant has a wide distribution and two or more centres of diversity (our major food crops, such as wheat, barley, pea, lentil, chickpea, flax, maize and lima bean); (v) non-centric – domesticant has a wide distribution, but no discernible centres of diversity (American beans, radish, sorghum, pearl millet, cole crops and bottle gourd).

While people in developed countries are familiar with only a few dozen oligocentric and non-centric crops, in reality, hundreds of endemic and semi-endemic crops were domesticated. In his book *Crops and Man*, Harlan (1992) provides what he calls a 'short list' of world crops, which encompasses 11 pages of text. Most of these crops are unknown outside their region of origin. In the publication *Lost Crops of the Incas* (Anon., 1989) over 30 species are described that were domesticated by the Andean peoples but are little grown outside South America (Table 7.7); many of these are restricted to specific elevational gradients (Fig. 7.7). Ethiopia provided the world with coffee, but also a unique assemblage of endemic crops, including the cereal teff, the oil crop noog, the mild narcotic chat, and enset, a relative of banana whose stem base rather than fruit is eaten (Harlan, 1992). Thousands of plant species have been utilized by somebody somewhere, but very few of them have attracted widespread attention.

Summary

The transition from hunter–gatherer to farmer was a gradual one that took thousands of years. The earliest grains and legumes were not domesticated until about 10,000 years ago, and it took another 5000 years for the rest of our major crops to emerge. The earliest group of domesticants were all herbaceous annual species and most were selfing, which provided quick harvests after planting and true-to-type seed. Once plants were domesticated, they were dramatically altered by humans through conscious and unconscious selection. The simple act of harvesting resulted in the selection of non-shattering traits, more determinate growth, more uniform ripening and higher seed production. The seedling competition that occurred after seed were scattered probably increased seedling vigour and rate of germination. Many of the traits associated with plant domestication are regulated by only a few genes, making rapid evolutionary change possible. Domestication often resulted in reduced levels of variability, but hybridization between crops and their wild progenitors occasionally increased their local adaptations and expanded their geographical range. Transfers of genes between wild and domesticated populations also led to the evolution of weeds that mimic the crop in such a way that their removal becomes difficult.

Table 7.7. Little-known plants domesticated by the Incas in South America (source – National Research Council, 1989, *Lost Crops of the Incas*, National Academy Press, Washington, DC).

Type of crop	Common name	Species	Distinctive properties
Roots and tubers	Achira	*Canna edulis*	Staple with unusually large starch grains
	Ahipa	*Pachyrhizus ahipa*	Crisp like apples, addition to salads
	Arracacha	*Arracacia xanthorrhiza*	Has flavours of celery, cabbage and chestnut
	Maca	*Lepidium meyenii*	Sweet tangy flavour; can be stored for years
	Mashua	*Tropaeolum tuberosum*	Starchy staple; very easy to grow
	Mauka	*Mirabilis expansa*	'Cassava' of the highlands
	Oca	*Oxalis tuberosa*	Second most important staple in the highlands (to potatoes)
	Potatoes	Many other than *Solanum tuberosum*	Most important staple; huge diversity
	Ulluco	*Ullucus tuberosus*	Very brightly coloured; staple in some areas
	Yacon	*Polymnia sonchifolia*	Sweet and juicy, but almost calorie free
Grains	Kaniwa	*Chenopodium pallidicaule*	Very nutritious and extremely hardy
	Kiwicha	*Amaranthus caudatus*	Protein quality is equivalent to milk
	Quinoa	*Chenopodium quinoa*	Excellent source of protein; extremely hardy
Vegetables	Basul	*Erythrina edulis*	Tree with edible seeds; staple in some areas
	Nuñas	*Phaseolus vulgaris*	Dropped into hot oil and popped
	Tarwi	*Lupinus mutabilis*	Extremely rich in protein and oil
	Peppers	Many *Capsicum*	Huge range in pungency and taste
	Squashes	Many *Cucurbita*	Unusually robust and productive
Fruits	Mora de Castilla	*Rubus glaucus*	Superior in flavour and size to other raspberries
	Ugni	*Myrtus ugni*	Sprightly flavour; blueberry relative
	Capuli cherry	*Prunus capuli*	A large, sweet black cherry
	Cherimoya	*Annona cherimola*	Has flavours of pawpaw, pineapple and banana
	Goldenberry	*Physalis peruviana*	Yellow fruits are excellent in jam; very hardy
	Highland pawpaws	*Carica* species	Unusually cold adapted
	Lucuma	*Pouteria lucuma*	Used as both a staple crop and a fresh fruit; a tree can feed a family
	Naranjilla (lulo)	*Solanum quitoense*	Like tomato; has particularly refreshing juice
	Pacay	*Inga* species	Long pods filled with soft white pulp (called ice-cream beans)
	Passion-fruits	*Passiflora* species	High quality and huge variabiliy
	Pepino	*Solanum muricatum*	Tastes like sweet melon
	Tamarillo	*Cyphomandra betacea*	A tree with a tomato-like fruit
Nuts	Quito palm	*Parajubaea cocoides*	Nut tastes like tiny coconuts

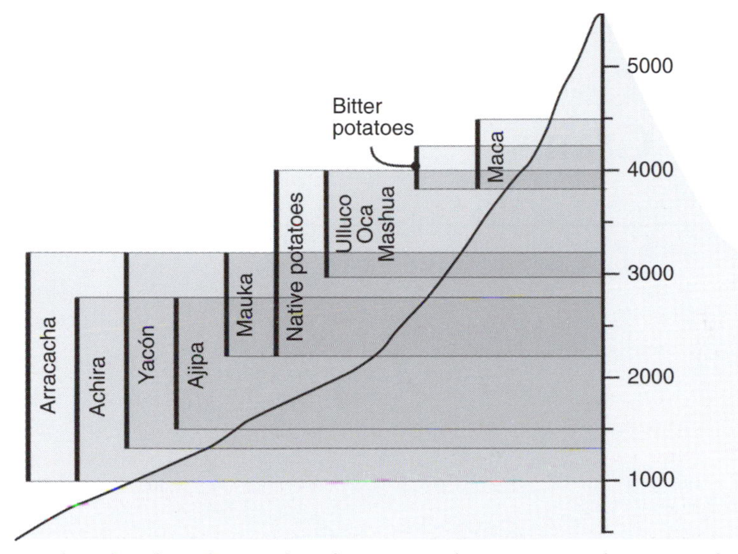

Fig. 7.7. Elevational gradients of Andean crops (from CIP Circular, September 1994, Centro Internacional de Papas, Lima, Peru).

Cereal Grains

8

The cereal grains represent a diverse group of species in the *Gramineae* (grasses) (Fig. 8.1). They are grown on all continents of the world except Antarctica and are the primary food source of most of the world. In fact, over two-thirds of our major food crops are cereals.

The quality of the information bearing on the evolutionary history of the grain species varies greatly. Some crops, such as barley, have relatively clear pasts due to good historical records and the presence of living ancestors. Others, such as rice and sorghum, are ambiguous due to spotty archaeological remains and a confusing array of existing taxa. Still others, such as maize, have missing or unclear progenitor species, making conjectures about their origin problematic. In this chapter, we review the available information on the evolutionary origin of each major grain species and tell as complete a story as possible.

Barley

Barley, *Hordeum vulgare* L., was certainly one of the earliest domesticants and it has a very simple evolutionary history (Zohary and Hopf, 1993). It was derived from what was traditionally considered a separate species, *Hordeum spontaneum* C. Koch, which is widely distributed in the initial area of cultivation (Fig. 8.2). The two taxa are so closely related that most investigators now consider them to be in the same species, *H. vulgare*, with the cultivated forms belonging to subsp. *vulgare* and the wild forms subsp. *spontaneum*. The genes separating wild and cultivated forms are easily transferred, and natural hybrids can be found with combinations of

Fig. 8.1. A comparison (left to right) of the inflorescences of rice, oats, barley, wheat and rye.

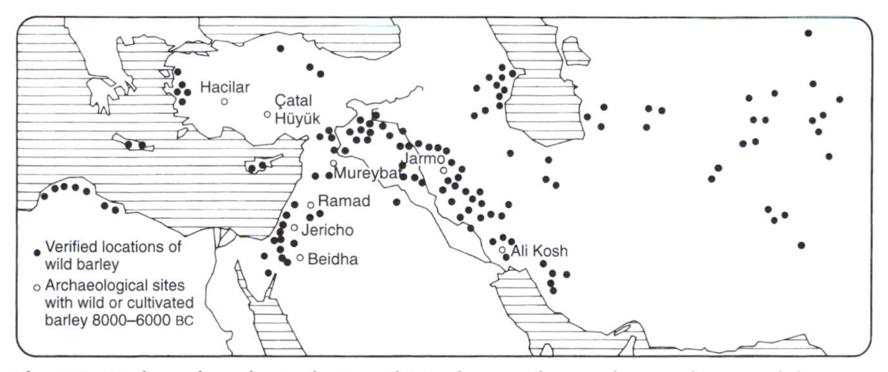

Fig. 8.2. Early archaeological sites of *Hordeum vulgare* subsp. *vulgare* and the distribution of its wild progenitor *H. vulgare* subsp. *spontaneum* (used with permission from J.R. Harlan and D. Zohary, © 1966, *Science* 153, 1074–1080, American Association for the Advancement of Science).

wild and domesticated characteristics (Harlan, 1976; van Bothmer *et al.*, 1990). Some of the brittle, six-rowed hybrids have even been given their own name, *Hordeum agriocrithon* Åberg (Zohary, 1973; von Bothmer *et al.*, 1999).

Only a few recessive alleles separate cultivated and wild barley (Nilen, 1971): (i) the wild forms have spikes that shatter easily at maturity, while the cultigens have a non-brittle spike that results from a mutation in either of two tightly linked 'brittle' genes, Bt_1 or Bt_2; (ii) the wild forms have tight glumes that are difficult to thresh, while the cultivated forms have naked grains that result from a recessive gene (n) that allows the husks to be easily removed by threshing; (iii) the domesticated forms have six- rather than two-row glumes due to two recessive mutations at the *vrs1* locus.

Cultivated barley possesses much of the isozyme diversity found in the native species suggesting that it had a broad ancestry (Ladizinsky and Genizi 2001); however, the cytoplasmic evidence indicates that the origin of barley cultivation was limited. When Clegg *et al.* (1984) examined cytoplasmic DNA (cpDNA) variation in 11 wild and nine cultivated accessions of barley, they found three chloroplast lineages in the wild populations, but only one in the cultivated types. In a much larger collection of over 300 wild and cultivated accessions, Neale *et al.* (1988) confirmed the findings of Clegg's group. They found the same three chloroplast families to be common in the wild, but almost all the cultivated forms belonged to the same chloroplast family. Most evidence points to a Near Eastern origin for barley cultivation (Zohary, 1999; Blattner and Bandani Mendez, 2001), but a secondary Moroccan origin cannot be eliminated. Weedy Moroccan populations show patterns of restriction fragment length polymorphism (RFLP) variation different from those from the Near East (Molina-Cano *et al.*, 1999) and sequence data from a DNA marker in close proximity to the *vrs1* locus indicate that mutations producing six-row glumes occurred at least twice near Turkmenistan and in the Mediterranean region (Tanno *et al.*, 2002).

People first began gathering wild, two-rowed forms with easily shattered spikes about 19,000 years ago and about 9500 before present (BP) were cultivating non-brittle types at Netiv Hagdud, north of Jericho and Abu Hureyra in what is now Syria. Barley was probably used first as a staple, but it did not take people very long before they were fermenting it into beer. Naked forms of two-row barley were being grown by about 8500 BP at Abu Hureyra and Tell Aswad, and the first six-rowed barley appeared at Ali Kosh at about the same time (Zohary and Hopf, 1993). Barley was a common companion to einkorn and emmer wheat in the earliest farming communities, but its importance as a staple crop gradually diminished over the next 5000 years. It came to be considered a poor man's grain, being grown on marginal sites that did not adequately support wheat. Today barley is grown across all temperate climates and is used primarily to make beer and feed livestock.

Maize

The history of maize has been difficult to trace as its fruiting cob and monoecious nature are unique in the *Gramineae*. In other grasses the sexual parts are in close proximity rather than isolated at different locations in the plant, and the grains are protected individually by glumes rather than being naked.

The closest relatives of maize are a few species of *Tripsacum* and a number of teosintes (Table 8.1), which were for a long time considered to be a separate genus, *Euchlaena*. The teosintes have a habit that resembles that of maize, but their tassels are not on a central spike like maize and their ears are much less complex (Figs 8.3 and 8.4). The *Tripsacum* species have pistillate and staminate flowers that are borne separately, but, unlike maize, they are adjacent on the spike. Their seeds are embedded in segments of the rachis and scatter when ripe.

Table 8.1. Maize and teosinte taxonomy according to Iltis and Doebley (1980).

Species and subspecies	Chromosome number (2n)	Synonyms
Cultivated maize		
Zea mays L. subsp. *mays*	20	*Z. mays*
Teosinte	20	*Zea* (or *Euchlaena*) *mexicana*
Z. mays subsp. *mexicana* (Shrader) Iltis	20	*Z.* (or *E.*) *mexicana*
Z. mays subsp. *parviglumis* Iltis & Doebley	20	*Z.* (or *E.*) *mexicana*
Z. mays subsp. *diploperennis* Iltis, Doebley & Guzmán	20	*Zea* (or *Euchlaena*) *diploperennis*
Z. mays subsp. *perennis* (Hitchc.) Reeves & Mangelsdorf	40	*Zea* (or *Euchlaena*) *perennis*
Zea luxurians (Durieu & Ascherson) Bird	20	*Euchlaena luxurians*

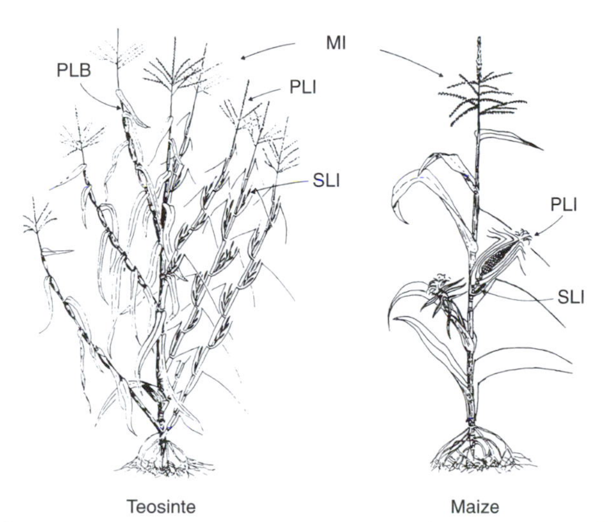

Fig. 8.3. Differences in fruiting structure of maize and teosinte. Note that tassels and ears are on the same fruiting stalk of teosinte, while on maize they are segregated on different spikes. MI, main inflorescence; PLI, primary lateral inflorescence; SLI, secondary lateral inflorescence; PLB, primary lateral branch. (Used with permission from J. Doebley *et al.*, 1990, *Proceedings of the National Academy of Sciences USA* 87, 9888–9892.)

Fig. 8.4. Grain-bearing inflorescences of maize and its relatives. Left to right: *Tripsacum dactyloides, Zea mexicana* and *Zea mays.*

All the native *Zea* species have very restricted ranges in Mexico and Central America, and they carry the same chromosome number, 2n = 20, except for *Z. perennis*, which is tetraploid. The *Tripsacum* species all have multiples of x = 18. Diploid teosinte crosses readily with maize, and reciprocal introgression may occasionally occur today in Mexico and Guatemala, where teosinte grows adjacent to cultivated maize, although evidence of modern gene flow is minimal (Doebley *et al.*, 1984; Doebley, 1990a). *Tripsacum* can be crossed with maize (Eubanks, 2001a,b) and a natural hybrid species has been identified (Talbert *et al.*, 1990), but the sterility barriers are sufficient to greatly limit natural introgression (Newell and deWet, 1973; James, 1979; deWet *et al.*, 1984a).

The high chromosome number of maize suggests that it might be an ancient polyploid formed from two diploids with 2n = 10. Several independent lines of evidence support this conjecture (Molina and Naranjo, 1987): (i) haploid maize shows many chromosome associations (Ting, 1985); (ii) normal taxa display secondary associations of bivalents (Vijendra Das, 1970); (iii) chromosomes of *Zea mays* form four subsets of five chromosomes in somatic metaphase cells rather than two sets of ten (Bennett, 1984); (iv) maize carries a high number of isozyme and restriction fragment length polymorphism (RFLP) duplications (Stuber and Goodman, 1983; Helentjaris *et al.*, 1988); (v) DNA sequence data have shown numerous duplications (Gaut and Doebley, 1997); and (vi) distant relatives *Coix* and *Sorghum* have a haploid number of 5.

The lack of fossil evidence of a prototype maize plant has led to three major hypotheses concerning the origin of maize: (i) maize, teosinte and *Tripsacum* were separate lineages that evolved from a common, unknown ancestor (Weatherwax, 1954); (ii) an interspecific hybridization between two or more native grasses produced maize (Mangelsdorf and Reeves, 1939; Eubanks, 2001a,b); and (iii) maize evolved directly from teosinte (Beadle, 1939; Galinat, 1973; Dorweiler and Doebley, 1997; Iltis, 2000).

In a complex scenario, Mangelsdorf and Reeves (1939) initially suggested that modern maize arose through a series of interspecific crosses, and in fact was the progenitor of teosinte. In their 'tripartite hypothesis' they envisioned that now extinct races of pod corn were introduced into Mexico and Central America and subsequently hybridized with *Tripsacum* to form teosinte. This new teosinte then hybridized with maize to produce superior races. Mangelsdorf (1974) later altered this hypothesis and considered teosinte to be a mutant derivative of maize.

In support of Mangelsdorf's hypotheses, pod corn has been found among the fossils of ancient communities that lived in New Mexico 4000 to 3000 BP and even older ears have been located in the Tehuacan Valley, Mexico, which appear to have traits of both popcorn and pod corn. Mangelsdorf (1958) also crossed modern races of popcorn and pod corn and obtained a hybrid that had a combination of grass and maize characteristics. However, there is no hint of where the pod corn came from.

Eubanks (2001a,b) has suggested that maize was derived from a hybridization between *Tripsacum dactyloides* and *Zea diploperennis*. Support for this hypothesis has come from her generating recombinant progeny that have maize-like flowering spikes and RFLP data where modern maize appears to carry a combination of fragments from a limited sample of native *Tripsacum* and *Zea*.

The most overwhelming support has been garnered for the teosinte hypothesis through molecular and isozyme studies (Bennetzen *et al.*, 2001; Smith, 2001). No intermediate forms between maize and teosinte have been found in the archaeological record and there is little evidence that humans ever cultivated teosinte (Galinat, 1973; Mangelsdorf, 1974, 1986), but a punctuated change could have occurred in the ear or tassel that led to a dramatically different crop (Iltis, 1983, 2000). In an extensive analysis of electrophoretic variation in native populations, one variety of *Z. mays* subsp. *parviglumis* was found to have a high genetic identity of 0.92 with maize, and the two were tightly grouped when the data were subjected to a principal component analysis (Doebley *et al.*, 1987). This suggests that they are directly related and are part of the same lineage. The close similarity between maize and *Z. mays* subsp. *parviglumis* has also been documented using complementary DNA (cDNA) restriction fragments (Doebley, 1990c) and ribosomal internal transcribed space (ITS) sequences (Buckler and Holtsford, 1996). Most recently, Matsuoka *et al.* (2002) were able to trace the origin of maize, using single sequence repeat

(SSR) markers, to a single domestication of subsp. *parviglumis* in southern Mexico about 9000 years ago. A paradox still remains in that the oldest evidence of maize cultivation at Guilá Naquitz Cave falls outside the current geographical range of *Z. mays* subsp. *parviglumis* (Benz, 2001; Piperno and Flannery, 2001). However, there is no assurance that this was indeed the first place that maize was domesticated, and the range of *Z. mays* subsp. *parviglumi* could have been very different when maize emerged (Smith, 2001).

Even though maize and teosinte are separated by numerous polygenic traits, the punctuated change in maize morphology was probably facilitated by there being only a few major loci involved in the evolutionary change. Iltis (1983) suggested that the emergence of maize may have been due primarily to a feminization of the tassel; however, John Doebley's laboratory has found several key quantitative trait loci (QTL) that separate teosinte from maize through regulation of glume toughness, naked kernels, sex expression and the number and length of internodes in both lateral branches and inflorescences of maize (reviewed in Chapter 7).

Maize became an integral part of a Mesoamerican crop assemblage that included beans and squash (Smith, 2001). From its early cultivation in south-western Mexico before 8000 BP, maize spread throughout Mexico, Central America and into the south-western USA over a period of 3000 years. It is likely that subsequent hybridization with teosinte played a role in the early development of maize, as the modern crop appears to contain as much as 77% of the landrace's diversity (Eyre-Walker *et al.*, 1998; Tenaillon *et al.*, 2001). Maize arrived in eastern North America through the south-western states about 2000 years ago (Smith, 1998). It appeared in South America by 6000 BP (Bush *et al.*, 1989) and was introduced from there into Florida via the Caribbean.

By the time the Europeans arrived in the Americas in the 1500s, maize was an important staple across a vast area from Argentina to Canada. Indians from all over North and South America had developed countless varieties of maize, many of which still exist in Mesoamerica (Doebley *et al.*, 1985; Bretting and Goodman, 1989). The primary forms that were developed were: (i) popcorn, which has extremely hard seeds that explode when heated; (ii) flint maizes, which are composed of hard starch; (iii) flour maizes, which have soft starch that can be ground into flour; (iv) dent maizes, which have soft starch at the top of the kernel and hard starch below; and (v) sweet corn, which is eaten as a sugary vegetable (Fig. 8.5). Within all these groups there exists tremendous variation for kernel colour, ear size, maturation dates and overall plant habit. The efforts of various primitive peoples represent a remarkable example of the changes possible under domestication. Heiser (1990) suggests that at least part of this diversity was produced by seeds being planted individually instead of being broadcast like the other grain species. This practice made people more aware of individual variation.

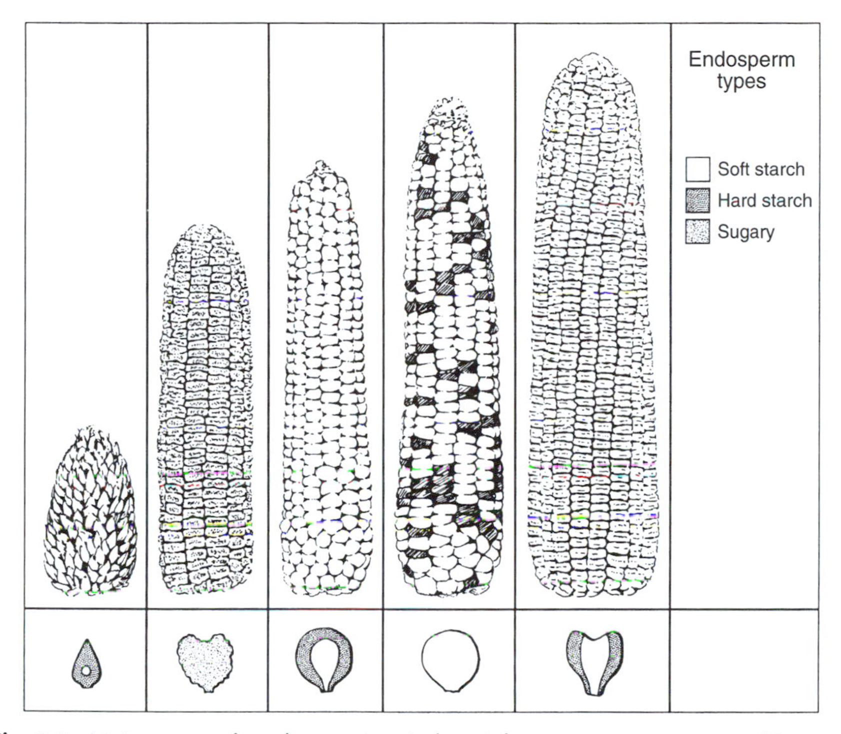

Fig. 8.5. Major types of modern maize. Left to right: popcorn, sweet corn, flint maize, flour maize, and dent maize. Endosperm types are depicted below each ear. (Redrawn from H.G. Baker, © 1970, *Plants and Civilization*, Wadsworth Publishing Company, Belmont, California.)

Millets

The millets represent a number of different genera in the grass tribe *Paniceae* (Fig. 8.6). *Panicum miliaceum* L. (proso or broom-corn millet), *Setaria italica* (L.) P. Beauv. (Italian or foxtail millet) and *Echinochloa frumentacea* (L.) Schum. (Japanese barnyard millet) are known collectively as small millets and represent several million hectares across the world, particularly in India. Other important millets are: *Pennisetum americanum* (L.) Schum. (bulrush, pearl or cat-tail millet), which is one of the principal grain crops in the driest areas of tropical Africa and India, and *Eleusine coracana* (finger millet), which is important in the highlands of Ethiopia and Uganda.

The evolutionary history of the small millets has not been extensively studied, but a few conjectures can be made about their origins. The Japanese millet has both tetraploid ($2n = 4x = 36$) and hexaploid races ($2n = 6x = 56$). The tetraploid race is very similar to the widespread weed, *Echinochloa crus-galli* (L.) Beauv., which is probably its progenitor. Two

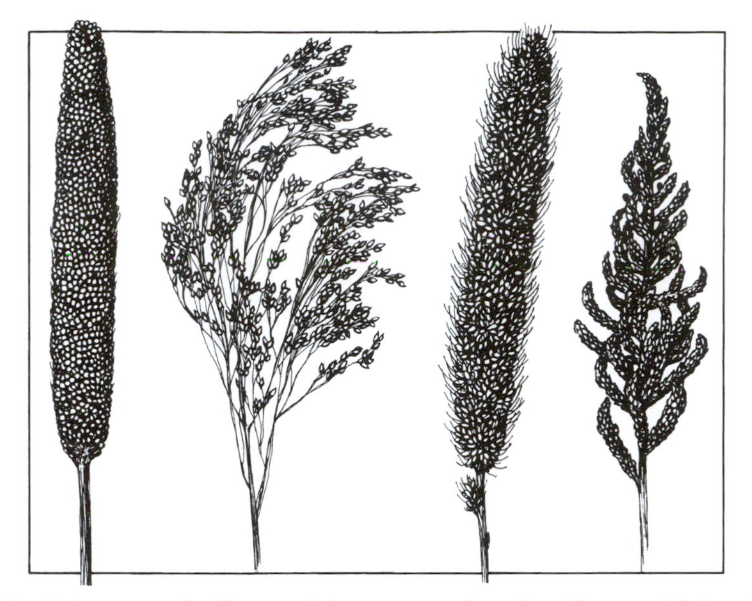

Fig. 8.6. Major types of millet: pearl, broom-corn, foxtail and barnyard (left to right).

hexaploid races have unique genomes: an Indian one, which was probably derived from the wild species *Echinochloa colonum*, and a Sino-Japanese one (*Echinochloa utilis*) which may have an eastern strain of *E. crus-galli* in its background (Yabuno, 1962). *P. miliaceum* (2n = 4x = 36) has been grown in China for at least 5000 years and in southern Europe for at least 3000 (deWet, 1995). Its closest wild progenitor is *P. miliaceum* var. *ruderale*, which is native to central China.

S. italica (2n = 2x = 18) probably arose from the common diploid weed *Setaria viridis* (L.) Beauv., which grows all over Eurasia (Ho, 1977; deWet *et al.*, 1979). The two are so similar that they can be considered subspecies of *S. viridis*, the cultivated forms being subsp. *italica* and the wild forms being subsp. *viridis* (Rao *et al.*, 1987). Conventional wisdom suggests that the crop originated in China by 6000 BP and then spread to Europe (Ho, 1975; Rao *et al.*, 1987), but the possibility of independent domestication in Europe cannot be excluded. In comparisons of isozyme variation, Jusuf and Pernes (1985) found the European cultivars to resemble most closely the French wild types, while the Chinese cultigens were more similar to native populations in Asia. These patterns could signal separate crop origins or a single domestication followed by introgressions with native populations (Doebley, 1989).

Pearl millet (2n = 2x = 14) belongs to a highly heterozygous group that contains both wild and cultivated forms. Stapf and Hubbard (1934) originally recognized 15 species of cultivated millets, but these taxa have more recently been lumped together into one species because of high interfertility.

Cultivation probably began in tropical West Africa about 4000 years ago (D'Andrea *et al.*, 2001) with the emergence of persistent spikelets and protruding grains instead of deciduous spikelets and enclosed grains. Pearl millet arrived in East Africa by 3000 BP and from there was taken on to India.

Finger millet (2n = 4x = 36) probably arose from *Eleusine indica*, a native species that has both diploid (subsp. *indica*) and allotetraploid (subsp. *africana*) races. The diploid is located throughout the tropics and subtropics, while the tetraploid is largely confined to eastern and southern Africa. Genomic *in situ* hybridization (GISH) has identified diploid *E. indica* and *Eleusine floccifolia* as the A and B genome donors of *E. indica* subsp. *africana* (Bisht and Mukai, 2001).

There are two general types of finger millet cultivars: (i) an African highland race that has a strong *E. indica* subsp. *africana* influence; and (ii) an Afro-Asian type that has strong *E. indica* subsp. *indica* influence. Most authors feel that the cultivars originally arose in Uganda from subsp. *africana* and then diverged through introgression with native populations to produce the modern cultivated races (Kennedy-O'Byrne, 1957; Mehra, 1963a,b; Purseglove, 1972). The origin of the crop from subsp. *africana* has been documented using a diverse array of molecular markers (Hilu and Johnson, 1992; Salimath *et al.*, 1995). Finger millet arrived in India, like the other millets, by 3000 BP, where it underwent substantial differentiation (deWet *et al.*, 1984b).

Oats

Oats present a confusing array of morphological and cytogenetic groups. Baum (1977, 1985a,b) divided the genus into 29 species using morphological patterns, while Ladizinsky and Zohary (1971) recognized only seven species based on the interfertility of the various taxa. The genus is composed of three ploidy groups: diploids (2n = 2x = 14), tetraploids (4n = 4x = 28) and hexaploids (2n = 6x = 42) (Table 8.2). Almost all the species are located in the Mediterranean basin and the Middle East. The cultivated species are represented by the widely grown hexaploid *Avena sativa* and the tetraploid *Avena abyssinica*, which is farmed regionally in Ethiopia. There are four distinct karyotypes (A, B, C, D) in the group and enough divergence in the A and C genomes for them often to be given separate subscripts (Rajhathy and Thomas, 1974; Leggett and Thomas, 1995). The various genomes have been distinguished using molecular markers (Fabijanski *et al.*, 1990; Solano *et al.*, 1992; Alicchio *et al.*, 1995), Giemsa C banding (Fominaya *et al.*, 1988; Jellen and Ladizinsky, 2000) and *in situ* hybridization (Linares *et al.*, 1998; Yang *et al.*, 1999). Evidence has grown that the A and D genomes are quite closely related; the D genome could actually have been derived from the A genome (Leggett, 1996; Li *et al.*, 2000).

Table 8.2. Some of the most important *Avena* species and their genomes (based on Rajhathy and Thomas, 1974; Baum, 1977).

Species	Chromosome number (2n)	Genome
A. clauda Dur.	14	CpCp
A. eriantha Dur.	14	CpCp
A. ventricosa Bal.	14	CvCv
A. canariensis Baum.	14	AcAc
A. hirtula Lag.	14	ApAp
A. longiglumis Dur.	14	A_1A_1
A. strigosa Schreb.	14	AsAs
A. wiestii Steud.	14, 28	AsAs
A. abyssinica Hochst.	28	AABB
A. barbata Pott.	28	AABB
A. valviloviana Malz.	28	AABB
A. insularis Ladiz.	28	AACC
A. maroccana Baum.	28	AACC
A. murphyi Ladiz.	28	AACC
A. fatua L.	42	AACCDD
A. sativa L.	42	AACCDD
A. sterilis L.	42	AACCDD

The existing karyotypic, crossability and molecular data suggest that the diploid, A-genome donor of the AC tetraploids and ACD hexaploids was *Avena canariensis* (Leggett, 1992; Leggett and Thomas, 1995; Li *et al.*, 2000). It is unclear which diploid species donated the C genome; however, either *Avena maroccana* or *Avena murphyi* may be the tetraploid progenitor of the hexaploid oats, with *A. murphyi* being perhaps the closest match. A newly described species, *Avena insularis*, also warrants attention (Ladizinsky, 1998, 1999).

Avena wiestii was thought to be the A genome donor of the AB tetraploids, based on karyotypic and interspecific hybridization data, but molecular studies suggest that they may be more closely related to *Avena strigosa* (Li *et al.*, 2000). It is still unknown where the B genome came from, as it is unique to the diploids. Either substantial differentiation occurred within the group after polyploidization to produce what now appears to be two distinct genomes (AB) or another diploid form is extinct or undiscovered (Rajhathy and Thomas, 1974).

The cultivated hexaploid oat, *A. sativa*, belongs to a genus of confusing morphological and cytogenetic groups, but its immediate progenitor is known, as wild forms of *A. sativa* are found all over western Asia and the Mediterranean region. These wild types have been variously referred to as *Avena sterilis* L. or *Avena fatua* L., depending on whether the spikelet has a single disarticulation point at the base of the floret. These taxa are interfertile with the cultivated forms and share the same karyotype, suggesting that they should be considered together as an *A. sativa* crop complex (Ladizinsky and Zohary, 1971).

Some authors have separated the cultivated oat hexaploids into three species: *A. sativa*, *Avena byzantina* and *Avena nuda* (Holden, 1976). The most widespread type, *A. sativa*, is separated from *A. byzantina* only by the fact that in *A. byzantina* the rachilla remains attached to the upper floret; in *A. sativa* the rachilla breaks at the base of the upper floret. The other race, *A. nuda*, is simply a free-threshing variant of *A. byzantina*. The subtleties of these differences, along with complete interfertility, suggest that these species should also be united under the umbrella *A. sativa*.

While oats were native to the ancient Near East, they are not thought to have been established as an independent crop until 3000 years ago in Central Europe (Helbaek, 1959). Oats were probably carried along as a weed or secondary crop in the group of cultivated plants that spread out from the Near East. They were not exploited individually until they reached their full potential in the cooler, moist climates of Central Europe. In fact, oats reached North America, Argentina and Australia during the colonization period, decades before they were grown as an individual crop in the Middle East.

Rice

The most commonly cultivated rice species, *Oryza sativa* (Asian or paddy rice), is grown primarily in the humid tropics and subtropics, with some cultivation on flooded upland sites, such as central California. Another less important rice, *Oryza glaberrima* (African rice), is grown in East Africa, but is being replaced by *O. sativa*.

The cultivated rices and their ancestors are considered to be diploid (2n = 24), although their high chromosome number indicates that they could be ancient, diploidized polyploids (Moringa, 1965). Ten genomes have been identified among the various sections of *Oryza*, based on chromosome pairing relationships (Vaughan, 1994), molecular markers (Wang *et al.*, 1992; Aggarwal *et al.*, 1999; Ge *et al.*, 2001) and sequencing (Ge *et al.*, 1999). The cultivated species and their closest relatives carry the A genome and form what is referred to as the *sativa* complex. The A genome is further divided with superscripts to denote small pairing aberrations and partial sterility among the various diploid species (Table 8.3). The two cultivated species are morphologically very similar, but they show some chromosomal divergence and their hybrids are sterile (Oka, 1974, 1975).

A wide array of cytological, morphological and molecular markers have confirmed that the wild progenitor of *O. sativa* was *Oryza rufipogan*, and *O. glaberrima* diverged from *Oryza breviligulata* (Oka, 1988; Joshi *et al.*, 2000; Bautista *et al.*, 2001; Ishii *et al.*, 2001), but, beyond this point, it is difficult to

Table 8.3. Genomes and distribution of the species of *Oryza* in section *Oryza* (based on Vaughn, 1994; Chang, 1995: Ge *et al*., 2001).

Species and complex	Chromosome number (2n)	Genome designation	Geographical distribution
O. sativa complex	24		
O. sativa L.	24	AA	Worldwide (cultivated)
O. glaberrima Steud.	24	A^gA^g	West Africa (cultivated)
O. barthii A. Chev.	24	A^gA^g	Africa
O. glumaipatula	24	$A^{gp}A^{gp}$	South and Central America
O. longistaminata	24	A^lA^l	Africa
O. meridionalis	24	$A^?A^?$	Australia
O. nivara	24	AA	Tropical and subtropical Asia
O. rufipogan Griff.	24	AA	Tropical and subtropical Asia
O. officinalis complex	24		
O. punctata Kotschy	24, 48	BB, BBCC	Africa
O. eichingeri A. Peter	24	CC	Africa
O. officinalis Wall.	24	CC	Tropical and subtropical Asia
O. rhizomatis	24	CC	Sri Lanka
O. minuta J. S. Presl.	48	BBCC	South-East Asia
O. alta Swallen	48	CCDD	South and Central America
O. grandiglumis Prod.	48	CCDD	South and Central America
O. latifolia Desv.	48	CCDD	South and Central America
O. australiensis Domin.	24	EE	Australia

develop a consistent phylogeny. Reproductive isolating barriers are not complete between most species, and their morphological differences are often obscured due to introgression (Harlan, 1969; Morishima and Oka, 1970). There is even debate over whether perennial and annual types were the progenitors of the cultivated forms or some intermediate annual/perennial population (Chang, 1995; Joshi *et al*., 2000). The progenitors of cultivated rice may only be conceptual taxa that do not now exist because of introgression between wild and cultivated taxa.

The origins of Asian rice cultivation are also clouded. Many authorities consider India to be its cradle, but strong cases have been made for much earlier origins in central China and South-East Asia (Higham, 1984; Maloney *et al*., 1989; Normile, 1997). From its beginnings somewhere in Central Asia around 10,000 BP, rice cultivation probably moved into Korea and Japan by 3000 BP. The cultivation of African rice probably began in the Niger delta about 3500 years ago and spread gradually across tropical East Africa. Asian rices arrived in Africa about 2000 BP (Chang, 1975). Rice found its way to the New World in 1647, when its cultivation was begun in the Carolinas.

O. sativa underwent substantial differentiation as its cultivation spread across the world. Three main races are now recognized based on ecocultural criteria that include both temperate (japonica) and tropical areas (indica and javanica, Table 8.4). Javanica may have arisen through hybridization between

Table 8.4. Major characteristics distinguishing races of *O. sativa* (based on Nayar, 1973).

Character	Japonica	Javanica	Indica
Primary area of cultivation	Japan and Taiwan	Indonesia	South-East Asia
Grain shape	Short	Large	Narrow
Length of second leaf blade	Short	Long	Long
Angle between second leaf and stem	Small	Small	Large
Texture of plant parts	Hard	Hard	Soft
Angle between flag leaf and stem	Medium	Large	Small
Flag leaf	Short, narrow	Long, wide	Long, narrow
Tiller number	Large	Small	Large
Tiller habit	Erect	Erect	Spreading
Leaf pubescence	None	Little	More
Glume pubescence	Dense	Dense	Sparse
Awns	Usually absent	Usually present	Usually absent
Shattering	Difficult	Difficult	Easy
Panicle length	Short	Long	Medium
Panicle branching	Few	Many	Intermediate
Panicle density	High	Moderate	Moderate
Panicle weight	Heavy	Heavy	Light
Plant height	Short	Taller	Tall

indica and japonica (Second, 1982, 1985; Glaszmann *et al.*, 1984). Tremendous diversity exists within all these groups, due to isolation and selection under diverse conditions, and sufficient divergence has occurred between races of the Japonica group to cause hybrid sterility (Engle *et al.*, 1969).

Rye

There are three clearly defined species of rye (Spencer and Hawkes, 1980): (i) *Secale cereale* L., the cultivated species, which also exists as a highly diverse annual weed in farms in Iran, Afghanistan and Transcaspia; (ii) *Secale montanum* Gussh., an outbreeding, widely distributed assemblage of perennial races located from Morocco east through the Mediterranean countries to Iraq and Iran; and (iii) *Secale sylvestre* Host., an annual inbreeder, which is widely distributed from Hungary to the steppes of southern Russia. One additional taxon, *Secale vavilovii* Grossh., may be sufficiently unique to warrant species status (Spencer and Hawkes, 1980; Zohary and Hopf, 1993).

Most authorities believe that *S. cereale* evolved from *S. montanum* Gussh. (Riley, 1955; Khush and Stebbins, 1961). These two species are similar cytologically, but vary by two reciprocal translocations involving three pairs of chromosomes. Stutz (1972) proposed that the translocations were accumulated gradually through two steps involving interspecies hybridization; however, the recent molecular information indicates that *S. cereale* is probably a direct derivative of *S. montanum* (Vences *et al.*, 1987; Murai *et al.*, 1989). It is possible that *S. cereale* arose by chance in an isolated population via parapatric speciation.

As with oats, rye developed as a secondary crop. It was probably picked up as a weed when the wheat–barley assemblage arrived in western Asia, where the native species are widely distributed. Like the other grain species, agronomic traits, such as rachis fragility, ear branching and growth habit, are determined by only a few genes. The precise origin of rye domestication is unknown, but it was being cultivated at several locations in the general area of Turkey, north-western Iran and Armenia by 6000 BP (Hillman, 1978; Evans, 1995). Rye arrived in Europe as a cultivated crop by 4000 BP (Khush, 1962). Because of its tough constitution, it may have performed better than wheat and barley in the cooler, nutrient-poor northern climates and therefore attracted human attention. In modern times, tetraploid and hexaploid wheats have been artificially hybridized with rye to form the new crop called *Triticale* (Larter, 1995).

Sorghum

The wild sorghums present a diverse array of morphological variability and have presented taxonomists with an interesting challenge. In the most complex taxonomy, Snowden (1936) recognized over 70 species, but most researchers now accept a much smaller number after multivariate analyses were employed to cluster groups via morphological traits (deWet, 1978; Doggett, 1988). The three main native species located in Asia and Africa are *Sorghum bicolor* subsp. *arundinaceum* (Desv.) deWet and Harlan, *Sorghum propinquum* (Kunth) Hitchc. and *Sorghum halepense* (L.) Pers. *Sorghum bicolor* subsp. *arundinaceum* is an allotetraploid ($2n = 2x = 20$) that is found in tropical Africa, while *S. propinquum* shares the same chromosome number and is located in Asia, Indonesia and the Philippines. *S. halepense* or Johnson grass is a segmental auto-allo-octoploid ($2n = 2x = 40$) and is located in an overlapping zone between the other two species. Molecular and cytogenetic data have strongly implicated *S. bicolor* as one of the progenitors of Johnson grass (Hoang-Tang and Liang, 1988; Hoang-Tang *et al.*, 1991). The other likely parent is *S. propinquum,* which is interfertile with *S. bicolor* and carries many of its molecular markers (Paterson *et al.*, 1995).

Cultivated *S. bicolor* subsp. *bicolor* was probably derived from *S. bicolor* subsp. *arundinaceum* through human selection for non-shattering heads, large seeds and heads, threshability and suitable maturities (Doggett and Rao, 1995). These traits are regulated by relatively few QTL (Paterson *et al.*, 1998). The oldest archaeological remains of domesticated sorghum have been found in India; however, sorghum could have been domesticated at several other locations, including Ethiopia (Doggett and Rao, 1995), West Africa and the savannah of central Africa (Harlan and deWet, 1972). RFLP and isozyme data indicate that modern cultivars are most closely related to wild populations in central–north-eastern Africa (Aldrich and Doebley, 1992; Aldrich *et al.*, 1992; Deu *et al.*, 1994). It is likely that sorghum cultivation spread from Africa to India through trade routes some 3000–2000 years ago, and it arrived in China at about the beginning of the Christian era.

Harlan and deWet (1972) recognized five main races of cultivated sorghum based on morphological patterns (Fig. 8.7): (i) Bicolor, which is widely distributed across the African savannah and Asia; (ii) Caudatum, found in central Sudan and the surrounding areas; (iii) Guinea, which is grown in eastern and western Africa; (iv) Durra, found primarily in Arabia and Asia Minor, but also grown in India, Burma and Ethiopia and along the Nile Valley; and (v) Kafir, which is cultivated primarily in southern Africa. These races overlap in many areas and freely hybridize among themselves and with the wild species to produce intermediate races. A weedy race of sorghum, *S. bicolor* subsp. *drummondii* (Steud.) de Wet, exists wherever cultivated and wild sorghums are sympatric (deWet, 1978).

The various cultivated types of today could have been derived from sev-

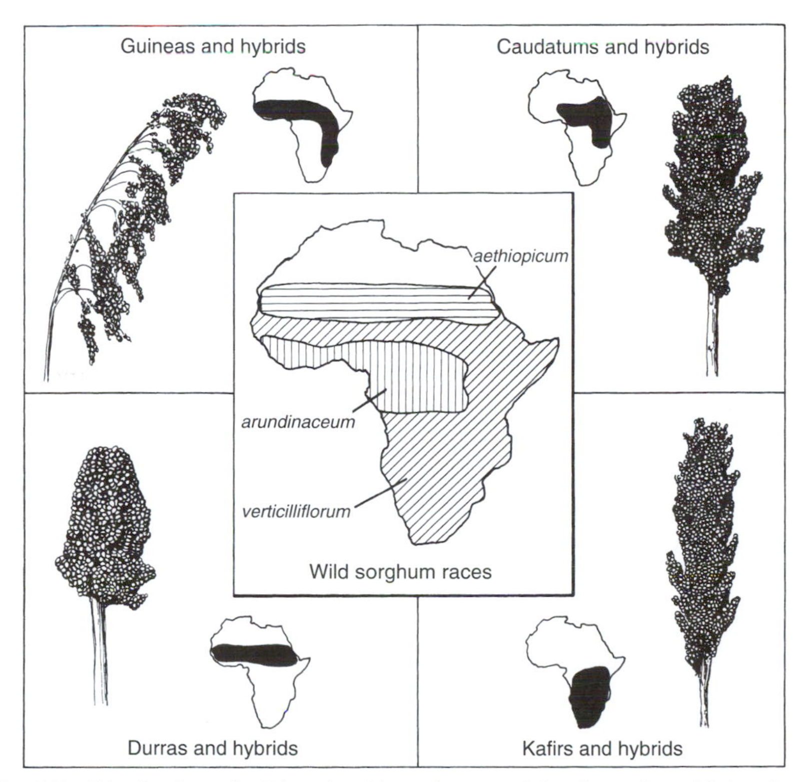

Fig. 8.7. Distribution of wild and cultivated races of *Sorghum*. An additional race, Bicolor, is widely scattered across the African savannah and in Asia from India to Japan and Indonesia.

eral wild races of *S. bicolor* subsp. *arundinaceum* that are located in different parts of Africa (Snowden, 1936; deWet and Huckabay, 1967). The West African sorghums probably came from var. *arundinaceum*, while the eastern–central group were developed from var. *verticilliflorum*. The north-eastern sorghums were developed from var. *aethiopicum*. It is not known whether these races were domesticated separately or were stimulated by adjacent cultivations. There was undoubtedly much hybridization occurring between wild and cultivated forms as agriculture spread throughout Africa. In fact, recent molecular analyses have indicated that most variation is more closely associated with geographical than with racial differentiation (Morden *et al.*, 1989; Menkir *et al.*, 1997; Djè *et al.*, 2000).

Wheat

The taxonomy of *Triticum* has undergone numerous reorganizations, which vary according to how much importance is given to levels of interfertility, ecological ranges and specific morphological traits (for an excellent review see http://www.ksu.edu/wgrc). We shall use the recent taxonomy of van Slageren (1994) and cite other relevant synonyms where warranted (Table 8.5).

Three ploidy levels of wheat were domesticated: diploid, tetraploid and hexaploid. The diploids are represented by the single cultivated species *Triticum monococcum* or einkorn, meaning one seed per spikelet. Einkorn is

Table 8.5. A classification of the wheat species *Triticum* L. (based on van Slageren, 1994).

Species	Subspecies	Status	Chromosome number	Genome
T. monococcum	*aegilopoides*	Wild	2n = 14	AA
	monococcum	Cultivated		
T. turartu		Wild		
T. turgidum	*carthlicum*	Cultivated	2n = 28	AABB
	dicoccoides	Wild		
	dicoccum	Cultivated		
	durum	Cultivated		
	turgidum	Cultivated		
	paleocolchicum	Archaeological		
	polonicum	Cultivated		
T. timopheevi	*armeniacum*	Wild	2n = 28	AAGG
	timopheevi	Cultivated		
T. aestivum	*spelta*	Cultivated	2n = 42	AABBDD
	macha	Cultivated		
	aestivum	Cultivated		
	compactum	Cultivated		
	sphaerococcum	Cultivated		

grown for a dark bread in some parts of the Middle East and southern Europe, but its glumes fit tightly around its seeds, making them hard to remove. For this reason, it is more commonly used to feed cattle and horses.

The tetraploid group is represented by a complex of several subspecies, with the most important cultivated ones being *Triticum turgidum* subsp. *dicoccoides* (emmer) and *T. turgidum* subsp. *durum* (durum). Emmer was widely grown in the Mediterranean region until Graeco-Roman times and makes good bread and pastry, but like einkorn has clinging glumes and delicate stems, which make it difficult to harvest and thresh. Emmer is now locally important as livestock feed in a diverse region, including the mountains of Europe and the Dakotas of the USA. Durum is grown widely in Italy, Spain and the USA. Its seeds are easily separated from its glumes, and it has a high gluten content that makes it sticky when wet. This makes it excellent for making spaghetti and macaroni. There is also another tetraploid species, *Triticum timopheevi*, but it is only cultivated in a restricted region of Russia.

The hexaploids are represented by the bread wheat, *Triticum aestivum*, which was the last domesticated wheat to appear but is now the most widely planted. It is a highly variable group that is divided into numerous subspecies (Table 8.5). With the exception of *spelta*, the bread wheats have tough inflorescence stems that do not shatter when harvested, and the seeds are easily threshed after gathering. Some of the hexaploid subspecies must be hulled (*spelta* and *macha*), while others are free threshing (*aestivum*, *compactum* and *sphaerococcum*). All hexaploid wheats are high in the protein gluten, which makes a fluffy, leavened bread. *T. aestivum* is grown throughout the world, but is particularly important in the continental climates of the Ukraine, central USA, Canada and Australia and the cool temperate climates of northern Europe, China and New Zealand.

Cultivated einkorn was probably derived from the wild subspecies, *T. monococcum* subsp. *aegilopoides* (*Triticum boeoticum* or *T. monococcum* subsp. *boeoticum*). The cultivated and wild subspecies look very similar and share numerous molecular markers (Asins and Carbonell, 1986; Hammer *et al.*, 2000). The native distribution of einkorns is in northern Syria, southern Turkey, northern Iraq, Iran and western Anatolia. The only substantial difference between the wild and cultivated forms is that the wild species has fragile spikes, which allow its seeds to scatter freely, while the cultivated variety has a tough rachis, which allows for more effective harvesting. In the cultivated forms, the mature ear breaks into individual spikelets through threshing and the kernels are wider.

The earliest archaeological evidence of people gathering brittle *T. monococcum* subsp. *aegilopoides* is found in remains at Tell Abu Hureyra, which are about 10,000 years old. The first cultivated einkorn is present by 8000 BP at several sites in the Near East, including Ali Kosh in Iran, Jarmo in Iraq and Cayönü and Can Hasan in Turkey (Helbaek, 1966; Renfrew, 1973). Brittle and non-brittle forms are found together at these locations, suggesting that gathering and farming were being employed simultaneously. Using amplified fragment length polymorphism (AFLP) molecular markers, Heun and co-workers (1997) have targeted the first domestication of einkorn to the Karacadağ Mountains in south-eastern Turkey.

The tetraploid wheat, *T. turgidum*, was derived from the natural hybridization of wild einkorn (AA) and another diploid species carrying the B genome (Fig. 8.8). Sequence data from plastid acetyl-coenzyme A (acetyl-CoA) carboxylase and 3-phosphoglycerate suggest that *Triticum urartu* is the diploid AA donor (Huang *et al.*, 2002). No diploid with the B genome has

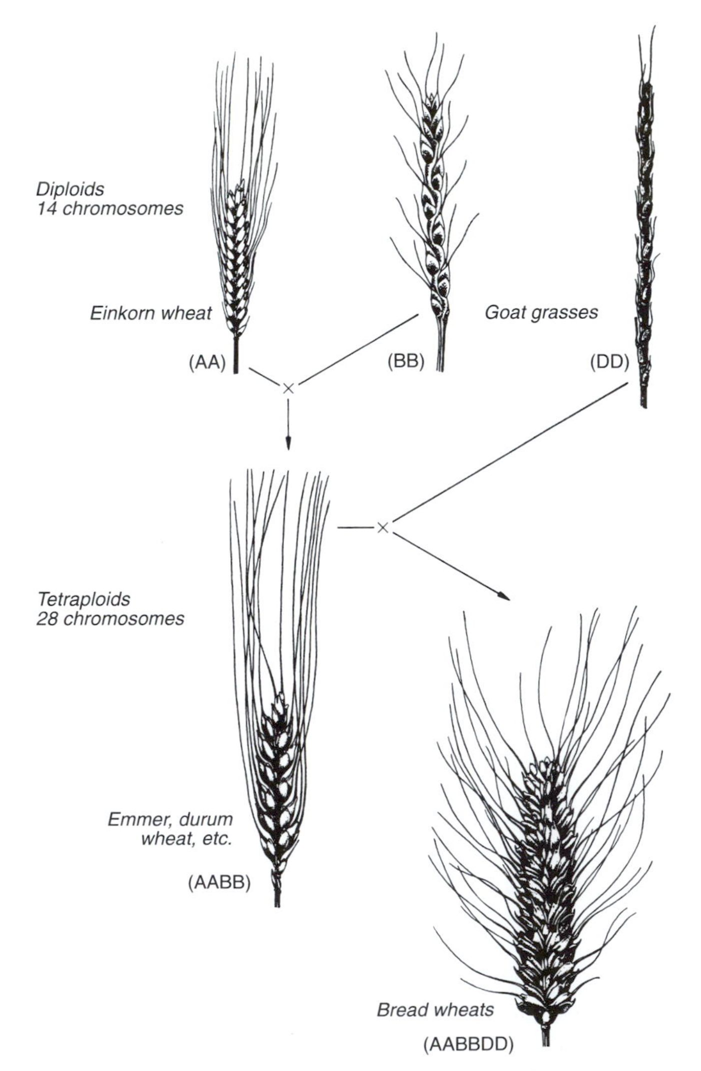

Fig. 8.8. Probable origin of domesticated wheats (adapted from P.C. Mangelsdorf, 1953).

been identified (Blake *et al.*, 1999); however, much evidence points to the S genome species *Aegilops speltoides* (*Triticum speltoides*), as being the closest living relative. Support for this hypothesis comes from cytological data (Riley *et al.*, 1958; Kerby and Kuspira, 1988; Friebe *et al.*, 1996), protein electrophoresis (Johnson, 1972), RFLP analysis of nuclear DNA sequences (Dvorak and Zhang, 1990; Talbert *et al.*, 1995; Badaeva *et al.*, 1996; Sasanuma *et al.*, 1996), mitochondrial and plastid RFLPs (Tsunewaki, 1989; Terachi *et al.*, 1990) and plastid sequence data (Huang *et al.*, 2002). *Aegilops speltoides* appears to be the plastid donor of all polyploid wheats (Wang, G.-Z. *et al.*, 1997), although more than one species could have contributed to the B genome through introgression (Zohary and Feldman, 1962).

The cultivated forms of emmer wheat were probably derived from wild populations of *Triticum dicoccoides*. Wild emmer is most common in the catchment area of the upper Jordan Valley, where it forms natural fields of grain with wild barley and oats. Further north it grows wild with barley and einkorn wheat. Wild forms of emmer wheat have been found at numerous prehistoric sites, with the oldest remains being at least 19,000 years old. The earliest cultivated forms have been discovered at Tell Aswad in Syria (van Zeist and Bakker-Heeres, 1985), Tell Abu Hureyra (Hillman, 1975) and Jericho (Hopf, 1983) in deposits around 9000 years old. Cultivated types have a non-brittle spike, which is regulated by a single major gene, and their grains are wider and thicker, with a rounder cross-section.

Emmer wheat spread rapidly across the Near East, even to areas where einkorn was not cultivated, and became the principal wheat in Neolithic times and the early Bronze Age. Emmer makes good bread and pastry, but its clinging glumes and delicate stem make it difficult to harvest and thresh. Durum seeds are easily separated from their glumes and, as previously mentioned, have a high gluten content that is sticky when wet and makes excellent pasta. They probably arose soon after emmer cultivation began (Zohary and Hopf, 1993). The free-threshing durum types probably evolved under cultivation through a gradual accumulation of genes reducing the toughness of glumes. This shift from hulledness to nakedness is regulated by a polygenic system. Scattered wild tetraploids also have the Q factor found in *T. aestivum* (see below), but these may have been picked up through hybridization with the bread wheats.

The hexaploid *T. aestivum* (AABBDD) is a relatively recent product of hybridizations between the cultivated tetraploid wheats *T. turgidum* subsp. *dicoccoides* (AABB) and the wild diploid goat grass *Aegilops tauschii* (*Aegilops squarrosa*) (Fig. 8.8; McFadden and Sears, 1946; Kihara, 1954; Lubbers *et al.*, 1991; Dvorak *et al.*, 1998). The geographical origin of the hexaploid wheats was probably outside the original Fertile Crescent, as the native range of *A. squarrosa* is in continental Central Asia rather than the Mediterranean Near East. A southern Caspian Sea–Transcaucasia origin for hexaploid wheat has been confirmed by isozyme (Jaaska, 1983) and ribosomal RFLPs (Dvorak *et al.*, 1998). The range of goat grass probably expanded as a weed in secondary, human-made habitats as farming spread.

The first evidence of hulled bread wheats (*T. aestivum* subsp. *spelta*) has been found at Arukhlo 1, Transcaucasia (Janushevich, 1984), dating from about 7000 BP, but, by this time, hulled forms were probably grown all across the Caspian belt. As in the other wheats, the genetics of the free-threshing trait is relatively simple, being regulated by two recessive genes, *q* on chromosome 5A, and *tg* (tenacious glumes) on chromosome 2D (Zohary and Hopf, 1993). All modern types are *tg/tg q/q*.

The addition of the D genome greatly expanded the range of wheats. The hexaploids spread rapidly and diverged into numerous races as new environments were met, mutations were accumulated and unique alleles were introgressed from native species. The extensive amount of interspecies hybridization now occurring naturally in the Middle East supports a polyphyletic origin for most of the polyploid species. Substantial chromosomal and genic modifications probably arose after the early hybridizations as the various taxa adapted to new environmental challenges, and their self-pollinating nature aided in the fixation of new coadapted complexes. This reorganization and diversification probably enhanced the adaptive potential of the genus, but has given great headaches to evolutionists trying to determine genomic origins.

9 Protein Plants

Wheat and barley were probably the first crops to be cultivated by our ancestors, but the legumes were not far behind. Lentils and peas were being harvested in the wild at about the same time that grain cultivation was beginning in the Near East, and it did not take people long to add the other pulses to their list of cultivated crops. Beans may have been domesticated even before maize in the New World.

In this chapter, we explore the origins of the most important legume crops. The evolutionary relationships of the legumes are often straightforward, because they have direct living relatives and simple ploidy relationships. They are also carbonized readily, like the grains, and are well represented in the archaeological record.

Chickpea

Chickpea (*Cicer arietinum* L.) belongs to a genus of 40 species that is located primarily in central and western Asia (van der Maesen, 1987). The majority of the *Cicer* species are perennial shrubs, but the group containing the cultivated forms are annual. Most species are diploid with 2n = 2x = 16. Three crossability groups have been identified among the species (Ladizinsky, 1995); the cultivated chickpea is in a group with *Cicer reticulatum* Davis and *Cicer echinospermum* Ladiz. These two diploid species closely resemble the cultivated chickpea, as they are annual and morphologically very similar and share many biochemical and molecular markers (Tayyar and Waines, 1996; Iruela *et al.*, 2002).

Of these two wild species, only *C. reticulatum* is fully interfertile with the cultivated forms and morphological, cytological and molecular studies have confirmed their close relationship (Labdi *et al.*, 1996; Ahmad, 2000; Iruela *et al.*, 2002). Therefore, *C. reticulatum* should probably be regarded as a subspecies, *C. arietinum* subsp. *reticulatum* (Ladiz) Cubero et Morreno. Crosses between the cultivated species and other wild species have generally failed (Ladizinsky and Adler, 1976a,b; Singh and Ocampo, 1993, 1997), although there are reports of successful hybridizations between cultivated chickpea and *Cicer pinnatifolium* Jaub et Spach, *Cicer judaicum* Boiss and *Cicer bijugum* Rech. (Singh *et al.*, 1994; Badami *et al.*, 1997).

Chickpea cultivation was probably associated with the emergence of the grain crops in the Near East. Interestingly, five closely related species coexist in that area, with similar growth habits and taste, but only *C. arietinum* was domesticated (Zohary, 1999). Carbonized seeds have been found at Cayönü, Turkey, and Tell Abu Hureyra, Syria (Zohary and Hopf, 1993; Ladizinsky, 1995). Both these locations are 9000–10,000 years old, but the seeds are very small and could represent cultivated material. The oldest remains of large seeds from clearly domesticated plants come from Bronze Age digs in Israel and Jordan, Jericho and Bab edh-Dhra. Chickpeas probably travelled to Europe along with the early grain crops. Seeds have been found in Greece from 8000 before present (BP) and southern France by 5000 BP. They arrived in India about 4000 years ago, and in Ethiopia from the Mediterranean about 3000 BP.

Over the course of domestication, the size of the seeds increased dramatically, from 3.5 to 6.0 mm (Fig. 9.1), and plant types that retained their pods and seeds were selected. The seed-coat also became smoother and thinner. As the crop was dispersed, two major morphological types emerged (Hawtin *et al.*, 1980): (i) plants with large, owl-shaped, light coloured and smooth seeds (Kabuli) with pale cream flowers; and (ii) plants with smaller, ram-shaped, dark and wrinkled seeds (Desi) with purple flowers. The Kabuli types are located primarily on the Mediterranean side of the chickpea distribution, while the Desi forms are on the eastern side. The Kabuli types are more distant from the wild types than the Desi, and this fits with their more distant geographical separation from their wild ancestor and their variant ecological range (Rowewal *et al.*, 1969).

Cowpea

The genus of the cowpea, *Vigna*, is a relatively large pantropical genus, with the majority of its species being found in Africa. All species have 22 chromosomes, with little cytogenetic divergence. The variability and complexity of morphological patterns within the group have led to the development and use of several different taxonomies (Lush and Evans, 1981; Ng, 1995); we shall use the most recent one of Pasquet (Table 9.1). In his treatment, *Vigna unguiculata* includes

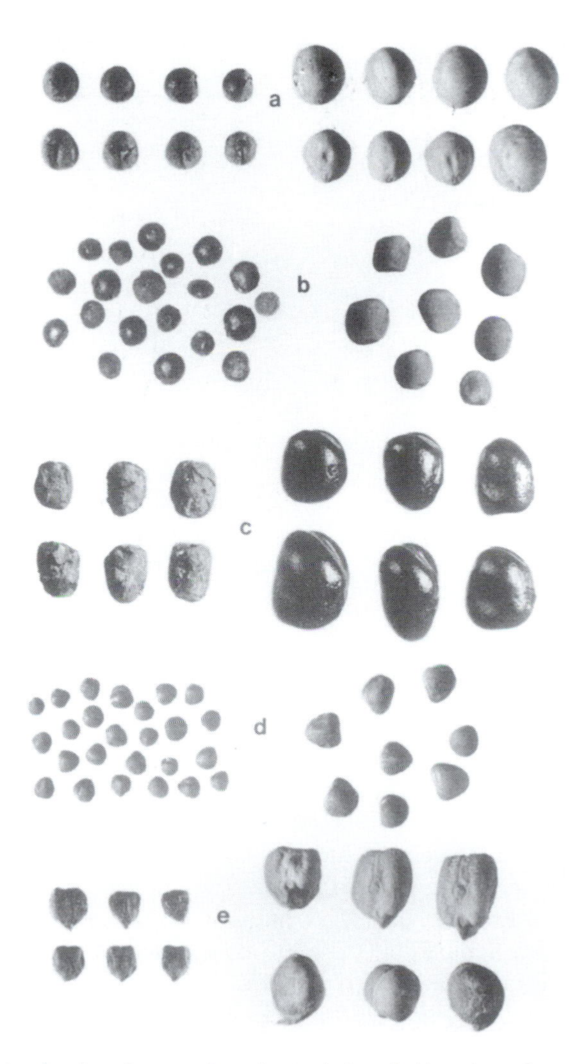

Fig. 9.1. Seeds of pulses from archaeological sites (left) and modern varieties (right): (a) pea, (b) lentil, (c) broad bean, (d) bitter vetch, and (e) chickpea (used with permission from M. Hopf, © 1986, Archaeological evidence of the spread and use of some members of the *Leguminosae* family. In: *The Origin and Domestication of Cultivated Crops,* Elsevier Science Publishers, New York).

the annual cowpeas and ten wild perennial subspecies. The closest wild subspecies to the cowpea are *pubescens*, *tenuis* and *alba*. The cultivars are further subdivided into five groups based on seed and pod characteristics (Ng and Maréchal, 1985; Pasquet, 1998): (i) 'Unguiculata', grown as a seed crop; (ii) 'Biflora' (catjang), used as fodder; (iii) 'Sesquipedalis' (yardlong or asparagus bean), grown as a green pod vegetable; (iv) 'Textilis', used for its peduncle fibres; and (v) 'Melanophthalmus' (black-eyed pea), grown as a seed crop.

Table 9.1. Subspecies and cultivar groups of the chickpea *V. unguiculata* (based on Pasquet, 1997, 1999).

Subspecies	Status	Cultivar group
alba	Wild	
baoulenis	Wild	
burundiensis	Wild	
letouzeyi	Wild	
pawekiae	Wild	
pubescens	Wild	
stenophylla	Wild	
tenuis	Wild	
unguiculata var. *spontanea*	Wild	
unguiculata var. *unguiculata*	Cultivated	'Biflora' (catjang)
		'Melanophthalmus' (black-eyed pea)
		'Sesquipedalis' (yardlong or asparagus bean)
		'Textilis'
		'Unguiculata'

There is conflicting evidence on where the crop was first domesticated (Vaillancourt and Weeden, 1992; Ng, 1995; Pasquet, 1999), probably due to the considerable amount of hybridization that has occurred between cultivated and wild races (Coulibaly *et al.*, 2002). In the most comprehensive molecular survey of wild populations, the cultivated cowpeas appear to be most closely related to western African populations of ssp. *ungiculata* var. *spontanea*, although enough variability exists for other locations not to be excluded (Coulibaly *et al.*, 2002). Regardless of its precise origin, cowpea was probably cultivated by 7000 to 6000 BP and arrived in India about 4000 years ago with the grain species (Ng, 1995).

Considerable divergence occurred in chickpea soon after its cultivation began. Ng (1995) believed cowpea was used as a fodder crop for cattle prior to domestication for human consumption. The original forms were probably spreading, short-day plants that readily scattered their seeds. Pod dehiscence and seed dormancy were most probably lost quickly in conjunction with domestication. Upright, day-neutral types may have first emerged after introgression with local wild relatives in the rain forests of Africa (Alexander and Coursey, 1969; Lush and Evans, 1981). The fodder and green pod vegetable were probably developed after the crop arrived in Asia.

Pea

The genus *Pisum* is represented by two, self-pollinating diploid species (2n = 2x = 14), *P. fulvum* Sibth. & Sm. and *P. sativum* L. (Davies, 1995).

P. fulvum is a wild species located in the eastern Mediterranean, while *P. sativum* (Fig. 9.2) is a complex aggregate of wild and cultivated races centred in the Mediterranean basin and Near East. The two species can be crossed with difficulty, although the hybrids are largely sterile (Ben-Ze'ev and Zohary, 1973). The interspecies cross using *P. sativum* as the maternal parent is more successful than the reciprocal, but the F_1 hybrids are weak and have many quadrivalents, trivalents and univalents at metaphase I. Regardless, hybrids are often found wherever the two species overlap (Davies, 1995).

The wild forms of *P. sativum* fall into two general morphological–ecological classes: (i) a tall, climbing type that was formerly called *Pisum elatius* Bieb. (= *P. sativum* subsp. *elatius*) and prefers mesic habitats; and (ii) a more xeric-loving, short type, formerly called *Pisum humile* Boiss. (= *P. sativum* subsp. *syriacum*), which often invades cereal farms. *P. sativum* subsp. *syriacum* was probably the ancestor of the cultivated types, as all subsp. *elatius* pea populations differ from the cultivated forms by a single chromosomal translocation,

Fig. 9.2. The pea plant (*Pisum sativum*) (used with permission from R.H.M. Langer and C.D. Hall, © 1983, *Agricultural Plants,* Cambridge University Press, New York).

and some subsp. *syriacum* populations in Turkey and Syria share the same chromosomal complement (Ben-Ze'ev and Zohary, 1973). Still, the high levels of introgression found in nature suggest that both native types could have contributed genes to the domesticated forms.

Wild and domesticated peas are difficult to distinguish in the archaeological record; however, the seed-coat appears to be diagnostic when it exists, as domesticated forms appear to have a smooth seed-coat, while the wild forms are rough coated (Zohary and Hopf, 1973). Smooth-coated forms are found with the earliest domesticated barleys and wheat in the Near East. Peas probably followed the spread of neolithic agriculture into Europe, and they were widespread in central Germany by 6000 BP (Zohary and Hopf, 1993). They arrived in the Nile Valley by 7000 BP and in India by 4000 BP.

As with most other seed crops, there was a general increase in size associated with pea cultivation along with the development of non-dehiscence of the pod, seed retention and the elimination of seed dormancy. The group has diverged substantially over the years, due at least in part to their self-breeding mechanism. Tremendous variation can be found in height, habit, flower colour and seed characteristics. Literally hundreds of landraces exist, and breeders have provided us with large numbers of varieties (Zohary and Hopf, 1993).

Lentil

Lentil belongs to the small genus *Lens*, with six annual diploid species (2n = 2x = 14) located in the Mediterranean basin and south-west Asia. *Lens* contains the cultivated *L. culinaris* subsp. *culinaris* (Fig. 9.3), its wild progenitor *L. culinaris* subsp. *orientalis* (= *Lens orientalis*) and five other wild species. Three crossability groups have been identified among these species: (i) *L. culinaris* and *Lens odemensis* Ladiz.; (ii) *Lens ervoides* (Bring.) Grand., *Lens nigricans* (M. Bieb.) Grand and *Lens lamottei* Czefranova; and (iii) *Lens tomentosus* Ladiz. (Ladizinksy *et al.*, 1984; van Oss *et al.*, 1997; Mayer and Bagga, 2002). Two integrating forms of cultivated lentil are recognized: (i) subsp. *microsperma*, with small grains and seeds (3–6 mm in diameter); and (ii) subsp. *macrosperma*, with large pods and seeds (Zohary, 1995). The species are primarily self-pollinated, like most other legume crops.

L. culinaris subsp. *culinaris* and subsp. *orientalis* have overlapping ranges, look very similar and share a high percentage of their molecular markers (Ladizinsky, 1999), although subsp. *orientalis* is much smaller and has pods that burst before harvest. There is considerable genetic divergence in subsp. *orientalis*, as some races share chromosome homology and crossability with the cultivated types, while others do not. Based on chloroplast DNA (cpDNA) restriction data, chromosome behaviour and crossability, Ladizinsky (1999) believes that the genetic stock from which lentil was

Fig. 9.3. The lentil plant (*Lens culinaris*) (reprinted with permission from M. Zohary, © 1972, Israel Academy of Sciences and Humanities).

domesticated is represented by three lines collected from Turkey, northern Syria and southern Syria. Further pinpointing of the origin of the lentil will take much more comprehensive collection and screening.

Lentil cultivation, like that of chickpea, was closely associated with the domestication of wheat and barley. It may even be one of the founder crops of Old World agriculture, since carbonized lentil seeds have been found in many of the ancient Near Eastern farming villages (Zohary and Hopf, 1993). Its pattern of migration across Asia and Europe closely matched that of the seminal grain species and other legume crops, and it arrived in Spain and Germany 6000–7000 years ago (Fig. 9.4). Lentils were grown in India by 4500 BP.

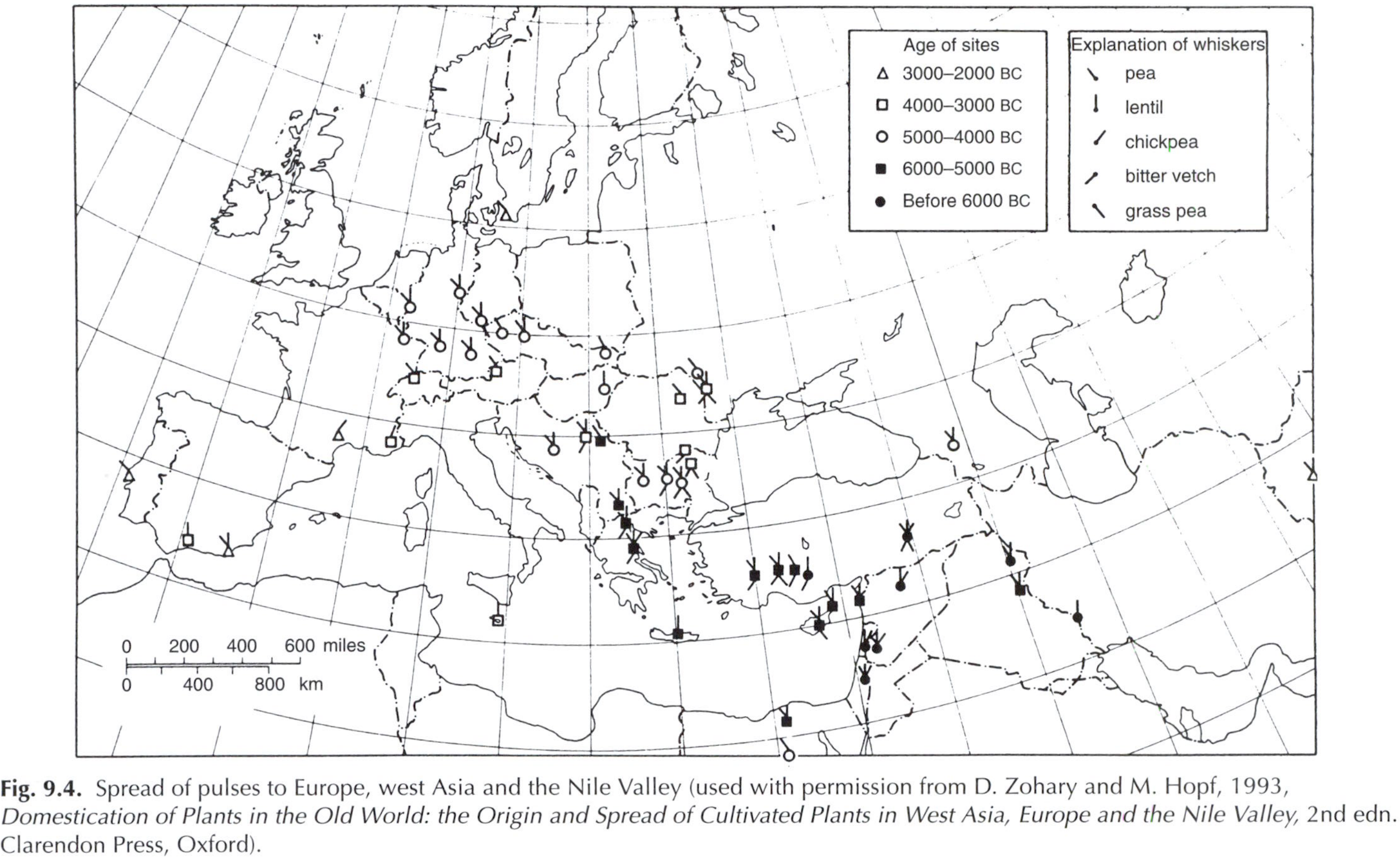

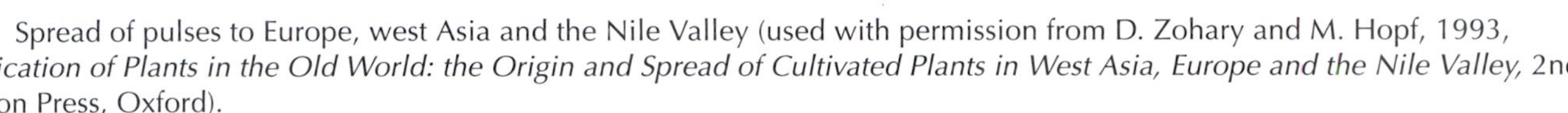
Fig. 9.4. Spread of pulses to Europe, west Asia and the Nile Valley (used with permission from D. Zohary and M. Hopf, 1993, *Domestication of Plants in the Old World: the Origin and Spread of Cultivated Plants in West Asia, Europe and the Nile Valley*, 2nd edn. Clarendon Press, Oxford).

Phaseolus Beans

There are dozens of *Phaseolus* species, all of American origin (Delgado Salinas *et al.*, 1999). Five of them are cultivated (Fig. 9.5): *Phaseolus vulgaris* L. (common bean, haricot, navy, French or snap bean), *Phaseolus coccineus* L. (runner or scarlet bean), *Phaseolus lunatus* L. (Lima, sieva, butter or Madagascar bean), *Phaseolus polyanthus* Greenm. (year bean) and *Phaseolus acutifolius* A. Gray (tepary bean). Until recently, *P. polyanthus* was considered to be a subspecies of *P. coccineus* (Schmit and Debouck, 1991).

All the cultivated species are diploid (2n = 2x = 22) and their direct progenitors are found in the wild. The wild and cultivated races of all species are fully to nearly fully compatible. *Phaseolus* beans inhabit a wide range of habitats, including the cool, humid uplands of Guatemala (*P. coccineus* and *P. polyanthus*), semi-arid regions in Guatemala, Arizona and Mexico (*P. acutifolius*), the tropics and subtropics of Central and South America (*P. lunatus*) and warm, temperate areas in Mexico–Guatemala (*P. vulgaris*) (Debouck, 2000). *P. polyanthus* and *P. coccineus* are outcrossed but can also self, while the other species are primarily self-pollinating.

The evolutionary relationships among the cultivated species have been studied at a number of different levels including crossability, seed proteins and molecular markers (Gepts, 1998; Debouck, 2000). *P. vulgaris*,

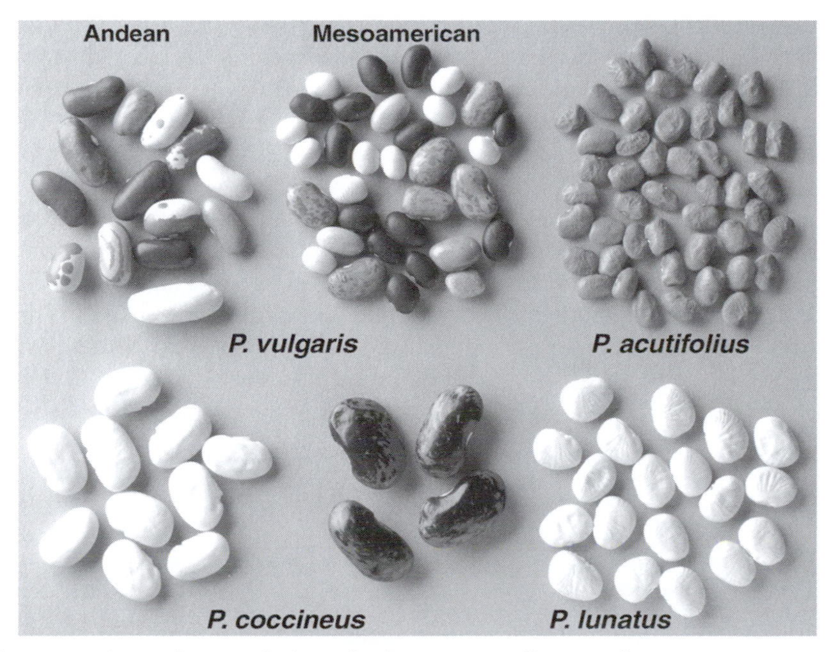

Fig. 9.5. Cultivated types of *Phaseolus* beans (Jim Kelly at Michigan State University provided representatives of the bean types).

P. polyanthus and *P. coccineus* are very closely related, with *P. acutifolius* in the middle and *P. lunatus* being the most distant. A number of other wild species share varying levels of compatibility with the cultigens, but they are not thought to be in their direct ancestry (Table 9.2).

Most investigators now believe that the common bean was domesticated independently in Mesoamerica and South America from two distinct wild taxa of *P. vulgaris*, var. *aborigineus* in South America and var. *mexicanus* in Mesoamerica (Gepts, 1998). Ancient seeds have been found in both Puebla, Mexico (2300 BP) and the Peruvian Andes (4400 BP) (Kaplin and Lynch, 1999), and landraces from the two regions have distinct floral structures, seed sizes, phytopathology, phaseolin seed proteins, allozymes and DNA markers (Gepts, 1990, 1998; Khairallah *et al.*, 1990). In addition, hybrids between plants from the two areas are mostly infertile. The combination of the dwarf lethal genes, DL_1 from Middle America and DL_2 from the Andean region, produces very weak F_1s.

There are not as many data on the origin of the other bean species, but the lima bean appears to have a polycentric origin like the common bean, with a large lima cultigen originating in the Andean region, and a small lima or sieva cultigen originating in Mesoamerica (Debouck, 1991; Caicedo *et al.*, 1999). The earliest evidence of domesticated lima beans comes from coastal Peru (5600 BP) (Kaplin and Lynch, 1999). Many investigators feel that *P. acutifolis* was originally domesticated in Central America (Debouck and Smartt, 1995), although an Aridoamerican origin (north-west Mexico–south-west USA) has been suggested (Pratt and Nabham, 1988). The earliest evidence of *P. acutifolius* in the archaeological record is from about 2500 BP in the Tehuacan Valley. Sketchy data indicate that *P. coccineus* was first tamed in Mexico about 1000 years ago (Kaplan and Lynch, 1999). *P. polyanthus* was probably domesticated in Guatemala in pre-Columbian times (Schmit and Debouck, 1991).

The common bean spread widely across North and South America over several thousand years and arrived in the Ohio Valley of the central USA by about 1000 BP. The Spanish explorers and traders took lima beans from Peru to Asia and Madagascar. The lima and common bean arrived in Africa via the slave trade. Remarkably, much of the variation in seed-coat, shape and growth habit found in South America has been maintained in the

Table 9.2. Gene pools of the various *Phaseolus* cultigens (adapted from Debouck, 2000).

Primary gene pool (GP1)	Secondary gene pool (GP2)			Tertiary gene pool (GP3)		
	Close	Intermediate	Distant	Close	Intermediate	Distant
P. vulgaris	*P. costaricensis*	*P. polyanthus*	*P. coccineus*		*P. acutifolius*	*P. filiformis*
P. polyanthus	*P. coccineus*	*P. costaricensis*	*P. vulgaris*			
P. coccineus	*P. polyanthus*		*P. vulgaris*			
P. acutifolius		*P. parvifolius*			*P. vulgaris*	
P. lunatus	*P. pachyrrhizoides*		*P. maculatus*	*P. jaliscanus*	*P. salicifolius*	

African landraces, even though the initial introductions were probably limited (Fig. 9.6; Martin and Adams, 1987a,b). *P. vulgaris* eventually reached Europe by the 16th century and was introduced back into eastern North America in the late 19th century.

In the last 10,000 years all the *Phaseolus* beans have undergone very similar changes under domestication, which are often regulated by only a few genes (Smartt, 1999). Some of the most dramatic changes associated with cultivation have been the shift from the perennial to the annual habit,

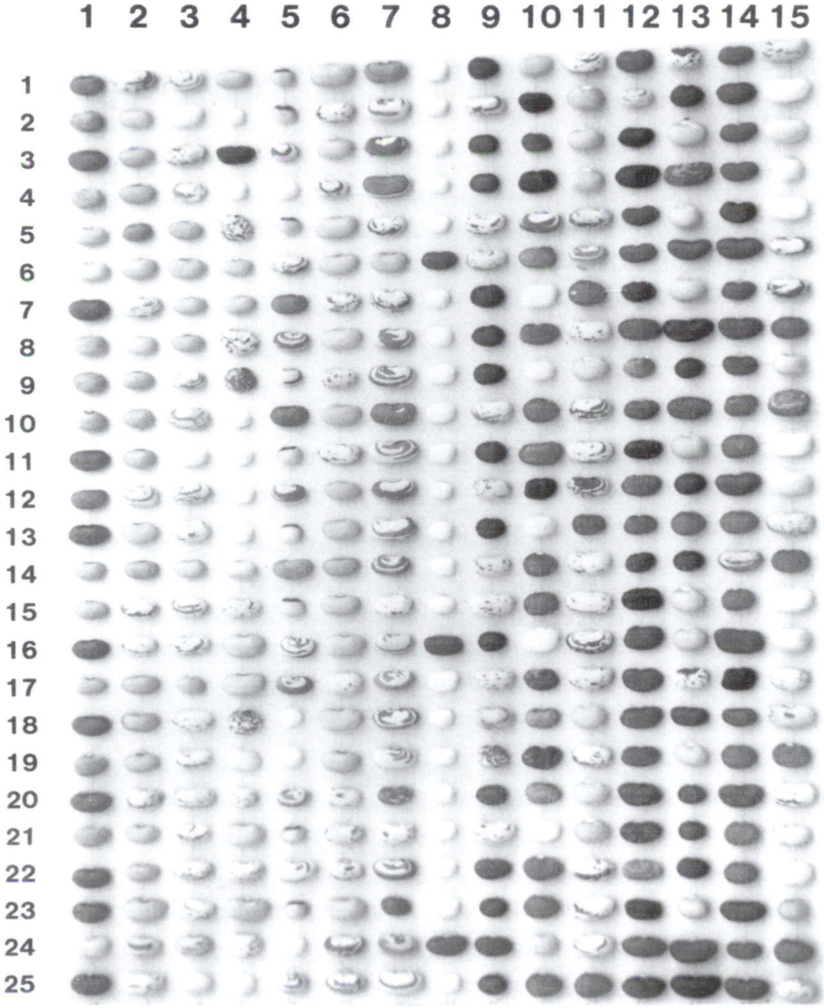

Fig. 9.6. *Phaseolus lunatus* seed types found in Malawi. Seeds are arranged in columns corresponding to 15 landraces. (Reprinted with permission from G.B. Martin and M.W. Adams, *Economic Botany* 41, 190–203, © 1987, New York Botanical Garden, New York.)

larger and softer seeds, a shift from short-day to day-neutral photoperiods and the development of more persistent pods. Considerable variation has also been developed in plant architecture, from indeterminate climbers to determinate bush types (Adams, 1974).

Faba Beans

Vicia faba L. (field or broad bean) is diploid (2n = 2x = 12) and it is outcrossed, although it has varying degrees of self-fertility. Several wild vetches show strong morphological similarities to the faba bean, such as *Vicia narbonensis* L. and *Vicia galiliea* Plitm. Et Zoh. (Schäfer, 1973; Zohary and Hopf, 1973), but most of them are 2n = 2x = 14 and interspecific hybridizations have failed (Ladizinisky, 1975). *V. faba* chromosomes are much larger than the wild species and contain much more DNA (Chooi, 1971), indicating that any evolutionary relationship with known species is remote (Bond, 1995). Either a 12-chromosome progenitor has yet to be discovered or the direct progenitor of the broad bean is extinct.

The evidence for a very early domestication of faba beans is scant, with the most ancient archaeological evidence coming from Israel, from 8000 to 8500 BP (Kislev, 1985). Most of the other findings are from several thousand years later and are restricted to areas east of Israel (Hanelt *et al.*, 1972). However, the crop probably originated in the Near East and spread west along the Mediterranean coasts until it reached Spain by 5000 to 4000 BP. Interestingly, faba beans did not arrive in China until the last millennium (presumably through the silk trade), even though China is now the leading producer in the world (Hanelt, 1972b). The Spaniards took faba beans to Mexico and South America, only a few hundred years ago.

Today, several different, intergrading subspecies of faba beans are recognized, based primarily on seed size: *minor*, *equina*, *faba* and *paucijuga* (Cubero, 1974). The smallest type, *minor*, is considered the most primitive, but all the subspecies contain varieties that carry primitive traits, such as shattering pods. *V. faba* subsp. *paucijuga* is the most self-fertile of all the subspecies and can set seed without bee activity.

Soybean

The soybean genus *Glycine* is divided into two subgenera, *Soya* and *Glycine*. The subgenus *Glycine* contains a confusing array of 15 wild perennial species whose centre of distribution is Australia; all carry 2n = 40 or 80 chromosomes, with some aneuploidy (Table 9.3). The subgenus *Soya* is composed of the annual cultivated species *Glycine max* and the wild species *Glycine soya*, which both carry 2n = 40 chromosomes. *G. soya* is distributed naturally in China, Japan, Korea, Taiwan and the former USSR (Hymowitz, 1995).

Table 9.3. Species of *Glycine* and their genomes (based on Hymowitz, 1995; Kollipara *et al.*, 1997).

Subgenus	Species	2n	Genome
Soya	*soya*	40	GG
	max	40	GG
Glycine	*albicans*	40	II
	arenaria	40	HH
	argyrea	40	A_2A_2
	canescens	40	AA
	cladestina	40	A_1A_1
	curvata	40	C_1C_1
	cyrtoloba	40	CC
	falcata	40	FF
	hirticaulis	80	H_1H_1
	lactovirens	40	I_1I_1
	latifolia	40	B_1B_1
	latrobeana	40	A_3A_3
	microphyta	40	BB
	pindanica	40	H_2H_2
	tabacina	40	B_2B_2
		80	Complex
	tomentella	38	EE
		40	DD
		78	Complex
		80	Complex

Much information has been developed on the evolutionary relationships within the subgenus *Glycine* (Doyle and Beachy, 1985; Singh and Hymowitz, 1995; Doyle *et al.*, 1990a,b). Numerous investigations on morphology, cytology, isozymes, restriction fragment length polymorphisms (RFLPs) and gene sequences have revealed a high number of nuclear genome groups and cytoplasms (Table 9.3, Fig. 9.7). As we discussed in Chapter 4, considerable diversity exists within each of these groups, and some of the polyploids have multiple origins. *Glycine tabacina* includes a diploid and two tetraploid types, which share one genome but differ at a second (Doyle *et al.*, 1999). *Glycine tomentella* represents a broad group of cytotypes, with 2n = 38, 40, 78 and 80 and carrying two or more genomes (Doyle *et al.*, 2002). Many of the races have more than one origin, and interbreeding between races has probably led to lineage recombination.

It is relatively clear that cultivated soybeans arose from *G. soya*. They are interfertile and have similar morphologies, distributions, isozyme banding patterns and DNA polymorphisms (Doyle and Beachy, 1985; Hymowitz and Singh, 1987). What is still unclear is who their diploid ancestor was, because *G. soya* and *G. max* have chromosome numbers that are too high to be anything but polyploids. Only a few remote relatives in the same sub-

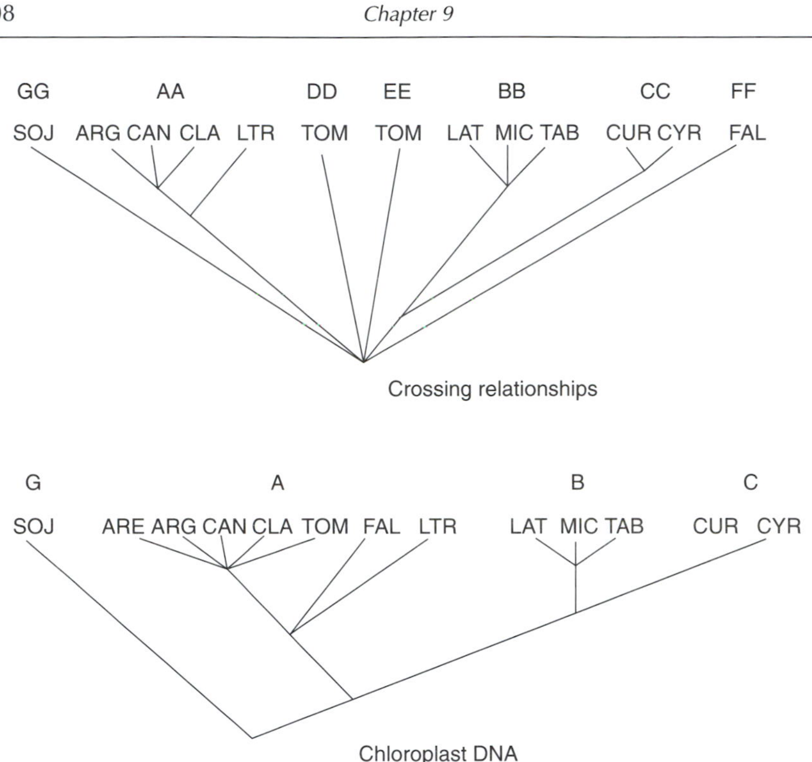

Fig. 9.7. Chloroplast DNA phylogeny of 13 of the *Glycine* species compared with their crossing relationships. Abbreviations are the first three letters of taxa listed in Table 9.3 except for SOJ = *soya* and LTR = latrobeana. AA to GG are genome designations based on crossing studies; A to G are chloroplast clades. (Used with permission from J.J. Doyle, J.L. Doyle and A.H.D. Brown, © 1990, *Evolution* 44, 371–389, Society for the Study of Evolution.)

tribe have 2n = 20, suggesting that the direct ancestor is either extinct or undiscovered (Kumar and Hymowitz, 1989). The answer to this question awaits further study and good luck.

Most investigators feel the eastern half of northern China was probably where soybean cultivation began about 3000–4000 years ago (Hymowitz, 1995). Based on RFLPs of chloroplast and mitochondrial DNA, the Yangtze River Valley of China appears to be the most likely origin of most domesticated soybeans, although another, more northern origin cannot be excluded (Shimamoto *et al.*, 1998, 2000). The major improvements instituted by people were the usual increased seed size, erect habit of growth and reduced seed shattering. Soybeans spread throughout China into Korea, Japan and South-East Asia during the expansion of the Ming dynasty 2000–3000 years ago. Europeans were aware of the soybean by 1712 and the Americans by 1765, but little planting was done before the 20th century (Hymowitz and Harlan, 1983). Soybeans were introduced into Brazil in 1882 to complete their entry into the New World (Hymowitz, 1995).

Starchy Staples and Sugars

10

Cereals and legumes may feed most of the world, but numerous other crops are rich sources of carbohydrates and are regionally important. Cassava, taro, yam, sweet potato and white potato are the primary foodstuffs of people in many parts of the world where grains do not do well, particularly in the tropics. Sugar cane and sugar beet are widely grown as a source of sugar for sweetening. The various starchy staples and sugars are not evolutionarily related, but most are propagated asexually and have far less protein than the legumes. In addition, most produce their edible parts under the ground.

Banana

There are two genera of wild bananas in the family *Musaceae*, *Musa* and *Ensete* (Simmonds, 1995a). Most of our domesticated bananas were derived from the wild, diploid species (2n = 2x = 22) *Musa acuminata* Colla (A genome) and *Musa balbisiana* Colla (B genome) in the section *Eumusa*. The diploids are all located in South-East Asia and the Pacific. As discussed in Chapter 4, the majority of the cultivars are triploids with the genomic constitutions of AAA, AAB and ABB, although there are a few cultivated diploids (AB) and tetraploids (AAAB, AABB, and ABBB) (Simmonds, 1995a). Isozymes and molecular markers have been used to confirm the genomic identities of many cultivated hybrids (Jarret and Litz, 1986; Osuji *et al.*, 1997; Pillay *et al.*, 2000).

Considerable variability exists among the cultivated bananas, both within their centre of origin in South-East Asia and in sub-Saharan Africa (Ortiz, 1997; Crouch *et al.*, 2000). Humans have taken great advantage of the diversity produced in bananas through interspecific hybridization and the accumulation of somatic mutations. Numerous inflorescence types (Fig. 10.1) and flavours abound. While only a few cultivars of bananas are grown on a large scale for export, many distinct clones are cultivated on the local scale in Asia, Africa and South America. The AAA types produce the sweetest fruit (bananas) and are eaten fresh; the various AB hybrids are more starchy (plantains) and are cooked or used to make beer.

Two other bananas are of minor importance. Manila hemp, *Musa textilis* Née (2n = 2x = 20), was once popular in the Philippine Islands and Borneo for its sheath fibres, which were made into rope. However, it is now

Fig. 10.1. Extremes in bunch and fruit shapes of bananas found in Asia. From left: 'Klue Teparod' (AABB), 'Ripping' (ABB), 'Pisang Seribu' (AAB), 'Pitogo' (ABB) and the wild, non-parthenocarpic *Musa acuminata* subsp. *malaccensis* (AA). (Drawn from a photograph in P. Rowe and F.E. Rosales (1996) Bananas and plantains. In J. Janick and J.N. Moore (eds) *Fruit Breeding*, John Wiley & Sons, New York.)

little grown (Simmonds, 1995a). *Ensete ventricosum* (Welw.) Cheesm. (2n = 2x = 18) is still grown in Ethiopia for its stems, which provide both starch and fibre. Both of these species have undergone minimal selection outside the wild.

All the varieties cultivated for fruit are parthenocarpic and seedless. Sexually produced bananas have large flinty seeds (Fig. 10.2), making edibility dependent on two characteristics: sterility and parthenocarpy. Seed set and zygotic viability vary greatly in both wild and cultivated populations of both *M. acuminata* and *M. balbisiana*, particularly among hybrids. Genetic female sterility is relatively common, and various types of meiotic errors lead to gametic sterility, including chromosomal rearrangements and numerical imbalances. Parthenocarpy is regulated by several complementary dominant genes that are found in wild populations of *M. acuminata* (Simmonds, 1995a).

Diploids were probably first domesticated in the Malay region of South-East Asia, after their fruit had been harvested from the wild for tens of thousands of years. The date of domestication is unknown, but certainly occurred thousands of years ago. At a very early stage, human beings must have selected individuals with both parthenocarpy and seed sterility to obtain an edible fruit. Edibility is thought to have first appeared in wild *M. acuminata*. Hybridization with *M. balbisiana* occurred as AA cultivars were taken out of their centre of origin, forming a classical polyploid complex (Fig. 10.3). The *balbisiana* types provided increased tolerance to cold and drought, while edibility in *M. balbisiana* was greatly enhanced by crosses with *M. acuminata* (Simmonds, 1995a).

The banana was probably introduced into Africa, India and the Polynesian Islands from Indonesia–Malaysia. It was being cultivated in Central Africa by 2500 years before present (BP) (Mindzie *et al.*, 2001). The banana was first brought to the attention of the Europeans by Alexander the

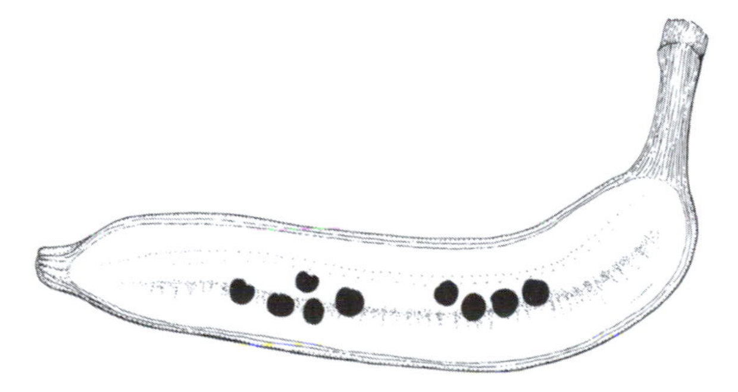

Fig. 10.2. Cross-section of a mature, diploid banana showing some fully developed seeds (drawn from a photograph in P. Rowe and F.E. Rosales (1996) Bananas and plantains. In: J. Janick and J.N. Moore (eds) *Fruit Breeding*, John Wiley & Sons, New York).

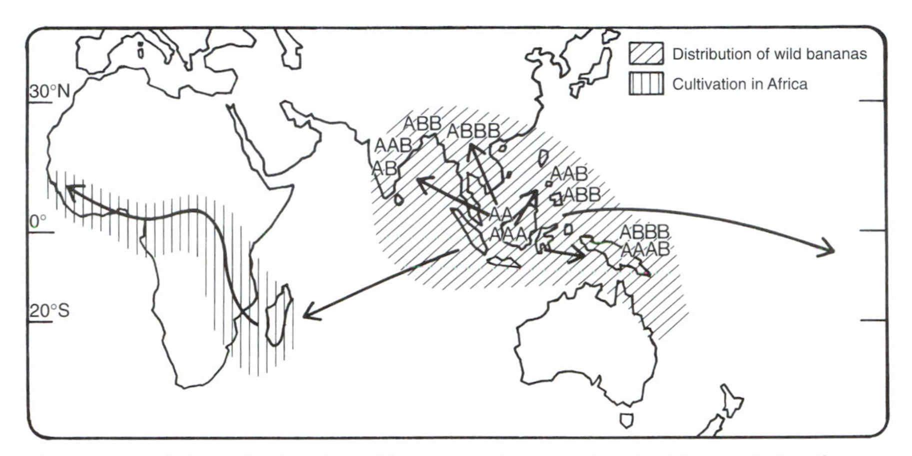

Fig. 10.3. Origin and migration of banana cultivation (used with permission from N.W. Simmonds, © 1985, *Principles of Crop Improvement*, Longman, London).

Great during his conquests. Portuguese travellers took the banana to the Canary Islands in the late 1400s and it reached Santo Domingo from there in 1516. Over the next century its cultivation spread throughout tropical America and the Caribbean region. Shipments to the USA began in the early 1900s from Central America and then Ecuador.

Cassava

Manihot esculenta Crantz (manioc, cassava or yuca) is a critical food source in tropical areas of South America, Africa, Asia and several oceanic islands (Fig. 10.4). It is grown primarily in wet lowlands, but produces well in somewhat arid environments and at elevations up to 2000 m. Cassava is a perennial shrub whose tuberous roots are higher in carbohydrate than rice or maize, although lower in protein (1–3%) (Heiser, 1990). New plantings are started by placing stem cuttings directly in the soil, and the plants often mature in less than a year. It is grown mostly by peasant farmers as a staple, but is also used to produce industrial starches, tapioca and animal feed. Cassava leaves are an excellent source of protein and are eaten in Africa as a pot-herb.

Humans domesticated both bitter and sweet forms of cassava. The bitter types have high levels of cyanogenic glycosides, which must be removed. This is done by peeling, grating, boiling and draining (Cock, 1982). The resulting pulp is then used directly to make a kind of bread, or it is dried and powdered for later consumption after frying. The grated manioc is also squeezed in porous tubes made of basketry and the resulting sap is used for sauces and alcoholic beverages. Tapioca is prepared by partially cooking small pellets of cassava starch.

Fig. 10.4. The manioc or cassava plant. Its total height can reach over 5 metres.

The genus *Manihot* is a member of the *Euphorbiaceae* (Rogers and Appan, 1973). There are about 100 species of *Manihot* and all have 36 chromosomes. Many of the species have been successfully intercrossed; most of the resulting hybrids have normal chromosome pairing at meiosis, with only a few univalents and tetravalents (Hahn *et al.*, 1990; Bai, 1992; Nassar *et al.*, 1995).

Even though cassava shows regular bivalent pairing, it is probably polyploid in origin due to its high chromosome number. Magoon *et al.* (1969) suggested that it was a segmental polyploid derived from two species with six similar and three dissimilar chromosomes. However, disomic inheritance has been documented across a wide range of random amplified polymorphic DNA (RAPD), microsatellite and isozyme loci (Fregene *et al.*, 1997). Since all existing species have the same chromosome number, polyploidization must have been an ancient event that occurred when the genera first emerged, long before domestication.

Until recently, manioc was believed to be a complex hybrid. Rogers and Appan (1973) felt that cassava arose from the hybridization of two closely related species. Likely candidates were thought to be *Manihot aesulifolia*, *Manihot isoloba*, *Manihot rubricaulis* and *Manihot pringeli*, as all are erect with tuberous roots like *M. esculenta*. Using patterns of variability in chloroplast DNA (cp DNA), nuclear ribosomal DNA and amplified fragment length polymorphisms (AFLPs), *M. esculenta* subsp. *flabellifolia* (or *Manihot tristis*) was determined to be one of the progenitors, but these markers left the possibility of other relationships unresolved (Fregene *et al.*, 1994; Roa *et al.*, 1997). Most recently, Olsen and Schaal (1999, 2001) used DNA sequence and microsatellite data to show that cassava was derived solely from wild *M. esculenta* subsp. *flabellifolia*. They found little evidence of recent hybridization, even with *Manihot pruinosa*, which is sympatric and interfertile (Olsen, 2002).

The location where cassava was first domesticated has also been long debated, with origins proposed in the dry scrublands of north-eastern Brazil, the savannahs of Columbia and Venezuela, the rain forests of the Amazon and the warm, moist lowlands of Mexico and Central America (Ugent *et al.*, 1986). The earliest archaeological evidence of cassava cultivation has been found in Peru and Chile from around 4000 BP (Ugent *et al.*, 1986; Rivera, 1991). The molecular data indicate that cassava was probably first domesticated along the southern border of the Amazon basin (Olsen and Schaal, 1999, 2001).

At the time of the arrival of the Europeans, only sweet forms of cassava were cultivated on the Peruvian coast and in Central America and Mexico. Cassava was used as an important garden crop, but not a staple (Sauer, 1993). In South America, sweet types were also grown as garden vegetables, but bitter forms were utilized as a staple crop. Cassava was restricted to the New World until the end of the 16th century, when sailors took it to the west coast of Africa. It had been used in the African slave-trade as a staple provision. Once established on the west coast of Africa, it spread rapidly among the islands of Réunion, Madagascar and Zanzibar and to the east coast, and finally moved into the interior from both coasts by the 19th century (Jennings, 1995). It found its way to India by 1800. Most of the introductions were by cuttings, but self-sown seedlings became established in most regions, resulting in high levels of local and regional diversity. Numerous landraces still exist, particularly among indigenous farmers, who maintain tremendous amounts of diversity (Boster, 1985; Salick *et al.*, 1997).

Potato

The edible, tuber-bearing *Solanum* species (Fig. 10.5) are a small part of a very large genus (Hawkes, 1990; Spooner and Hijmans, 2001). Potatoes belong to section *Petota* (*Tuberarium*), subsection *Potato* (*Hyperbasarthrum*) of *Solanum*. Their basic chromosome number is x = 12, and they are found at several ploidy levels, including diploid, triploid, tetraploid, pentaploid and hexaploid. Both autopolyploids and allopolyploids are present among the higher ploidies. The diploids have an S-allele incompatibility system and are primarily outcrossed, while the polyploids are self-compatible and generally self-pollinated (Simmonds, 1995b). The most important cultivated species, *Solanum tuberosum,* is considered by most to be autopolyploid because it displays tetrasomic inheritance (Martinez-Zapater and Oliver, 1984; Douches and Quiros, 1988b).

Tetraploid *S. tuberosum* subsp. *tuberosum* is by far the most widely grown taxon, but diploids, triploids and pentaploids are still grown by native South Americans (Table 10.1). The related tetraploid, *S. tuberosum* subsp. *andigena*, is also cultivated at some locations in cool, temperate Chile. There is so much morphological overlap among the various landraces of potato that Human and Spooner (2002) have suggested that all the landraces should be referred to as a single species, *S. tuberosum*, with eight cultivar groups arranged along the previously recognized species borders.

The majority of native species are located in South America and 75% of them are diploid (Hawkes, 1990; Ochoa, 1990; Spooner and Hijmans, 2001). In the frost-free Andean valleys are found the *Solanum* diploids *S. brevicaule* Bitt, *S. multidissecum* Hawkes, *S. bukasovii* Juz., *S. canasense* Hawkes, *S. soukupii* Hawkes, *S. leptophytes* Bitt., *S. multiinterruptum* Bitt, *S. abbottianum* Juz., *S. liriunianum* Card. et Hawkes, *S. ochoae* Vargus, *S. spegazinii* Bitt. and *S. vidaurrei* Card., which constitute what is called the *brevicaule* complex. Most cross easily, and natural hybrids have been reported in most combinations, although the hybrids are often poorly adapted (Dodds, 1965; Hawkes, 1979). Regardless, substantial interspecific hybridization is still going on within wild and cultivated populations, and many hybrid swarms have been described (Brush *et al.*, 1981; Johns and Keen, 1986). Because of this promiscuity and morphological continuity, van den Berg *et al.* (1998) have suggested that all the South American species in the *brevicaule* complex should be considered as one species, *S. tuberosum*.

The tetraploid potato has a confusing ancestry, based to a large extent on the high degree of hybridization occurring in the genus. Several types of morphological, cytoplasmic and molecular data have been accumulated on the origin of tetraploid *S. tuberosum*, and they do not all agree on a com-

Table 10.1. Cultivated South American potato species (*Solanum*).

Species	Chromosome number (2n)	Location
S. ajanhuiri Juz. et Buk.	24	High altitudes around Lake Titicaca
S. goniocalyx Juz. et Buk.		Widespread in humid–temperate maritime climates
S. phureja Juz. et Buk.		Northern Andean valleys
S. stenotomum Juz. et Buk.		Northern high Andes
S. vernei Bitt. et Wittm.		North-west Argentina
S. × *chaucha* Juz. et Buk.	36	Andean valleys from Ecuador to Bolivia
S. × *juzepczukii* Buk.		Central high Andes
S. tuberosum L.	48	
subsp. *tuberosum* L.		Worldwide
subsp. *andigena* Juz. et Buk.		Southern South America
S. *cutilobium* Juz. et Buk.	60	Central high Andes

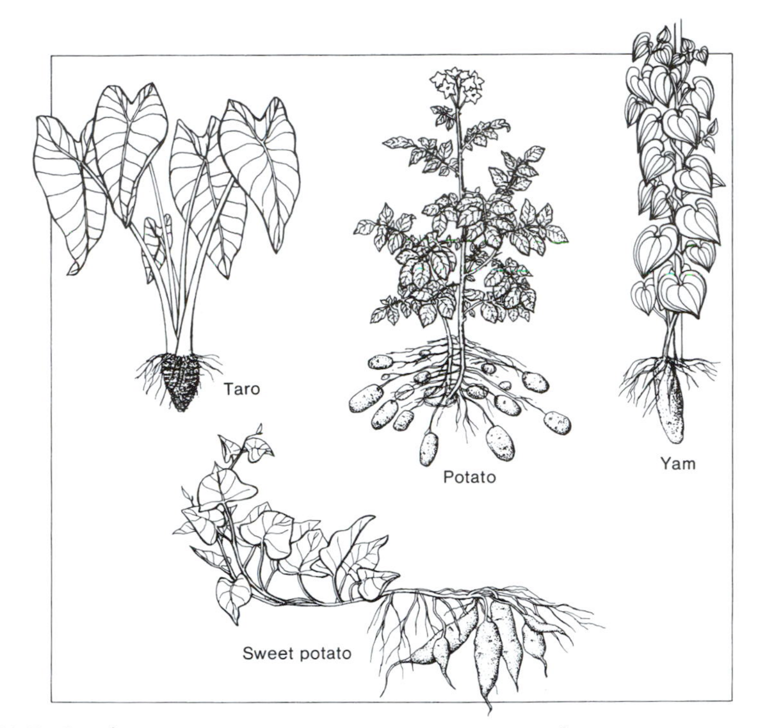

Fig. 10.5. Starchy root crops: potato, sweet potato, taro and yam.

mon set of ancestors; this probably reflects a complex origin. Using morphological data, at least four diploids have been implicated in the background of *S. tuberosum* including *S. leptophytes*, *S. canasense, S. soukupii* and *Solanum sparsipilum* Juz. et Buk. (Simmonds, 1995b). *Solanum stentomum* and *S. sparsipilum* contain many of the isozyme alleles found in cultivated tetraploid potatoes, with *S. stentomum* showing the greatest similarity (I = 0.95 versus 0.71) (Oliver and Martinez-Zapater, 1984). Nuclear restriction fragment length polymorphism (RFLP) data indicate that *S. stentomum* and *S. canasense* are the diploids most closely related to the tetraploid forms (Debener *et al.*, 1990). *S. tuberosum* subsp. *andigena* appears to have a mixed cytoplasmic parentage (Hosaka, 1995), while *S. tuberosum* subsp. *tuberosum* has a unique cytoplasm that may have evolved from *Solanum chacoense* Bitt. through a deletion (Hosaka and Hanneman, 1988). *S. chacoense* and some populations of *Solanum maglia* Schlechtd. have cytoplasmic sterility factors similar to those of subsp. *tuberosum* (Grun, 1990). All the diploid species produce relatively high proportions of unreduced gametes (Mok and Peloquin, 1975; Iwanaga and Peloquin, 1982), so phylogenetic borders are likely to be blurred by interspecific mixing at the polyploid level.

The cultivated potato, *S. tuberosum* subsp. *tuberosum*, made its first appearance in Europe in the mid-1800s, and considerable debate has revolved around its origin. It has been suggested that subsp. *tuberosum* was developed independently in Europe from South American subsp. *andigena* (Salaman, 1949; Hawkes, 1967), but it is more likely that European subsp. *tuberosum* was a direct import from Chile (Grun, 1979). *S. tuberosum* subsp. *andigena* was first brought into Europe in the 1500s from the Andes and was an important crop until the fungus disease late blight almost eliminated it in the 1840s. A clone of resistant *S. tuberosum* subsp. *tuberosum* ('Rough Purple Chili') was then introduced in the mid-1800s from Chile and filled the void left by subsp. *andigena* (Goodrich, 1863; Grun, 1990). This one clone played an important role in the subsequent development of most North American and European potatoes.

Even in South America, it is unlikely that *S. tuberosum* subsp. *tuberosum* evolved exclusively from subsp. *andigena* (Fig. 10.6). The cytoplasmic sensitivity sterility factors commonly found in subsp. *tuberosum* have not been found in subsp. *andigena* (Grun *et al.*, 1977), and the restriction patterns of cpDNA and mitochondrial (mtDNA) from the two subspecies differ greatly (Hosaka, 1986; Hosaka and Hanneman, 1988; Grun and Kao, 1989) as well as their nuclear microsatellites (Raker and Spooner, 2002).

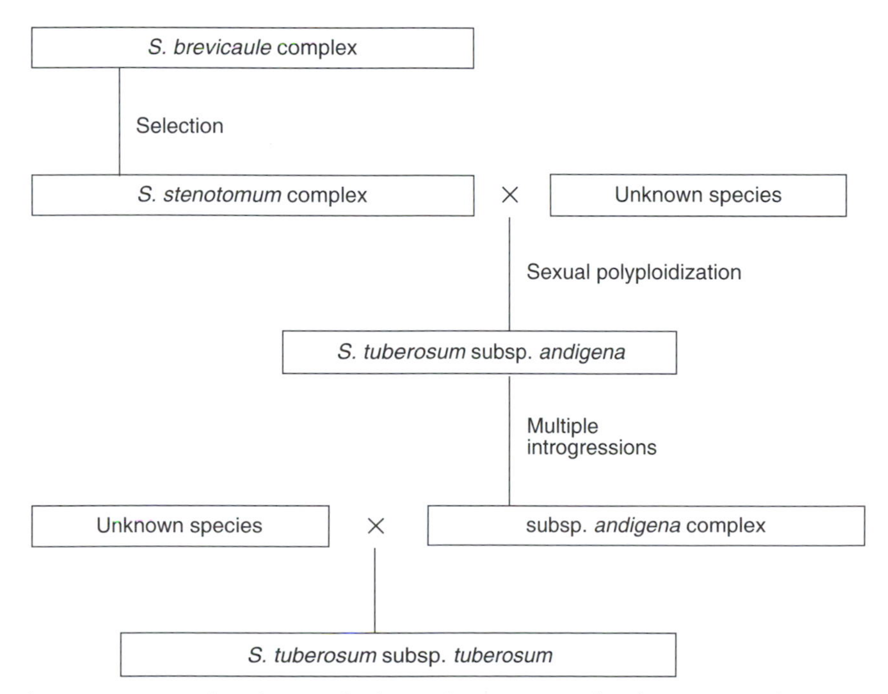

Fig. 10.6. Proposed evolution of cultivated potatoes (used with permission from P. Grun, © 1990, *Economic Botany* 44, 39–55, New York Botanical Garden, New York).

Also, some allozymes found in subsp. *tuberosum* are absent in subsp. *andigena*. Grun (1990) suggests that there may have been an extinct species at the southern edge of the subsp. *andigena* range that had the unique characteristics of subsp. *tuberosum,* or such a species may exist but remain undiscovered.

Probably the first step in South American potato cultivation was the collection of clones that were edible. Wild tubers are generally very bitter to the taste and can contain toxic amounts of alkaloids. The exact period when people began this activity is unknown, but the oldest evidence of potato use dates to 13,000 BP (Ugent *et al.*, 1982). The original area of domestication was probably in the high plateau of Bolivia–Peru. Potato cultivation spread throughout highland South America as a complex of diploid, triploid and tetraploid forms. The potato was first taken to Europe by Spanish invaders in 1537 and was spread throughout Europe by the end of the century. It arrived in North America from Europe in 1621.

Sugar Cane

There are six commonly recognized species of *Saccharum*, the genus of sugar cane (Table 10.2). They are all largely cross-pollinated by wind and suffer inbreeding depression. All are polyploids with very high chromosome numbers and considerable amounts of aneuploidy. Polysomic inheritance predominates within the group, although it is not complete (Al-Janabi *et al.*, 1993, 1994; Da Silva *et al.*, 1993). All sugar cane species can be successfully crossed, and they have limited interfertility with numerous other genera, including *Narenga*, *Erianthus*, *Sorghum*, *Schlerostachya*, *Imperata* and *Zea* (Stevenson, 1965).

A generally consistent phylogeny of *Saccharum* has emerged using morphological, cytological and molecular data (Daniels and Roach, 1987; Jannoo *et al.*, 1999; Nair *et al.*, 1999). *Saccharum spontaneum* is an autopolyploid, but the other species originated from interspecies hybridization. *Saccharum robustum* represents a very diverse population of introgres-

Table 10.2. Sugar cane species (*Saccharum*) and their chromosomal numbers.

Species	Status	Chromosome number (2n)	Location
S. barberi Jeswiet	Cultivated	82–92	Northern India
S. edule Hassk.	Cultivated	60, 70 and 80	New Guinea
S. officinarum L.	Cultivated	80	Worldwide tropics
S. robustum Brandes & Jeswiet	Wild	60 and 80	Borneo – New Guinea
S. sinense Roxb.	Cultivated	82–124	South-East Asia, China
S. spontaneum L.	Wild	40–180	Northern Africa to China

sants and may have arisen as an intergeneric hybrid of *S. spontaneum* and several other species. *Saccharum edule* may be an allopolyploid of *Saccharum officinarum* and *S. robustum*. *Saccharum barberi* probably resulted from the natural hybridization of *S. spontaneum* and *S. officinarum*. *Saccharum sinense* is closely related to *S. barberi,* but can be distinguished from it based on its horticultural characteristics (Daniels *et al.*, 1991).

Sugar cane was probably first domesticated in or near New Guinea approximately 10,000 years ago. Because of its morphology and similar chromosome number, *S. robustum* is thought to be the progenitor of *S. officinarum*. Cultivation probably moved northward towards continental Asia, where *S. officinarum* ('noble' canes) hybridized with *S. spontaneum* to produce *S. sinense* ('thin' canes). These hybrids were less sweet and less robust than the noble canes, but were hardier and could be grown successfully on subtropical mainlands (Sauer, 1993). This new type eventually became established in the India–China area. At the same time, *S. officinarum* was moving eastward across the Pacific. The other cultivated species, *S. barberi* and *S. edule*, also had early origins, but their spread was much more limited.

Sugar manufacture first began about 3000 years ago in India; before that time the canes were grown as garden plants for chewing (Sauer, 1990). Europeans did not become aware of the crop until the explorations of Alexander the Great in Asia. Columbus is credited with introducing sugar cane into the New World in the late 14th century. Today the various sugar cane species are found all across the tropics and are the source of approximately 50% of the world's sugar.

Sugar Beet

The cultivated beet, *Beta vulgaris* L. subsp. *vulgaris,* belongs to the goosefoot family, *Chenopodiaceae*. There are numerous types of domesticated beets: (i) leaf beets, which are used as a leafy vegetable and do not have a swollen hypocotyl; (ii) garden beets, whose swollen hypocotyl is eaten as a salad vegetable; (iii) mangels, whose swollen hypocotyl is used primarily as a fodder crop; (iv) sugar beets, whose root is an important sugar source; and (v) fodder beets, which are used to feed livestock and are hybrids of mangels and sugar beets (Fig. 10.7). By far the most widely grown type is the sugar beet, which predominates in Europe, the former USSR and North America; it is the source of about half of the world's sugar production.

All the cultivated types were derived from the subsp. *maritima* (L.) Thell., which occurs naturally on seashores in cool, temperate parts of Europe and Asia. Both morphological and molecular evidence are consistent with this possibility (Coons, 1975; Mikami *et al.*, 1984; Kishima *et al.*, 1987). *B. vulgaris* subsp. *vulgaris* is mostly outcrossed and anemophilous (Dark, 1971). It is primarily a diploid ($2n = 2x = 18$), although autopolyploids arise periodically via unreduced gametes. Tetraploid cultivars have

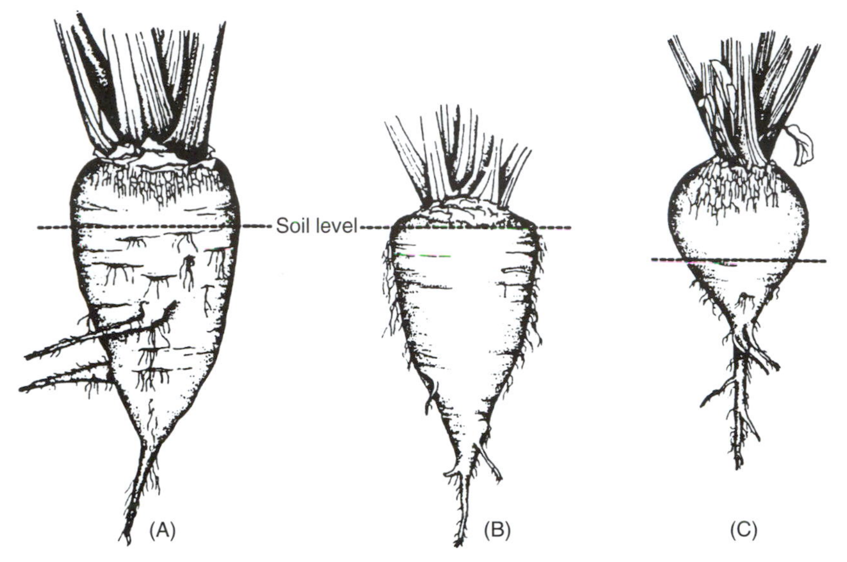

Fig. 10.7. Various forms of beets: (A) fodder beet, (B) sugar beet and (C) beetroot. Note differences in soil level. (Used with permission from R.H.M. Langer and C.D. Hill, © 1982, *Agricultural Plants,* Cambridge University Press, New York.)

been produced artificially with little commercial success, but triploid types have become important in Europe. Cytoplasmic male sterility has been exploited to produce commercial hybrid seed.

Domestication probably began in the eastern Mediterranean, where the leaves were used as pot-herbs and animal feed (Ford-Lloyd and Williams, 1975). Leaf beets were described in the ancient writings of Aristotle and Theophrastus. Romans used the beet rather extensively, and it found its way from Italy to Europe with the 'barbarian' invaders. Beets were introduced into the USA in 1800.

Beet was first recognized to have high sugar levels in the 17th century (Deerr, 1949). Marggraf discovered sugar in the sap of the root in 1747, and his student Achard was given a grant by the King of Prussia to develop a commercial industry. The first sugar factory was built in 1801 at Kunern, Silesia. The sugar beet industry in the USA was established at the end of the 19th century. The early sugar beet cultivars contained approximately 6% sugar, while the modern types have over 20%.

Sweet Potato

Ipomoea is a pantropical genus with more than 50 species recognized. It belongs to the family *Convolvulaceae* and displays a polyploid series of 2n = 30, 60 and 90 (Austin, 1978). The domesticated sweet potato, *Ipomoea batatas* (L.) Lam (Fig. 10.5), is the only hexaploid in the section.

There has been considerable debate about the phylogenetic organization of the sweet potato and its close allies. Based on morphological data, Ting and Kehr (1953) felt the sweet potato was an allotetraploid derived from unknown species, while Nishiyama and his group (1975) suggested it was an autohexaploid of the diploid *Ipomoea trifida* and was part of a polyploid complex (Fig. 10.8). Nishiyama and Temamura (1962) found what they thought was a wild hexaploid type in Mexico, but it was later shown to be a feral sweet potato (Jones, 1967). Jones (1967) and Martin and Jones (1972) felt that *I. trifida* and *I. batatas* evolved separately from an unknown, but common ancestor. Recent molecular data have indicated that *I. batatas* is much more closely related to *I. trifida* than any other diploid species, strongly supporting an autopolyploid origin for the sweet potato (Huang and Sun, 2000). Diploid *I. trifida* produces 2n pollen (Orjeda *et al.*, 1990; Jones, 1990) and the chromosomes of 2x forms associate readily with 6x types (Shiotani and Kawase, 1989). *I. trifida* is the only diploid species that is self-incompatible, similar to the sweet potato (Diaz *et al.*, 1996).

The place where sweet potato was first domesticated is also disputed. The oldest archaeological proofs of sweet potatoes are found in Peru (10,000 to 8000 BP), but these may be the remains of gathering (Ugent *et al.*, 1981) and a Middle American origin cannot be excluded (Heiser, 1990). Zhang *et al.* (2000) found the highest amount of AFLP diversity in Central America, leading them to propose that it is the centre of origin. Other ancient remains have been found in Hawaii, New Zealand and across Polynesia (Yen, 1982). It is not clear how the crop reached Polynesia from America before the voyages of the Europeans; however, a number of explanations have been presented, ranging from early transoceanic movements of humans to natural dispersal by drifting capsules (Sauer, 1993).

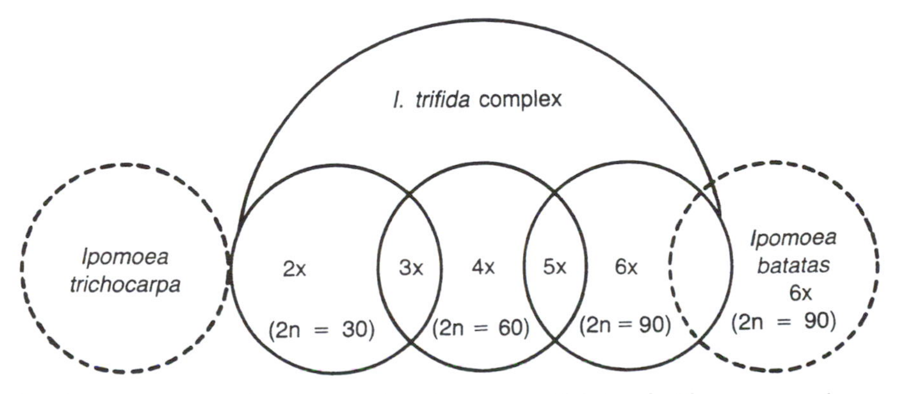

Fig. 10.8. Proposed polyploid complex of *Ipomoea trifida* (used with permission from K. Kobayashi, © 1984, *Proceedings of the Sixth Symposium of the International Society for Tropical Fruit Crops*).

One thing is clear, Columbus discovered the sweet potato in the West Indies and was the first to take it back to Europe (Yen, 1982). It was then spread along the so-called 'batata line' to Africa, Brazil and India by Portuguese explorers, who used it to prevent scurvy (batata was the West Indian Arawak's name for sweet potato). Spanish trading galleons dispersed sweet potatoes along the 'camote line' between Mexico and the Philippines (the word camote was derived from the Mayan word camotli). The sweet potato was introduced into southern China by 1594 and Japan by 1674.

Taro

Taro or dasheen (*Colocasia esculenta*) (Fig. 10.5) belongs to the *Araceae* or aroid family. It is the primary starch source of millions of tropical peoples, most notably those located in the Pacific and Caribbean islands and West Africa. It is grown for its underground corms, and is farmed in flooded or swampy land that will support little else that is edible. Wild populations of *C. esculenta* are found throughout south-central Asia and exist at two ploidy levels, 2n = 2x = 26 and 2n = 2x = 42 (Kuruvilla and Singh, 1981). The triploids are probably autopolyploid (Tahara *et al.*, 1999; Ochiai *et al.*, 2001). Major lineages of diploid and triploid taros have been identified based on chromosomal (Coates *et al.*, 1988) and RAPD variation (Irwin *et al.*, 1998). Taro is reproduced almost exclusively asexually, which has resulted in its landraces being extremely variable, both morphologically and cytogenetically (Kuruvilla and Singh, 1981; Sreekumari and Mathew, 1991).

Many anthropologists believe that taro was the first irrigated crop (Plunckett, 1979) and its cultivation may be more ancient than that of rice at many locations (Taiwan, Philippines, Assam and Timor). Domestication probably began in Indo-Malaya 4000–7000 years ago (Spier, 1951; Plunckett *et al.*, 1970), where it spread east to Asia and the Pacific islands and west towards Arabia. It has been concluded from isozyme data that Oceania taros originated from Indonesia (Lebot and Aradhya, 1991). Traders and explorers took taro to West Africa approximately 2000 years ago, and it arrived several hundred years ago in the Caribbean islands and tropical America via slave ships. Taro was brought to Japan in prehistory, from either Taiwan or the Yangtse River area of China (Sasaki, 1986).

A number of other aroids are important food sources, including the Asian types *Alocasia indica*, *Alocasia macrorrhiza* and *Cyrtosperma chamissonis*, and the South American types *Xanthosoma atrovirens*, *Xanthosoma sagittifolium* and *Xanthosoma violaceum* (Table 10.3). Based on RAPD and isozyme data, *Alocasia* appears to be more closely related to *Colocasia* than to *Xanthosoma*, which fits their overlapping geographical distributions (Ochiai *et al.*, 2001). The origins of these aroids is generally clouded, although *Xanthosoma* or cocoyam cultivation is known to predate Columbus in Hispaniola, Central and South America and the Caribbean

Table 10.3. Major species of edible aroids (based on Plunckett, 1979).

Location	Species	Chromosome number (2n)
Asia	*Alocasia indica* Schott.	28
	Alocasia macrorrhiza Schott.	26, 28
	Colocasia esculenta (L.) Schott.	28, 42
South America	*Cyrtosperma chamissonis* (Schott.) Merr.	?
	Xanthosoma atrovirens Koch.	26
	Xanthosoma sagittifolium Schott.	26
	Xanthosoma violaceum Schott.	26

(Sauer, 1993). Cocoyam was introduced into West Africa in the colonial period as slave provisions (Sauer, 1993) and in many regions is now an important crop. In Nigeria, taro (*Colocasia*) is called 'old cocoyam' and *Xanthosoma* is referred to as 'new cocoyam' to distinguish the original African aroid from the imported one (Valenzuela and DeFrank, 1995).

Yam

Yams (Fig. 10.5) are represented by hundreds of Old and New World species of *Dioscorea* in the family *Dioscoreaceae* (Fig. 10.9). They are an extremely important staple in West Africa and Nigeria, and are also widely grown in South-East Asia, Oceania, the Caribbean and tropical America. Their production is mostly limited to subsistence farming (Valenzuela and DeFrank, 1995), but they are rich sources of steroidal sapogenins, which are the starting compounds in the synthesis of cortisones and sex hormones (Putz and Mooney, 1992). Tremendous diversity exists among the cultivated yams for both morphological and biochemical traits. Some of the cultivated types contain toxic alkaloids that can harm or kill people, but these compounds are rendered harmless by peeling and boiling.

Probably the most important species worldwide is now *Dioscorea rotundata* Poir. (white yam), but numerous other species are cultivated (Table 10.4). Polyploidy is common in all the geographical assemblage of yams, with counts as high as 2n = 100 being reported in some cultivated types (Hahn, 1995). Aneuploidy is also common, and single individuals may display variable chromosome numbers. Most cultivated species have chromosome numbers based on either 9 or 10.

Yam domestication is thought to have taken place separately in South-East Asia, Japan, Africa and tropical America, with different species involved in each region (Table 10.4). The evolutionary relationships of these species are still ambiguous. Cultivated forms of *Dioscorea esculenta* (Chinese yam), *Dioscorea bulbifera* (aerial yam), *Dioscorea cayenensis* (yellow yam) and

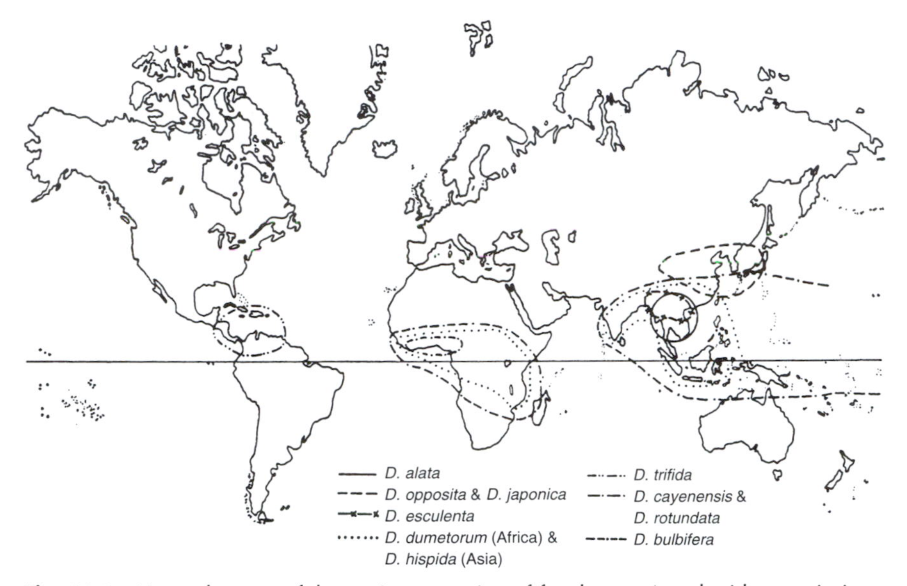

Fig. 10.9. Natural range of the various species of food yams (used with permission from D.C. Coursey, © 1967, *Yams*, Longman Group UK, London).

Table 10.4. Most important cultivated yam species (*Dioscorea*) and their chromosome numbers.

Location	Species	Chromosome number (2n)
South-East Asia	*D. alata* L.	20, 30, 40, 50, 60, 70, 80
	D. esculenta (Lour.) Burk.	30, 40, 60, 90, 100
	D. bulbifera L.	30, 40, 50, 60, 70, 80, 100
	D. hispida Dennst.	40, 60
Japan	*D. opposita* Thunb.	40
	D. japonica Thunb.	40
Africa	*D. cayenensis* Lam.	36, 54, 60, 63, 66, 80, 120, 140
	D. rotundata Poir.	40, 80
	D. dumetorum (Kunth) Pax.	36, 40, 45, 54
America	*D. trifida* L.	54, 72, 81

Dioscorea trifida (cush-cush yam) appear to have been derived directly from the native populations that are still found in the wild today. However, *Dioscorea alata* (water yam) may be of hybrid origin and could have been selected by humans after the hybridization of several candidate Asian species. African *D. rotundata* and *D. cayenensis* are also thought to be products of interspecific hybridization, but it is not clear whether *D. cayenensis* was a parent of *D. rotundata* (or vice versa) or they evolved separately. These two species are so similar that many consider them to be subspecies

(Ayensu and Coursey, 1972) or part of a polyploid complex (Hamon and Bakary, 1990). Molecular data have been collected to elucidate species relationships, but it has not yet clarified the phylogenetic picture (Terauchi *et al.*, 1992; Ramser *et al.*, 1997; Mignouna *et al.*, 1998; Choi *et al.*, 2002).

Some have postulated that the development of modern yam cultivation was stimulated 5000 years ago by grain-crop agriculture. However, it is likely that its roots go much deeper. Coursey (1979) suggested that yam cultivation began at least 10,000 years ago in the north-eastern parts of the South-East Asian peninsula and spread throughout Melanesia 3500 years ago. African cultivation may have begun 11,000 years ago when people began to move into the West African savannah (Coursey, 1975). An early centre of domestication probably also arose in Japan and central–coastal China. New World yam domestication lagged behind the Old World, possibly due to the earlier cultivation of cassava as a primary food source. It began at the border between Brazil and Guyana and spread to the Caribbean (Ayensu and Coursey, 1972). Intercontinental distribution of the yam species did not begin until the last 500 years with the advent of long-distance ocean travel and the establishment of trade routes. Yams served as a good source of nutrition on long sea voyages.

11

Fruits, Vegetables, Oils and Fibres

Even though the grains, legumes and starchy roots are of primary importance in feeding the world, there are numerous other crops that play an important role in human diets. As was already mentioned, Harlan (1992) lists over 250 food crops and suggests there are hundreds more. These crops are sometimes a major source of calories, but they are more commonly a supplemental source of nutrients that break up the tedium of limited crop diets. They are also used as drugs, medicine and fibre. In this chapter, we shall discuss 12 of the most frequently consumed fruits, vegetables, oils and fibres.

Fruits

Apples

The genus of apples, *Malus*, belongs to the subfamily *Pomoideae* of the *Rosaceae* family. Another important fruit tree, pear (*Pyrus*), belongs to the same subfamily. There are over 30 primary species of apple and most can be readily hybridized (Korban, 1986; Way *et al.*, 1991). The cultivated apple is probably the result of interspecific hybridization and is most appropriately called *Malus* × *domestica* (Korban and Skirvin, 1984). Its primary wild ancestor is *Malus sieversii* whose range is centred at the border between western China and the former Soviet Union (Harris *et al.*, 2002). Apples are the main forest tree there and display the full range of colours, forms and tastes found in domesticated apples across the world (Forsline *et al.*, 1994; Hokanson *et al.*, 1997b). Other species of *Malus* which contributed to the genetic background of the apple include: *M. orientalis* of Caucasia,

M. sylvestris from Europe, *M. baccata* from Siberia, *M. mandshurica* from Manchuria and *M. prunifolia* from China. It is likely that these species hybridized with domesticated apples as they were spread by humans.

The bulk of the apple species are 2n = 2x = 34 (Table 11.1), although higher somatic numbers of 51, 68 and 85 exist; several of the cultivated types are triploid (Chyi and Weeden, 1984). It is possible that the high chromosome number of apple represents an ancient genomic duplication, since there are several other rosaceous fruit species with lower haploid chromosome numbers of n = 8 and 9. Allopolyploidy is indicated, as apples generally display disomic inheritance at what appear to be duplicated isozyme loci (Weeden and Lamb, 1987). Apples are largely self-incompatible and many are apomictic. They are propagated vegetatively, usually as composites with a separate rootstock and scion.

Apples were certainly among the earliest fruits to be gathered by people, and their domestication was probably preceded by a long period of unintentional planting via rubbish disposal. It is difficult to determine exactly when the apple was first domesticated, but the Greeks and Romans were growing apples at least 2500 years ago. They actively selected superior seedlings and were budding and grafting by 2000 before present (BP) (Janick *et al.*, 1996). The most likely beginning of cultivation was in the region between the Caspian and Black Seas (Vavilov, 1949–1950); apple cultivation had reached the Near East by 3000 BP (Zohary and Hopf, 1993). The Romans spread the apple across Europe during their invasions and it was dispersed

Table 11.1. Distribution of selected apple species (*Malus*) in subsection *Pumilae* and their chromosome numbers (adapted from Way *et al.*, 1991).

Species	Chromosome number (2n)	Distribution
M. asiatica Nakai	34	North and north-east China, Korea
M. baccata (L.) Borkh.	34, 68	North and north-east China
M. × domestica	34, 51, 68	Worldwide
M. floribunda (Siebold) ex Van Houte	34	Japan
M. halliana Koehne	34	Japan
M. hupehensis (Pampan.) Rehder	51	Central China
M. mandshurica (Maxim.) V. Komarov	34	Manchuria
M. micromalus Makino	34	South-east China, Korea
M. orientalis Uglitzk. ex Juz.	?	Caucasia
M. prunifolia (Willd.) Borkh.	34	North and north-east China, Korea
M. pumila Miller	34	Europe
M. sieversii (Lodeb.) M. Roemer	?	North-west China
M. spectabilis (Aiton) Borkh.	34, 68	China
M. sikkimensis (Wenzig) Koehne ex C. Schneider	51	Himalayas
M. sylvestris Miller	34	Europe

to the New World by European settlers during the 16th century. Jonathan Chapman (Johnny Appleseed) is credited with spreading the apple throughout eastern and central North America during the 18th century.

Citrus

The genus *Citrus* contains a confusing array of diversity (Fig. 11.1) and anywhere from three to 145 species have been recognized, depending on the

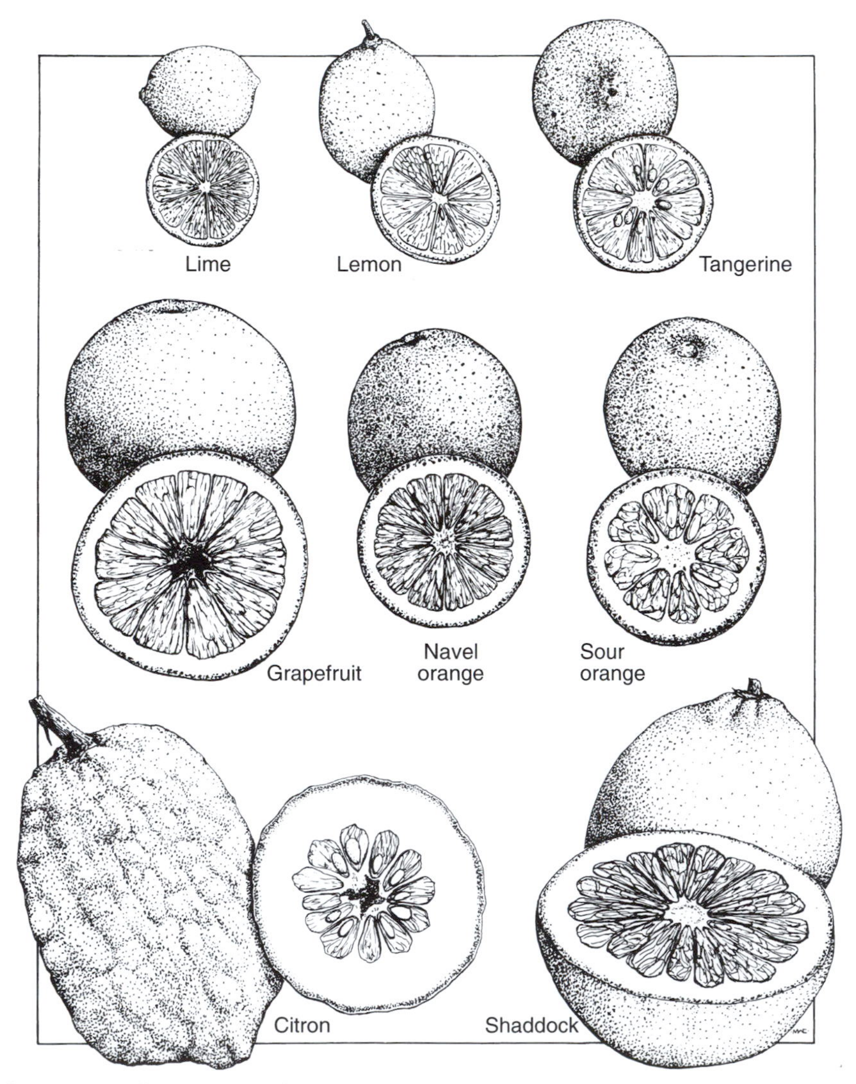

Fig. 11.1. Different types of citrus crops.

authority (Tanaka, 1954; Hodgson, 1961; Swingle, 1967). Probably the most comprehensive attempt at citrus taxonomy was undertaken by Barrett and Rhodes (1976). They studied variation patterns in 147 morphological characters, using numerical methods, and came to the conclusion that there are only three clearly distinct taxa: (i) *Citrus medica* (citron); (ii) *Citrus grandis* (shaddock or pummelo); and (iii) *Citrus reticulata* (mandarin). Recent molecular data have generally supported this conclusion (Fang *et al.*, 1998; Federici *et al.*, 1998; Nicolosi *et al.*, 2000). The difficulty in assorting more groups stems from the fact that many of the commercial species are morphologically similar and highly interfertile. Apomixis is also rampant in the genus, and many of the types probably arose from a common ancestor via asexual means (Roose *et al.*, 1995). The only simple part about *Citrus* taxonomy is that polyploidy is rare and most species share a basic chromosome number of x = 9.

Morphological and molecular data have provided clues about the origin of many hybrid citrus taxa (Scora, 1975; Barrett and Rhodes, 1976). Early types of shaddock and mandarin probably hybridized to produce the sweet and sour oranges. Citrons contributed the largest part of the lemon genome, with smaller contributions from mandarin and shaddock (Gulson and Roose, 2001a,b). Shaddock provided the cytoplasm of oranges and many lemons (Green *et al.*, 1986; Gulson and Roose, 2001a,b). The lime was a hybrid of citron and one or more generally unspecified taxa. Mandarin was the donor of at least some lime cytoplasms.

The origins of most citrus taxa are clouded, but some conjectures can be made (Sauer, 1993). Citron may have originated in India and was spread prehistorically to the Near East and China. It arrived in Greece by 2500 BP. Shaddock and mandarin cultivation probably began in tropical South-East Asia and spread to China by 2500 BP. Sweet and sour oranges are likely to have been generated repeatedly where shaddocks and mandarins were grown together, although the complex and stable karyotype of sweet orange cultivars suggests a monophyletic origin (Pedrosa *et al.*, 2000). Oranges were first recorded in Chinese writings about the same time as shaddock and mandarin. The lime probably had a South-East Asian origin. The origin of the lemon is unknown, but it probably arose repeatedly where citron and lime cultivation overlapped. The grapefruit is of very modern origin, having emerged as a hybrid of shaddock and sweet orange on the island of Barbados in 1750.

The citrus species found their way to Europe and ultimately to the New World through a variety of routes. The shaddock moved through India to North Africa and arrived in Spain about 800 years ago. Lemon may have been spread from India to Rome via the Near East. Mandarin did not arrive in Europe before the 19th century and the lime did not reach temperate Asia or Europe until almost modern times. Columbus introduced a variety of citrus species to the New World in 1497, and these were spread all over the world by the Portuguese and Spanish during their expeditions in the 16th century.

Grape

As with citrus, the traditional grape taxonomy is confusing and somewhat artificial (Alleweldt *et al.*, 1991). About 60 species of *Vitis* have been described across the temperate world and all of them can be successfully intercrossed. There is considerable geographical overlap among species and interspecific hybrids are common (Olmo, 1995). In fact, interspecific hybridization played an important role in the spread of grape cultivation. The domesticated grape, *Vitis vinifera* L., was derived directly from native populations of *V. vinifera* subsp. *sylvestris*, but as grape cultivation entered new regions, the cultigens hybridized with local types to produce better adapted material. In spite of this genic mixing, wild and cultivated grapes still show considerable amounts of geographical and ecological divergence (Lamboy and Alpha, 1998; Sefc *et al.*, 2000).

The genus belongs to the family *Vitaceae*. It is primarily dioecious, although most cultivated types are hermaphroditic. *Vitis* is unique in the family in that most of its species have 38 chromosomes, rather than the more common $2n = 40$, and its chromosomes are much smaller. The muscadine grape, *Vitis rotundifolia* Michaux, is one of the few 40-chromosome species, and some have put it in a separate genus, *Muscadinia*. The high chromosome number of *Vitis* may represent ancient polyploidy, but the group has regular bivalents at meiosis (Patel and Olmo, 1955) and appears to have undergone considerable gene silencing (Weeden *et al.*, 1988; Parfitt and Arulsekar, 1989).

Cultivation of the grape had certainly begun in Middle Asia by 6000 BP, where wild populations of *V. vinifera* still exist (Fig. 11.2). There is ample archaeological evidence that wine grapes were being grown in a large area spanning the Aegean, Mesopotamia and Egypt by 4000 BP (Zohary and Hopf, 1993). From the very beginning, grapes were propagated by cuttings or layering. Grape growing began to spread out from Asia Minor and Greece about 3000 years ago, and by 2500 BP it had reached France. Its later distri-

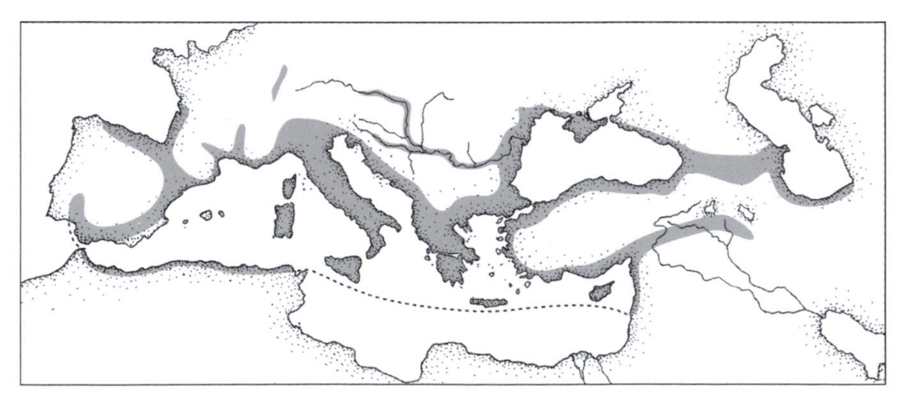

Fig. 11.2. Distribution of the wild grape, *Vitis vinifera* subsp. *sylvestris* (adapted from Zohary and Hopf, 1993).

bution was strongly associated with the use of wine in the consecration of the Christian mass.

Grapes were initially introduced to the New World by Portuguese and Spanish explorers (Olmo, 1995). European wine grapes were established in the Mediterranean-like climates of California in the middle 1800s. The first successful grape grown widely outside California was 'Concord', which appeared in eastern North America as either a mutant of a native bunch grape (*Vutis labrusca*) or a hybrid with *V. vinifera*. It was used primarily for fresh fruit and juice, and is still grown today. *V. rotundifolia* was domesticated for both wine and eating by the colonists in the Carolinas in the late 1700s. Native North American species found their way back to Europe in the late 1800s when it was discovered that they could be used as rootstocks to combat the devastation wrought by the root aphid, *Phylloxera* (Alleweldt *et al.*, 1991; Olmo, 1995). Recent problems with mildews and *Phylloxera* have further stimulated the use of wild species in breeding programmes.

Peach

The peach (*Prunus persica* (L.) Batsch) is the most widely grown species in a very important genus containing the plum (*Prunus domestica* L.), apricot (*Prunus armeniaca* (L.) Kostina), almond (*Prunus amygdalus* Batsch), sweet cherry (*Prunus avium* L.) and sour cherry (*Prunus cerasus* L.) (Table 11.2).

Table 11.2. Native peach species (*Prunus*) and their domesticated relatives (adapted from Scorza and Okie, 1991).

Relatives	Common name	Chromosome number (2n)	Distribution
Wild			
P. davidiana (Carr.) Franch	Mountain peach, Shan tao	16	North China
P. ferganensis (Kost. & Rjab) Kov. & Kost.	Xinjiang tao	16	North-east China
P. kansuensis Rehd.	Wild peach, Kansu tao	16	North-west China
P. mira Koehne	Tibetan peach, Xizang-tao	16	West China and Himalayas
P. persica (L.) Batsch	Peach, Maotao	16	China
Domesticated			
P. domestica L.	European plum	48	West Asia, Europe
P. salicina Lindl.	Japanese plum	16	China
P. armeniaca (L.) Kostina	Apricot	16	Asia
P. amygdalus Batsch	Almond	16	South-west Asia
P. avium L.	Sweet cherry	16	West Asia, south-east Europe
P. cerasus L.	Sour cherry	32	West Asia, south-east Europe

Peach belongs to the family *Rosaceae* and the subgenus *Amygdalus*. It is largely outcrossed due to self-incompatibility. There are at least 77 wild species of *Prunus* and most of them are found in Central Asia. While polyploidy is common in the genus *Prunus*, the cultivated peach is diploid and has a chromosome number of 2n = 2x = 16.

Five species of peach are generally recognized: *P. persica*, *Prunus davidiana* (Carr.) Franch, *Prunus mira* Koehne, *Prunus kansuensis* Rehd. and *Prunus ferganensis* (Kost. & Rjab) Kov. & Kost. All are found in China. The domesticated peach can be readily hybridized with native populations of *P. persica* and all the other wild species. Successful hybrids have also been produced between peach and almond, apricot, plum and sour cherry (Parfitt *et al.*, 1985; Scorza and Okie, 1991). In most cases, these wide hybrids are largely sterile, although F_1s of almond and peach are highly fertile (Armstrong, 1957) and can be employed as rootstocks for both peach and almond.

Peach cultivation probably originated in western China from wild populations of *P. persica* (Hedrick, 1917; Scorza and Okie, 1991). The peach is mentioned in 4000-year-old Chinese writings, and most of the known variation in cultivated peaches is found in Chinese landraces. Peaches arrived in Greece through Persia about 2500 BP and in Rome 500 years later. The Romans spread the peach throughout their empire. The peach came to Florida, Mexico and South America in the mid-1500s via Spanish and Portuguese explorers. It became feral in the south-eastern USA and Mexico, and was further spread throughout North America by Indians.

Strawberry

The genus of the strawberry is *Fragaria*, which belongs to the *Rosaceae* and the subfamily *Rosoideae*. The major cultivated species, *Fragaria* × *ananassa*, is an octoploid (2n = 8x = 56) of interspecific origin. It originally appeared as an accidental hybrid in Europe in about 1750, when plants of *Fragaria chiloensis* from Chile were planted next to *Fragaria virginiana* from the Atlantic seaboard of North America (Darrow, 1966). Both wild species are octoploid and predominantly dioecious, but this trait was selectively removed from the breeding stock in the last 75 years. Gender is regulated by three alleles with differing levels of dominance (female > hermaphrodite > male) (Ahmadi and Bringhurst, 1991).

The evolutionary history of the octoploids is clouded and subject to speculation. Diploid, tetraploid and hexaploid species are found in Europe and Asia (Table 11.3), but octoploids are restricted to the New World and perhaps the Iturup Island north-east of Japan (Staudt, 1989). Only one diploid species, *Fragaria vesca*, is located in North America. The genomic complement of the octoploids is AAA′A′BBB′B′ (Bringhurst, 1990), with *F. vesca* being the only known progenitor, although numerous Asian species have not been investigated (Hancock, 1999). The most likely scenario is that

Table 11.3. Native strawberry species (*Fragaria*) and their distribution.

Species	Chromosome number (2n)	Distribution
F. daltoniana J. Gay	14	Himalayas
F. gracilosa Lozinsk	14	China
F. iinumae Makino	14	Japan
F. mandshurica Staudt	14	Manchuria
F. nilgerrensis Schlect.	14	South-eastern Asia
F. nipponica Lindl.	14	Japan
F. nubicola Lindl.	14	Himalayas
F. pentaphylla Lozinsk	14	China
F. vesca L.	14	North America, northern Asia and Europe
F. viridis Duch.	14	Europe and Asia
F. yesoensis Hara.	14	Japan
F. moupinensis (French.) Card	28	Southern China
F. orientalis Losinsk	28	Northern Asia
F. × *bringhurstii* Staudt	35, 42	California
F. moschata Duch.	42	Northern and Central Europe
F. chiloensis (L.) Miller	56	Pacific Coast North America, Hawaii and Chile
F. iturupensis Staudt	56	Japan
F. virginiana Miller	56	Central and eastern North America
F. × *ananassa* Duchesne in Lamarck.	56	Cultivated worldwide

the octoploids originated in north-eastern Asia when *F. vesca* combined with other unknown diploids, and the polyploid derivatives then migrated across the Bering Strait and dispersed across North America. The Chilean *F. chiloensis* was most probably carried by birds from North America. The octoploids are completely diploidized, as they show disomic inheritance at their isozyme loci (Arulsekar *et al.*, 1981).

F. vesca was probably cultivated by the ancient Romans and Greeks, and by the 1300s *F. vesca*, *Fragaria viridis* and *Fragaria moschata* were being grown all across Europe (Darrow, 1966). Strawberries are asexually reproduced by stolons and can easily be transplanted from the wild to gardens. *F. chiloensis* was domesticated at least 1000 years ago by the Mapuche Indians of Chile (Hancock *et al.*, 1999). Imported *F. chiloensis* and *F. virginiana* rose in importance during the 1700s in Europe, ultimately to be replaced by the hybrid *F.* × *ananassa* in the late 1700s. North American production began about 1800 with native selections of *F. virginiana*, which were supplanted in the mid-1800s with European hybrid varieties of *F.* × *ananassa*. In the last century, strawberry cultivation has spread throughout all the temperate regions of the world.

Vegetables

Cole crops

The amazingly diverse genus of *Brassica* has given us the oil-seed rape or canola (*B. napus* L. and hybrids with *B. campestris*), the turnip and Chinese cabbage (*B. campestris* L.), rutabaga or swede (*B. napus* L.), black mustard (*B. nigra* Koch), brown mustard (*B. juncea* (L.) Czern) and Ethiopian mustard (*B. carinata*). The single *Cruciferae* species, *Brassica oleracea*, has by itself yielded a remarkable array of vegetables, including the cabbage, kale, Brussels sprouts, cauliflower, broccoli and kohlrabi (Fig. 11.3).

An overall diagram of genomic relationships in the genus *Brassica* was first presented by U in 1935 (Fig. 11.4). He suggested that brown mustard (leaf mustard) was a hybrid of black mustard and turnip, rutabaga and oil-

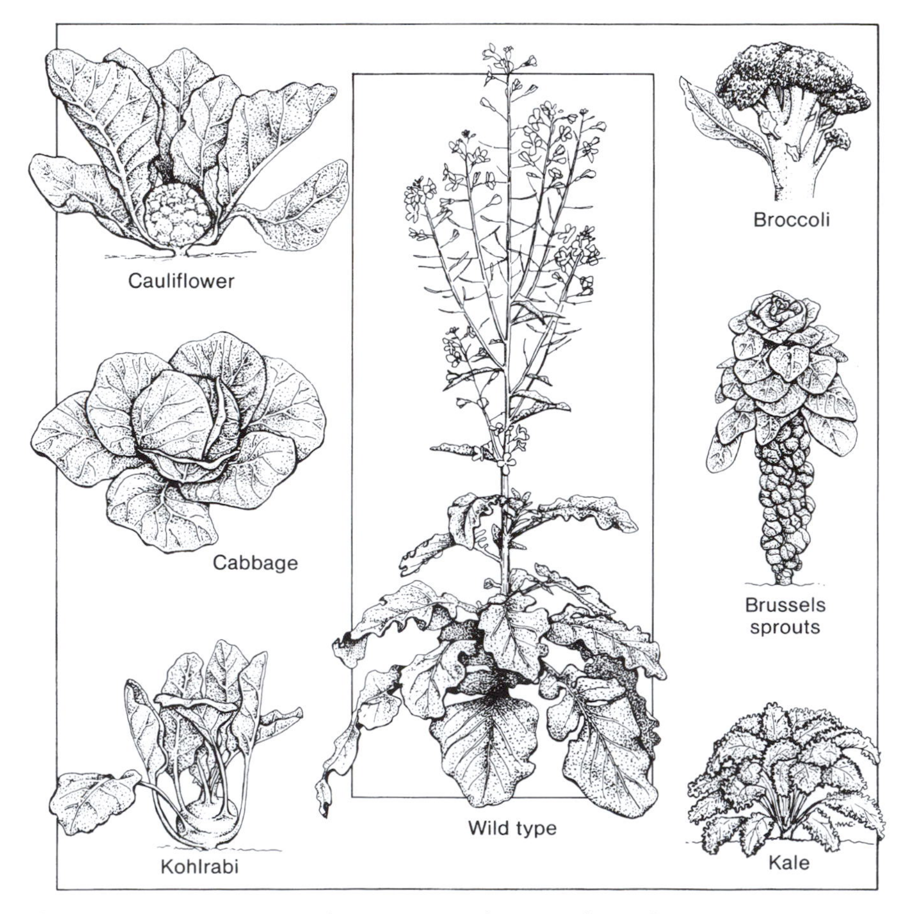

Fig. 11.3. Crops originating from *Brassica oleracea* subsp. *oleracea*.

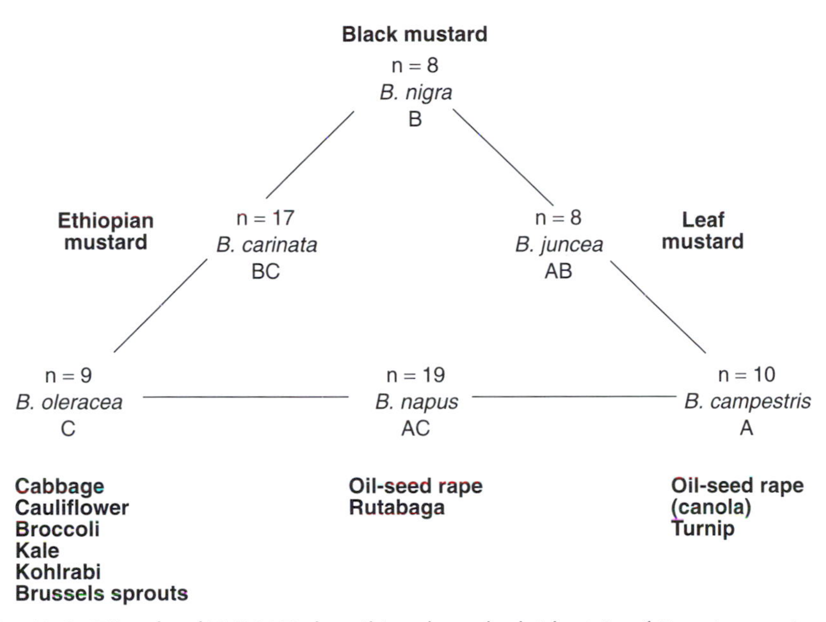

Fig. 11.4. Triangle of U (1935) describing the polyploid origin of *Brassica* species.

seed rape were derived from turnip and kale, and Ethiopian mustard was a hybrid of black mustard and kale. This scheme has been supported by many different types of evidence, including artificial synthesis of the hybrids, electrophoretic studies, molecular analysis of chloroplast and nuclear DNA and genomic *in situ* hybridization (Song *et al.*, 1990; Langercrantz and Lydiate, 1996; Snowdon *et al.*, 2002). The progenitor of the A, B and C genomes is thought to be an unknown hexaploid that was reorganized into the eight, nine and ten chromosome sets of *B. nigra*, *B. oleracea* and *Brassica rapa* through chromosomal fusions and rearrangements.

The first cultivated *Brassica* species was probably *B. campestris*, which was grown initially for its oil-seed (Thompson, 1979). Another species, *B. napus,* was also domesticated for its oil-seeds, but not until the Middle Ages. There is only meagre evidence of *B. campestris* beginnings as an oil crop, but it is likely to have been domesticated repeatedly about 4000 years ago from wild populations in an area spanning the Mediterranean to India. The turnip form of *B. campestris* emerged 3000 years later in northern Europe. Both *B. campestris* and *B. napus* probably first caught farmers' attention as weeds of wheat and other crops. These two species freely intercross and their hybrids are now our most important oil-seed crops.

The leafy kales may have been the second *Brassica* species to be cultivated, as the Greeks recorded growing them at least 2500 years ago (Helm, 1963). They were derived from *Brassica oleracea* subsp. *oleracea,* found naturally across the coast of the Mediterranean from Greece to England. An early stage of their domestication must have been a reduction of the bitter-tasting glucosinolates, which are found at high levels in the wild species (Josefsson, 1967). The other types of *B. oleracea* emerged much later, as humans began to actively select for the enlargement of different plant parts (Gray, 1982). Early types of cabbages were first grown in ancient Rome and Germany over 1000 years ago. Broccoli, cauliflower and Brussels sprouts are a more recent development, appearing within the last 500 years – cauliflower from northern Europe and broccoli from the eastern Mediterranean. Brussels sprouts first appeared as a spontaneous mutation in France in 1750.

The various types of *B. oleracea* were developed primarily by disruptive selection on the polymorphisms already available in wild *B. oleracea*, except for Brussels sprouts. Buckman demonstrated this in 1860 at the Royal Agricultural College in southern England by selecting broccoli-like cultigens from wild plants in only a few generations. Cauliflower is most closely related to broccoli, while cabbage is most closely related to the kales (Song *et al.*, 1990). As each new *Brassica* type emerged, it spread rapidly throughout Europe and the Mediterranean countries, and hybridization with wild congeners undoubtedly played a role in the development of crop types. Helm (1963) provided evidence that *Brassica cretica* contributed heavily to the development of cauliflower. Palmer *et al.* (1983) found chloroplast DNA in two populations of *B. napus* that was probably the result of recent introgressive hybridization with *B. oleracea* and *B. campestris.*

Black, brown (Indian) and Ethiopian mustards were also very early domesticants, but probably followed oil-seed rape and the leafy kales. All the mustards are used primarily as spices, with the exception of brown mustard, which is also used for its oil-seed and as greens (Hemingway, 1995). Black mustard is referred to in the early written literature of both Babylonia and India and is found native in Asia Minor and Iran. Brown mustard is found naturally from Central Asia to the Himalayas and was probably domesticated separately in India, China and the Caucasus. Ethiopian mustard arose in Ethiopia where the range of its parents – weedy *B. nigra* and cultivated *B. oleracea* – overlapped.

Squash and gourds

The cucurbit genus (*Cucurbita*) has probably made as important a contribution to our list of crops as the *Brassica*. The individual species are not as diverse as *B. oleracea*, but a large group of them have been domesticated (Table 11.4). Pumpkins and squashes are *Cucurbita pepo*, *Cucurbita argyrosperma* (= *Cucurbita mixta*), *Cucurbita moschata* and *Cucurbita maxima*. The cucumber belongs to *Cucumis sativus*, while the musk melons are in *Cucumis melo* and the water melons belong to *Citrullus lanatus*. Gourds are in both *Lagenaria siceraria* and *Cucurbita ficifolia*. In general, these taxa are reproductively isolated and are distinct for a number of molecular markers (den Nijs and Visser, 1985; Perl-Treves and Galun, 1985; Jobst *et al.*, 1998).

Table 11.4. Major cucurbit crops and their centre of origin.

Species	Common name	Chromosome number (2n)	Centre of origin
Cucumis sativus L.	Cucumber	14	India
Cucumis melo L.	Musk melon, cantalope	24	Africa, India
Citrullus lanatus (Thunb.) Mansf.	Water melon	22	South Africa
Lagenaria siceraria Standl.	White-flowered gourd	22	Africa, the Americas
Cucurbita pepo L.	Summer squash, pumpkin, marrow	40	Mexico
Cucurbita argyrosperma Hort. (= *Cucurbita mixta*)	Winter squash	40	Mexico, Central America
Cucurbita moschata Duchesne	Winter squash	40	Mexico, Central America
Cucurbita maxima Duchesne	Winter squash, pumpkin	40	Northern South America, Central America
Cucurbita ficifolia Bouché	Fig-leaf gourd	40	Mexico, Central America, northern South America

The cucumber is 2n = 14, while all the other species are 2n = 22 or 40. The high chromosome number of many of the cucurbit species suggests that they are of polyploid origin and this has been confirmed by examining isozyme segregation data (Weeden and Robinson, 1990).

Recent studies indicate that each cultivated *Cucurbita* arose from a different wild taxa in the New World, but all their ancestors are not clear (Nee, 1990; Decker-Walters, 1990; Sanjur *et al.*, 2002). South American *Cucurbita andeana* was probably the ancestor of *C. maxima*, and *Curcurbita sororia* Bailey from Mexico was the probable progenitor of *C. argyrosperma*. Domesticated *C. pepo* appears to contain two lineages with separate origins: subsp. *pepo* and *ovifera*. *C. pepo* subsp. *ovifera* shares the same mitochondrial DNA (mtDNA) with several wild taxa including *C. pepo* subsp. *fraterna*, *C. pepo* subsp. *ovifera* var. *texana* and *C. pepo* subsp. *ovifera* var. *ozarkana*. *C. pepo* subsp. *pepo* carries a unique mtDNA haplotype that is not associated with any other known wild species. *C. moschata* shares some sequence homology with *C. sororia* but it is an unlikely progenitor, while *C. ficifolia* is distinct from all known species.

The *Cucurbita* species were among the earliest domesticants. Evidence for *C. pepo* cultivation comes from about 10,000–9000 BP (Mesoamerica), followed by *C. argyrosperma* (southern Mexico – 7000 BP), *C. moschata* (southern Mexico – 7000 BP), *C. ficifolia* (Peru – 5000 BP) and *C. maxima* (Peru – 4000 BP) (Merrick, 1995). There is also evidence of the domestication of *Cucurbita ecuadorensis* in Ecuador from 10,000 to 12,000 years ago (Piperno and Stothert, 2003). Many of the species were first domesticated for their edible seeds and were gradually changed into thick-fleshed containers through artificial selection. The squashes and gourds were important components of the diets of the Aztecs, Incas and Mayas and the American Indians carried them throughout North America. While most of the cucurbits were introduced into what is now the USA, an independent origin of domesticated *C. pepo* in eastern North America is likely (Decker, 1988). The various *Cucurbita* species were taken to Europe and Asia after the discovery of the New World.

All the other cucurbit crops had Old World origins. There is evidence that the cucumber, *C. sativus* (2n = 14) was domesticated in India and China 3000–4000 years ago, although its progenitors are unclear (Bates and Robinson, 1995). One possible candidate, *C. sativa* var. *hardwickii*, is found in the wild in Asia, but it might be a feral derivative. Another Chinese species, *Cucumis hystrix* Chakr., is morphologically similar to cucumber but has different chromosome numbers and distinct isozyme frequencies (Chen *et al.*, 1997). The musk melon, *C. melo* (2n = 24), came from eastern tropical Africa, and the water melon, *C. lanatus* (2n = 22), originated in Central Africa. Both were domesticated in recent history from existing wild species of the same name. The bottle gourd, *L. siceraria,* does not appear to have a direct wild relative but was presumably domesticated first in Africa, where all the other wild *Lagenaria* species are found. Interestingly, the bottle gourd was being cultivated in the Americas by 15,000–9000 BP and by 6000–5000 BP in Asia, even though its roots lay in Africa (Heiser, 1979). Oceanic drifting is the most likely explanation for this disjunctive distribution (Decker-Walters *et al.*, 2001).

Chilli peppers

The peppers belong to the family *Solanaceae*. There are five cultivated and over 20 wild species that share the chromosome number 2n = 2x = 24 (Heiser and Smith, 1953; Eshbaugh, 1980). Restriction fragment length polymorphism (RFLP) and random amplified polymorphic DNA (RAPD) data have supported the traditional morphometric and cytogenetic classifications (Prince *et al.*, 1992, 1995; Rodriguez *et al.*, 1999). The most widely cultivated species is *Capsicum annuum* L., which contains a diverse array of types representing both sweet and hot peppers. It is grown worldwide. The rest of the cultigens, *Capsicum baccatum* L., *Capsicum frutescens* L., *Capsicum chinense* Joeg. and *Capsicum pubescens* Ruiz et Pav, are used primarily as spices, and their culture is largely confined to South America (Fig. 11.5) and, to a limited extent, Africa.

A major portion of *Capsicum* evolution appears to have occurred in south–central Bolivia (MacLeod *et al.*, 1982). Primitive types probably migrated out of this region into the Andes and lowland Amazonia and speciated along the way. Cytogenetic studies indicate that some of the species differentiation was associated with small chromosomal rearrangements (Kumar *et al.*, 1987), although hybrids with varying levels of fertility can be obtained

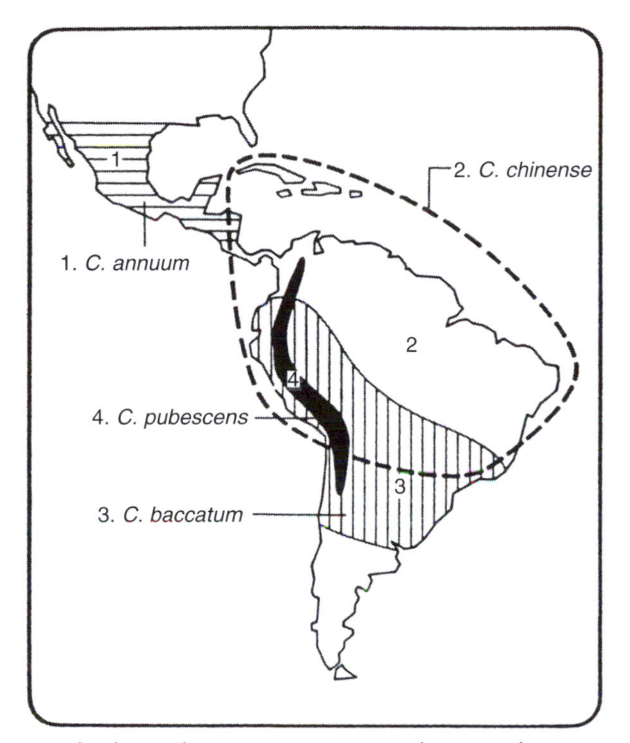

Fig. 11.5. Range of cultivated *Capsicum* peppers at the time of European discovery (used with permission from N.W. Simmonds, 1985, *Principles of Crop Improvement*, Longman, London).

among most of the cultivated species and many of the wild types (Fig. 11.6). *C. pubescens* is the species most strongly isolated from the rest. All the cultigens are self-compatible, even though many of the wild species are outcrossed through either heterostyly or self-incompatibility (Heiser, 1995a).

Chilli peppers were originally cultivated at several independent locations in the Americas (Pickersgill, 1989). Several-thousand-year-old remains of *C. annuum* have been found at Tehuacan, Mexico, but this material may have been wild collected plants; the oldest definite domesticants are over 4000 years old from Mexico and South America (Brücher, 1988). *C. baccatum* was probably domesticated in Bolivia, with the oldest remains being found along coastal Peru dating from 5000 to 4000 BP. *C. chinense* originated in lowland Brazil and was spread to Peru by 3000–2000 BP. It is not known if each of these domestications was totally independent or stimulated by other domestications.

At one time, the five cultivated forms were thought to have originated from one progenitor species, but now most researchers feel there were at least three evolutionary lines leading to the cultivated taxa (Eshbaugh *et al.*, 1983; Moscone *et al.*, 1996). *C. annuum*, *C. baccatum* and *C. pubescens* are much too different to have emerged from the same progenitor in the last

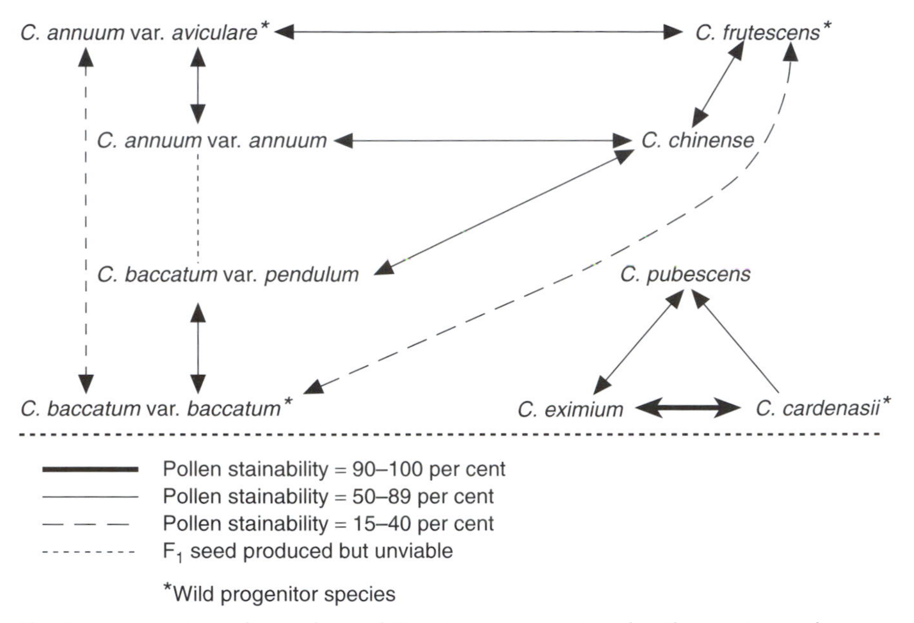

Fig. 11.6. Crossing relationships of *Capsicum* species (used with permission from W.H. Eshbaugh, © 1975, *Bulletin of the Torrey Botanical Club* 102; 396–403).

2000–3000 years (MacLeod *et al.*, 1979, 1982). There is a wild form of *C. baccatum* subsp. *baccatum* that is very similar to cultivated *C. baccatum* var. *pendulum*. Cultivated *C. pubescens* shows a close relationship to the wild species *Capsicum eximium*, *Capsicum cardenasii* and *Capsicum tovari* (Heiser, 1995a). The cultigens *C. annuum* var. *annuum*, *C. chinense* and *C. frutescens* all appear to have a common ancestor, which may be *C. annuum* var. *aviculare*. Most of the changes associated with domestication involved fruit size and colour, along with a shift from outcrossing to selfing.

Like so many other New World crops, Europeans became aware of chilli peppers after the voyage of Columbus and they gained almost immediate acceptance (Heiser, 1995a). Peppers came to the widespread attention of people in North America and Africa in the last couple of centuries.

Tomato

The genus of tomato, *Lycopersicon*, belongs to the family *Solanaceae*. There are nine wild species distributed throughout Central and South America (Table 11.5). A wild form of *Lycopersicon esculentum* var. *cesasiforme*, found in Mexico, Central America and South America, is probably the progenitor of the domesticated species (Jenkins, 1948; Rick, 1995). There do not appear to be any chromosomal structural differences separating the various species,

Table 11.5. Native species of tomato, *Lycopersicon* and *Solanum*, and their distribution. All species are 2n = 2x = 24.

Species	Distribution
L. cheesmanii Riley	Galapagos Islands
L. chilense Dun	Southern Peru and northern Chile
L. chmielewskii	Inter-Andean Peru
L. esculentum var. *cesasiforme* (Dun.) A. Gray	Andes, Brazil, Colombia, Central America and Mexico
L. hirsutum Humb. & Bonpl.	South–central Peru to northern Ecuador
L. parviflorum	Inter-Andean Peru
L. peruvianum (L.) Mill.	Peru and northern Chile
L. pimpinellifolium (Justl.) Mill	Coastal Peru and Ecuador
S. pennellii Corr	Central Peruvian Andes

and *L. esculentum* can be hybridized with all the other species with varying levels of success. Most of the species are self-incompatible, with the exception of *L. esculentum* and *Lycopersicon pimpinellifolium*, which are self-fertile and show substantial variation in rates of outcrossing (Rick *et al.*, 1977).

Mexico is generally accepted as the place where the tomato was first domesticated, although the evidence is not conclusive (Rick, 1995). Isozyme data indicate that var. *cesasiforme* could have originated in the area of Ecuador–Peru and then spread to Mexico, or the Andes may have been a secondary area of domestication (Rick and Fobes, 1975; Rick and Holle, 1990). There is also evidence that the early Mexican domesticants may have found their way to South America, where they introgressed with native races of *L. pimpinellifolium* (Rick *et al.*, 1977).

The initial date of domestication is unknown, but tomato cultivation was well established when the European explorers arrived in Mexico and the Americas. The North American and European tomato cultivars probably came from Mesoamerica, as they are more similar to the Mexican types than to the South American ones. Fears of toxicity initially slowed the spread of the tomato in Europe because it belongs to the deadly nightshade family, and it was not until the early 1800s that it was widely used for food. Tomato became established in North America a few decades later.

Fibres and Oils

Cotton

The genus *Gossypium* belongs to the family *Malvaceae*. It contains 39 diploid species (2n = 2x = 26) and six allotetraploids (2n = 4n = 52) (Fryxell *et al.*, 1992). Four species of cotton are cultivated: the diploids, *Gossypium herbaceum* and *Gossypium arboreum*, which carry the A genome, and the tetraploids, *Gossypium hirsutum* and *Gossypium barbadense,* which carry the A and D genomes (Table 11.6). The high ploidy number of the so-called diploids, along with substantial levels of enzyme multiplicity and DNA duplication, suggest an ancient polyploid beginning (Endrizzi *et al.*, 1985; Wendel, 2000).

The origin of the various cotton species is still under some debate, but recent molecular data have greatly clarified the picture. *G. arboreum* has long been thought to be a derivative of *G. herbaceum*, but the two species are so different genetically that it is more likely that they diverged before domestication (Wendel, 1989). The ancestor of cultivated *G. herbaceum* could be *G. herbaceum* subsp. *africanum*; the progenitor of *G. arboreum* remains unknown.

Extensive cytological and genetic data indicate that the progenitors of the tetraploids (AADD) are closely related to *Gossypium raimondii* (DD) and *G. herbaceum* (AA) (Phillips, 1966; Wendel, 1989; Small and Wendel, 1998). However, these two species are located on different continents, leaving much speculation about how their ancestors could have hybridized. Several hypotheses have been proposed: (i) the diploids were sympatric on the supercontinent of Pangaea and formed the tetraploids before the separation of South America and South Africa; (ii) the origin occurred in Polynesia before the establishment of the Pacific Ocean in its present form; (iii) people carried an Old World form to America or vice versa and the hybridization occurred; or (iv) the polyploids arose in South America after an A genome propagule drifted across the ocean from Africa or Asia. The latter hypothesis is the most popular, since the genetic data indicate a monophyletic origin for the allopolyploids, and the divergence pattern fits a time period between Pangaea and the emergence of farming. Wendel and Albert (1992) suggest that the A genome originally evolved in Africa and Asia, and then dispersed to the Americas across the Pacific, rather than the Atlantic. This makes the most sense, as the D genome species are located on the western side of the New World. They probably arose in north-western Mexico and then spread to Peru.

Table 11.6. Selected *Gossypium* species and their distribution.

Species	Chromosome number (2n)	Genome content	Distribution
G. herbaceum L.	26	AA	Wild: savannah of southern Africa Cultivated: Africa and Asia
G. arboreum L.		AA	Cultivated: Asia and Africa
G. raimondii Ulb.		DD	Wild: Peru
G. barbadense L.	52	AADD	Wild: western South America Cultivated: Arabia, Syria, Asia and South America
G. hirsutum L.		AADD	Wild: Central America and Mexico Cultivated: worldwide

The diploid species have a long history of cultivation that goes back at least 5000 years (Phillips, 1979). *G. herbaceum* was first domesticated in Arabia and Syria, while *G. arboreum* cultivation probably began in India. *G. arboreum* became dominant all across Africa and Asia, until it was replaced by the tetraploids in the last 100 years. Currently, *G. arboreum* is grown only to a limited extent in India, and *G. herbaceum* cultivation has only a spotty distribution in Africa and Asia.

Evidence of tetraploid cultivation also dates to antiquity. Cotton bolls and fibres of domesticated *G. barbadense* have been found in central–coastal Peru dating back to 4500 BP. The oldest cultivated remains of *G. hirsutum* have been located in Mexico from 5500 BP, although they are not thought to represent the earliest domestications. RFLP diversity patterns support the Yucatan peninsula as the primary site for *G. hirsutum* domestication (Brubaker and Wendel, 1994).

The Spanish and Portuguese colonists and traders spread the tetraploids to Spain, Africa and India via Cuba and Brazil. *G. hirsutum* (upland cotton) was introduced into the south-eastern USA in the mid-1700s from Mexico. *G. barbadense* (sea-island cotton) arrived in the Carolinas and Georgia from South America in the late 1700s and flourished in the south for several decades until the boll weevil prevented its successful cultivation. *G. hirsutum* now comprises about 95% of the world's crop, while *G. barbadense* accounts for all but 1% of the rest.

Groundnut

The genus *Arachis* (legume family) is restricted to South America and contains dozens of species in seven sections. It is grown throughout the world in warm temperate and tropical regions. The various species are mostly self-pollinated, but a high proportion of them can be successfully intercrossed. Their centre of diversity falls in Bolivia, like the peppers. Most of the species are diploids with $2n = 2x = 20$, but there are at least two allotetraploids with $2n = 2x = 40$, including cultivated *Arachis hypogea* L. and wild *Arachis monticola* Krop. et Rig. (Smartt *et al.*, 1978). Two distinct sets of chromosomes are found in groundnut, one marked by a pair of chromosomes significantly smaller than the others (A chromosomes) (Fig. 11.7) and a set that carries a pair with a secondary constriction (B chromosomes) (Husted, 1936). Some multivalents are formed during meiosis, suggesting segmental allopolyploidy, but the tetraploid species are mostly diploidized.

The cultivated groundnut was probably derived from the wild tetraploid, *A. monticola*, which is native to north-western Argentina, and freely hybridizes with it. For a long time, the most likely diploid progenitors were thought to be *Arachis cardenasii* and *Arachis batizocoi*, but more recent molecular and cytological data indicate that the progenitors are more closely related to *Arachis duranensis* and *Arachis ipaensis* (Kochert *et al.*, 1996; Raina and Mukai, 1999). *A. duranensis* carries the A genome, while *A. ipaensis* possesses the B genome.

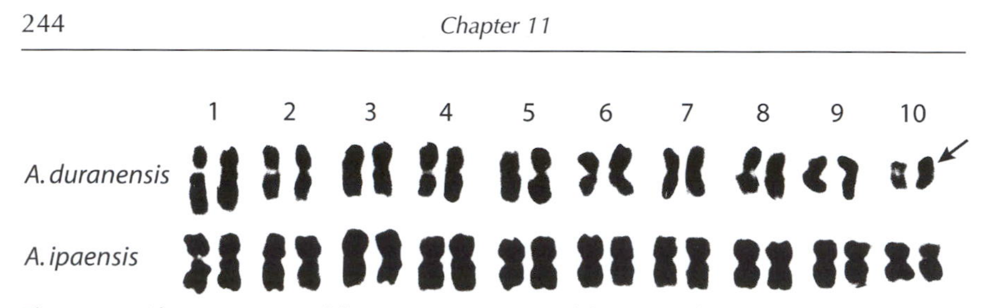

Fig. 11.7. Chromosomes of the two progenitors of the groundnut (*Arachis*). The small A chromosome is arrowed (used with permission from G. Kochert *et al.*, © 1996, *American Journal of Botany* 83, 1282–1291).

The oldest archaeological records of groundnuts are from Peru dated 5000–4000 BP (Gregory *et al.*, 1980; Singh, 1995). However, their ancestry is probably much older and their cultivation may have actually begun in Bolivia when more robust types with less fragile pods were selected from *A. monticola*. By 1500, groundnuts were widely distributed throughout South America, the Caribbean and Mexico. Spanish travellers took them from Mexico to eastern Asia, and Portuguese sailors distributed them to Africa from Brazil. India and North America received their first plants from Africa in the 1600s.

Sunflower

Two *Helianthus* species are cultivated, the sunflower, *H. annuus* L., and the Jerusalem artichoke, *H. tuberosum* L. The sunflower is widely grown as an oil crop, while the Jerusalem artichoke has a more limited use as a tuber crop. The genus *Helianthus* is a member of the *Asteraceae* family, which is divided into four sections (Schilling and Heiser, 1981). *H. annuus* belongs to the section *Annui*, whose 13 species are located in the western USA. They are all described as diploids, but cytological evidence suggests they are probably ancient polyploids; premeiotic treatment of microspores with colchicine induces quadrivalent formation (Jackson and Murry, 1983). Wild and cultivated *Helianthus* show considerable variability in the size of their ribosomal genes (Choumane and Heizmann, 1988) and their total DNA content (Sims and Price, 1985). Sunflowers can be crossed with most other species in their section, although hybrids generally display reduced fertility. *H. tuberosum* belongs to a different section (*Diaricati*) from that of the sunflower and is a hexaploid.

Sunflower was probably gathered by the earliest people inhabiting the western USA. It is a camp-following weed and was introduced by nomadic tribes into central USA where it was first fully domesticated (Heiser, 1976). Biochemical and molecular evidence suggests that our cultivars arose from relatively few genotypes (Riesenberg and Seiler, 1990; Burke *et al.*, 2002), and only a few major quantitative trait loci (QTL) are associated with the major morphological differences between wild and domesticated forms (Cronn *et al.*, 1997). The first types selected were probably rare single-headed or monocephalic plants (Fig. 11.8). The exact time of origin is

Fig. 11.8. Wild *Helianthus annuus* next to a monocephalic cultivar.

unknown, but a reasonable estimate is 2000–3000 years ago (Heiser, 1995b). Sunflower cultivation may have predated the arrival of maize, beans and squash from Mexico.

By the time Europeans came to the New World, the sunflower was being raised in a broad band from Mexico to southern Canada (Heiser, 1995b). The first European introductions were from Mexico to Spain in the 16th century, although sunflowers were carried to Europe from all over North America. Sunflower's popularity as a food source grew very slowly in Europe until it arrived in Russia, where it quickly became an important oil crop. Its popularity may have exploded because it was not on the list of oily foods forbidden on certain holy days (Heiser, 1995b).

12

Postscript: Germ-plasm Resources

Introduction

Within this book, we have described the evolutionary forces that produced crop species and have followed the changes associated with domestication. Genetic variation is the foundation on which all the crops were shaped, both before and after human intervention. Our species, *Homo sapiens*, was presented with a plenitude of potential crops and all we had to do was decide on how to most effectively exploit them.

As our population has continued to grow, we have found ourselves with reduced space and natural resources (Fig. 12.1). There are over 6 billion people in the world, and they all must find something to eat and a place to live. At current rates of growth, this number will double in the next few decades. Associated with this crush of people is the loss of wild plant populations. These are being rapidly eliminated by the increasing sprawl of human activities and home sites. Vast areas of native vegetation have been eliminated, and many more are threatened by the plough, slash-and-burn farming, logging trucks and developers.

Modern breeding methods have greatly expanded the productivity of our major crops, but continued high yields hang on a thread of genetic vulnerability. We have developed most of our cultivars from a very narrow genetic base (Tanksley and McCouch, 1997), and as a result they are very susceptible to environmental perturbations. The decimation of huge acreages by the potato late blight famine in Ireland in 1845 and by the maize blight in central USA in 1970 serve as vivid reminders that genetic variability is the key to continued productivity (Plucknett *et al.*, 1987). Biotic and abiotic stresses can at any time decimate a crop with too narrow a base of variability.

Years ago	Cultural stage	Area populated	Assumed density per square kilometre	Total population (millions)
1,000,000	Lower Palaeolithic	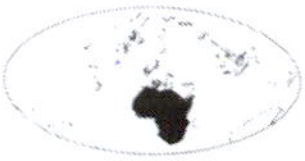	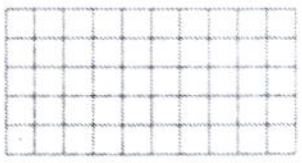 0.00425	0.125
300,000	Middle Palaeolithic		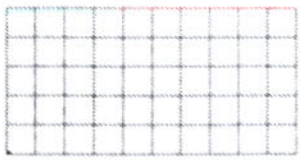 0.012	1
25,000	Upper Palaeolithic		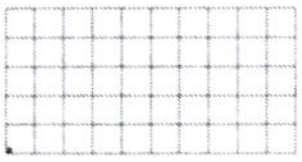 0.04	3.34
10,000	Mesolithic		 0.04	5.32
6000	Village farming and early urban		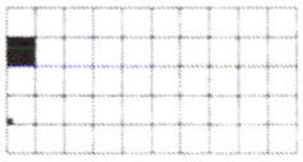 1.0[a] 0.04	86.5
2000	Village farming and urban		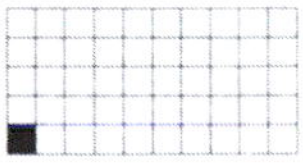 1.0	133
Year AD	**Cultural stage**	**Area populated**	**Assumed density per square kilometre**	**Total population (millions)**
1650	Farming and industrial		 3.7	545
1750	Farming and industrial		 4.9	728
1800	Farming and industrial		 6.2	906
1900	Farming and industrial		11.0	1610
1950	Farming and industrial		 16.4	2400
2000	Farming and industrial		 46.0	6079

Fig. 12.1. Changes in world population numbers and density over the last million years (used with permission from L.L. Cavalli-Sforza and W.F. Bodner, © 1971, *The Genetics of Human Populations*, W.H. Freeman, San Francisco).

Continued yield and quality improvements also depend on the acquisition of new genes. The Green Revolution, which provided some of the less developed countries with new, high-yielding types of rice and wheat, was based on the identification and transfer of novel genes for dwarfing and other desirable traits (Doyle, 1985). In addition, most breeders have found the solution to yield plateaux or new pest challenges to be the incorporation of unique germ-plasm.

Concentration on a narrow genetic base means that the genetic variability necessary to feed our growing numbers is being eliminated by the expansion of that same exploding population. As natural populations are destroyed, we are losing a storehouse of allelic diversity. Wild populations carry a vast array of genetic variability that is available to breeders who want to improve crop quality and yield or disease and stress tolerance. This makes the preservation of our native germ-plasm sources absolutely critical to our future breeding successes and perhaps the continued survival of our race.

Ex situ Conservation

There has been a growing effort to collect wild germ-plasm and store it in repositories (Shands, 1990; Guarino *et al.*, 1995; Engles *et al.*, 2002). This *ex situ* conservation allows genotypes to be catalogued and made readily available to interested individuals. Probably the most extensive collections are those coordinated by the International Board for Plant Genetic Resources (headquartered in Rome, Italy) and the US National Germplasm System (headquartered in Beltsville, Maryland). An extensive list of international germ-plasm collections can be found in Plucknett *et al.* (1987) and Dearing and Guarino (1995).

Such centralized collections are extremely critical as the production of new cultivars continues and the number of public breeders maintaining collections in industrialized countries declines (Brooks and Vest, 1988; Knight, 2003). Moreover, limits to storage space have demanded that collection sizes be kept to a minimum (Goodman, 1990). There is only so much material that can be stored in a viable state, particularly when it is clonally propagated and must be maintained as whole plants in the field or greenhouse. This limitation leaves a high likelihood that important germ-plasm will be left uncollected or poorly maintained.

The space problem has stimulated much research on improving methods of plant storage. Seeds and pollen of many species have been found to survive years or even decades under proper conditions and require less space than whole plants, although they can only be used when cultivar purity is not important (Breese, 1989). Preservation of plants in tissue culture or as frozen buds has shown high promise as a way to save space and maintain specific genotypes (Forsline *et al.*, 1998; Reed, 2001), but these methods are costly and require specialized equipment. Clearly, no matter how efficient we become at storage, someone will still have to make a decision as to what stays and what goes.

Deciding on what germ-plasm should be collected and stored is a great challenge. The general goal in any sampling strategy is to obtain as much of the genetic diversity as possible, based on space and time limitations. High priority must be given to genes that solve existing problems, but other genes with unknown benefits should also be collected for future unforeseen circumstances. There is simply no reliable way to predict which alleles will be useful in future varieties until a need arises or they are incorporated.

Several different strategic points must be taken into consideration before collecting germ-plasm. A decision must be made on what constitutes the crop and where it is located (Brown and Marshall, 1995). Important related questions are the following: (i) What is the natural range of the species? (ii) Are there one or more relatives that should be collected? (iii) Is the crop sexually or asexually propagated? (iv) What is the best maintenance strategy? Decisions on these parameters are critical before effective sampling can even begin.

The germ-plasm collector must also face the big question of how many individuals to collect. Marshall and Brown (1975) and Brown and Marshall (1995) have devised equations that predict the sample size needed to adequately sample the common alleles in a population under different evolutionary models. In general, it takes 15–80 samples to adequately represent the common alleles. They conclude that localized common alleles should be the major target of a sampling strategy, since rare alleles will be acquired essentially at random except in the most comprehensive sampling efforts. Based on this assumption, the best strategy to use with no a priori knowledge of variation patterns is to collect 50–100 random individuals from as many diverse sites as possible.

Of course, this model assumes that each population and each species has the same level of genetic variation, which we know is not true. Each species has its own peculiar evolutionary history, which must be taken into consideration in developing the most effective sampling strategy. For this reason, thorough studies of the ecological genetics of species are warranted before extensive collections are made. Important questions must be faced: (i) Are the species broken into environmentally and genetically distinct populations? (ii) Are the populations significantly substructured or even in their distribution? (iii) Are they outcrossed or inbred? Isozyme and DNA marker variability has proved instructive in determining variation patterns (Brown, 1988; Clegg, 1990; Hamrick and Godt, 1997), but it does not always accurately represent morphological patterns (Harrison *et al.*, 1997). Nothing can replace extensive field experience.

In situ Conservation

Probably the simplest solution to the storage problem is to maintain the wild populations themselves as genetic storehouses. This is called *in situ* conservation. We may not know exactly what genes are contained in the native populations, but at least they will be there for later searching when a need arises. This approach depends less on the whims of individual breeders and repository directors and makes the loss of potentially valuable genes due to accident less likely. Also, the populations are still subject to the sieve of natural evolution, making it possible that new adaptations may arise (Ingram and Williams, 1984).

A blanket preservation of all plant populations is, of course, impossible but efforts should be made to at least maintain large representative populations of the ecological range found in the progenitors of each crop species and its allies. A much larger sample of naturally segregating variability can be maintained in nature than can ever be held in even the largest repository. Effective *in situ* conservation requires knowledge of the variability patterns of the crop species and their ranges of adaptation, but this information is not unique to this type of conservation strategy. As we previously discussed, extensive evolutionary information is also needed to effectively collect native material for repositories.

While such surveys may seem like an impossible task, many crop species and their environments have already been censused for their variation patterns and this information is available (Auricht *et al.*, 1995; Pendergast, 1995). It is hoped that the growing awareness of the fragility of our natural environment will result not only in the initiation of new ecological studies, but also in the extensive cataloguing of natural populations. The germ-plasm repositories could then concentrate on maintaining cultivars and natural materials that are in danger of eroding *in situ*. The repositories will still need to maintain a base collection of natural variability for ready deployment, but the emphasis would be placed on exploration and information transmittal, rather than the storage of huge numbers of genotypes.

References

Abbott, R.J. (1992) Plant invasions, interspecific hybridization and the evolution of new plant taxa. *TREE* 7, 401–405.

Adams, M.W. (1974) Plant architecture and physiological efficiency in the field bean. *Plant Architecture and Physiological Efficiency in the Field Bean*. CIAT, Cali, Colombia, pp. 266–278.

Adams, W.T. and Allard, R.W. (1977) Effect of polyploidy on phosphoglucose isomerase diversity in *Festuca microstachys*. *Proceedings of the National Academy of Sciences USA* 74, 1652–1656.

Aggarwal, R.K., Brar, D.S., Nandi, S., Huang, N. and Khush, G.S. (1999) Phylogenetic relationships among *Oryza* species revealed by AFLP markers. *Theoretical and Applied Genetics* 98, 1320–1328.

Ahmad, F. (2000) A comparative study of chromosome morphology among nine annual species of *Cicer* L. *Cytobios* 101, 37–53.

Ahmadi, H. and Bringhurst, R.S. (1991) Genetics of sex expression in *Fragaria* species. *American Journal of Botany* 78, 504–514.

Ahn, S. and Tanksley, S.D. (1993) Comparative linkage maps of rice and maize genomes. *Proceedings of the National Academy of Sciences USA* 90, 7980–7984.

Ainsworth, C. (2000) Boys and girls come out and play: the molecular biology of dioecious plants. *Annals of Botany* 86, 211–221.

Albrigio, A., Spettoli, P. and Cacco, G. (1978) Changes in gene expression from diploid to autotetraploid status of *Lycopersicon esculentum*. *Plant Physiology* 44, 77–80.

Aldrich, P.R. and Doebley, J. (1992) Restriction fragment variation in the nuclear and chloroplast genomes of cultivated and wild *Sorghum bicolor*. *Theoretical and Applied Genetics* 85, 293–302.

Aldrich, P.R., Doebley, J., Schertz, K.F. and Stec, A. (1992) Patterns of allozyme variation in cultivated and wild *Sorghum bicolor*. *Theoretical and Applied Genetics* 85, 451–460.

Alexander, D.E. (1960) Performance of genetically induced corn tetraploids. *American Seed Trade Association* 15, 68–74.

Alexander, J. and Coursey, D.G. (1969) The origins of yam cultivation. In: Ucko, P.J. and Dimbleby, G.W. (eds) *The Domestication and Exploitation of Plants and Animals*. Longman, London, pp. 405–425.

Alicchio, R., Araci, L. and Conte, L. (1995) Restriction fragment length polymorphism based phylogenetic analysis of *Avena*. *Genome* 38, 1279–1284.

Al-Janabi, S.M., Honeycutt, R.J., McClelland, M. and Sobral, B.W.S. (1993) A genetic linkage map of *Saccharum spontaneum* L. *Genetics* 134, 1249–1260.

Al-Janabi, S.M., Honeycutt, R.J. and Sobral, B.W.S. (1994) Chromosome assortment in *Saccharum*. *Theoretical and Applied Genetics* 89, 959–963.

Allard, R.W. (1960) *Principles of Plant Breeding*. John Wiley & Sons, New York.

Allard, R.W. (1988) Genetic changes associated with the evolution of adaptedness in cultivated plants and their wild progenitors. *Journal of Heredity* 79, 225–238.

Allard, R.W. (1990) The genetics of host–pathogen coevolution: implications for genetic resource conservation. *Journal of Heredity* 81, 1–6.

Allard, R.W. and Kahler, A.L. (1971) Allozyme polymorphisms in plant populations. *Stadler Symposia* 3, 9–24.

Allard, R.W., Babbel, E., Clegg, M. and Kahler, A. (1972) Evidence for coadaptation in *Avena barbata*. *Proceedings of the National Academy of Sciences USA* 69, 3043–3048.

Alleweldt, G., Speigle-Roy, P. and Reisch, B. (1991) Grapes (*Vitis*). In: Moore, J.N. and Ballington, J.R. (eds) *Genetic Resources of Temperate Fruit and Nut Crops*. International Society of Horticultural Science, Wageningen, pp. 289–328.

Ammerman, A.J. and Cavalli-Sforza, L.L. (1984) *The Neolithic Transition and the Genetics of Populations in Europe*. Princeton University Press, Princeton, New Jersey.

Anderson, E. (1948) Hybridization of the habitat. *Evolution* 2, 1–9.

Anderson, E. (1949) *Introgressive Hybridization*. John Wiley & Sons, New York.

Anderson, E. (1954) *Plants, Man and Life*. Melrose, London.

Anderson, E. (1961) The analysis of variation in cultivated plants with special reference to introgression. *Euphytica* 10, 79–86.

Anderson, E. and Hubricht, L. (1938) Hybridization in *Tradescantia* III. The evidence for introgressive hybridization. *American Journal of Botany* 25, 396–402.

Anonymous (1989) *Lost Crops of the Incas*. National Academy Press, Washington, DC.

Antonovics, J. (1971) The effects of a heterogeneous environment on the genetics of natural populations. *American Naturalist* 59, 593–599.

Antonovics, J. and Bradshaw, A.D. (1970) Evolution in closely adjacent plant populations VIII. Clinal patterns at a mine boundary. *Heredity* 27, 349–362.

Appenzeller, T. (1998) Art: evolution or revolution. *Science* 282, 1451–1454.

Aragoncillo, C., Rodriguez-Loperen, M.A., Salcedo, G., Carbonero, P. and Garcia-Olmedo, F. (1978) Influence of homologous chromosomes on gene-dosage effects in allohexaploid wheat (*Triticum aestivum* L.). *Proceedings of the National Academy of Sciences USA* 75, 1446–1450.

Armstrong, D.L. (1957) Cytogenetic study of some derivatives of the F_1 hybrid *Prunus amygdalus* × *P. persica*. PhD thesis, University of California, Davis.

Armstrong, J.A., Powell, J.M. and Richards, A.J. (1982) *Pollination and Evolution*. Royal Botanical Gardens, Sydney.

Arnold, M.L. (1993) *Iris nelsonii* (Iridaceae): origin and genetic composition of a homoploid hybrid species. *American Journal of Botany* 80, 577–583.

Arriola, P.E. and Ellstrand, N.C. (1997) Fitness of interspecific hybrids in the genus *Sorghum*: persistence of crop genes in wild populations. *Ecological Applications* 7, 512–518.

Arulsekar, S., Bringhurst, R.S. and Voth, V. (1981) Inheritance of PGI and LAP isozymes in octoploid strawberries. *Journal of the American Society for Horticultural Science* 106, 679–683.

Asins, M.J. and Carbonell, E.A. (1986) A comparative study on variability and phylogeny of *Triticum* species. 2. Interspecific relationships. *Theoretical and Applied Genetics* 72, 559–568.

Auricht, G.C., Reid, R. and Guarino, L. (1995) Published information on the natural and human environment. In: Guarino, L., Rao, V.R. and Reid, R. (eds) *Collecting Plant Genetic Diversity*. CAB International, Wallingford, UK.

Austin, D.F. (1978) The *Ipomoeae* complex – I. Taxonomy. *Bulletin of the Torrey Botanical Club* 105, 114–129.

Ayala, F.J. (1975) Genetic differentiation during the speciation process. *Evolutionary Biology* 8, 1–78.

Ayala, F.J. (1982) *Population and Evolutionary Genetics: a Primer*. Benjamin/ Cummings Publishing Company, Menlo Park, California.

Ayensu, E.S. and Coursey, D.G. (1972) Guinea yams. *Economic Botany* 26, 301–318.

Babcock, E.B. and Stebbins, G.L. (1938) *The American Species of* Crepis. *Their Interrelationships and Distribution as Affected by Polyploidy and Apomixis*. Publication No. 504, Carnegie Institute of Washington, Washington, DC.

Badaeva, E.D., Friebe, B. and Gill, B.S. (1996) Genome differentiation in *Aegilops*. 1. Distribution of highly repetitive DNA sequences on chromosomes of diploid species. *Genome* 39, 293–306.

Badami, P.S., Mallikarjuna, N. and Moss, J. (1997) Interspecific hybridization between *Cicer arietinum* and *C. pinnatifidum*. *Plant Breeding* 116, 393–395.

Baetcke, K.P., Sparrow, A.H., Neumann, C.H. and Schwemmer, S.S. (1967) The relationship of DNA to nuclear and chromosome volumes and radiosensitivity (LD50). *Proceedings of the National Academy of Sciences USA* 58, 533–540.

Bai, K.V. (1992) Cytogenetics of *Manihot* species and interspecific hybrids. In: Roca, W. and Thro, A.M. (eds) *Proceedings of the 1st International Scientific Meeting of the Cassava Biotechnology Network*. Centro Internacional de Agricultura Tropical, Cali, Columbia, pp. 51–55.

Bailey, C.H. and Hough, L.F. (1975) Apricots. In: Janick, J. and Moore, J.N. (eds) *Advances in Fruit Breeding*. Purdue University Press, West Lafayette, Indiana. pp. 367–384.

Bailey, D.C. (1983) Isozyme variation and plant breeders' rights. In: Tanksley, S.D. and Orton, T.J. (eds) *Isozymes in Plant Genetics and Breeding*. Elsevier, New York, pp. 425–440.

Baker, H.G. (1970) *Plants and Civilization*. Wadsworth Publishing Company, Belmont, California.

Ballington, J.R. and Galletta, G.J. (1978) Comparative crossability of 4 diploid *Vaccinium* species. *Journal of the American Horticultural Society* 103, 554–560.

Balter, M. (1998) Why settle down? The mystery of communities. *Science* 282, 1442–1445.

Balter, M. (2001) In search of the first Europeans. *Science* 291, 1722–1725.
Barakat, A., Matassi, G. and Bernardi, G. (1998) Distribution of genes in the genome of *Arabidopsis thaliana* and its implications for the genome organization of plants. *Proceedings of the National Academy of Sciences USA* 95, 10044–10049.
Barber, H.N. (1970) Hybridization and the evolution of plants. *Taxon* 19, 154–160.
Barrett, H.C. and Rhodes, A.M. (1976) A numerical taxonomy study of affinity relationships in cultivated citrus and close relatives. *Systematic Botany* 1, 105–136.
Barrett, S.C.H. (1983) Crop mimicry in weeds. *Economic Botany* 37, 255–282.
Bates, D.M. and Robinson, R.W. (1995) Cucumbers, melons and water-melons: *Cucumis* and *Citrullus* (Cucurbitaceae). In: Smartt, J. and Simmonds, N.W. (eds) *Evolution of Crop Plants*. Longman Scientific and Technical, Harlow, UK.
Baum, B.R. (1977) *Oats: Wild and Cultivated. A Monograph of the Genus* Avena *L. (Poaceae)*. Biosystematics Research Institute, Canada Department of Agriculture, Ottawa.
Baum, B.R. (1985a) *Avena atlantica*: a new diploid species of the oat genus from Morocco. *Canadian Journal of Botany* 63, 1057–1060.
Baum, B.R. (1985b) A new tetraploid species of oat. *Canadian Journal of Botany* 63, 1379–1385.
Bauman, U., Juttner, J., Bian, X. and Langridge, P. (2000) Self-incompatibility in the grasses. *Annals of Botany* 85, 203–209.
Baumgartner, B.J., Rapp, J.C. and Mullet, J.E. (1989) Plastid transcription activity and DNA copy number increase in early barley chloroplast development. *Plant Physiology* 89, 1011–1018.
Baur, E. (1924) Untersuchungen Ober das Wesen, die Entstehung und die Vererbung von Rassenunterscheiden bei *Antirrhinum majus*. *Bibliotheca Genetica* 4, 1–170.
Bautista, N.S., Solis, R., Kamijima, O. and Ishii, T. (2001) RAPD, RFLP and SSLP analysis of phylogenetic relationships between cultivated and wild species of rice. *Genes and Genetic Systems* 76, 71–79.
Bazzaz, F.A., Levin, D.A., Levy, M. and Schmierbach, M.R. (1982) The effect of chromosome doubling on photosynthetic rates in *Phlox*. *Photosynthetica* 17, 89–92.
Beadle, G.W. (1939) Teosinte and the origin of maize. *Journal of Heredity* 30, 245–247.
Beck, C.B. (1976) *Origin and Early Evolution of Angiosperms*. Columbia University Press, New York.
Bennett, M.D. (1972) Nuclear DNA content and minimum generation time in herbaceous plants. *Proceedings of the Royal Society, London, Ser. B* 181, 109–135.
Bennett, M.D. (1976) DNA amount, latitude and crop plant distribution. *Environmental and Experimental Botany* 16, 93–108.
Bennett, M.D. (1984) The genome, the natural karyotype and biosystematics. In: Grant, W.F. (ed.) *Plant Biosystematics*. Academic Press, New York.
Bennett, M.D. (1987) Variation in genomic form in plants and its ecological implications. *New Phytologist* 106, 177–200.
Bennetzen, J.L. (2000a) Transposable element contributions to plant gene and genome evolution. *Plant Molecular Biology* 42, 251–269.
Bennetzen, J.L. (2000b) Comparative sequence analysis of plant nuclear genomes: microcolinearity and its many exceptions. *Plant Cell* 12, 1021–1029.

Bennetzen, J., Buckler, E., Chandler, V., Doebley, J., Dorweiler, J., Gaut, B., Freeling, M., Hake, S., Kellogg, E., Poethig, R.S., Walbot, V. and Weeler, S. (2001) Genetic evidence and the origin of maize. *Latin American Antiquity* 12, 84–86.

Benz, B.F. (2001) Archeological evidence of teosinte domestication from Guilá Naquitz, Oaxaca. *Proceedings of the National Academy of Sciences USA* 98, 2104–2106.

Ben-Ze'ev, N. and Zohary, D. (1973) Species relationships in the genus *Pisum*. *Israel Journal of Botany* 22, 73–91.

Bernstrom, P. (1952) Cytogenetic intraspecific studies in *Lamium*. I. *Hereditas* 38, 163–220.

Berry, M.A. (1985) The age of maize in the greater Southwest: a critical review. In: Ford, I. (ed.) *Prehistoric Food Production in North America*. Museum of Anthropology, University of Michigan, Ann Arbor, pp. 279–308.

Bever, J.D. and Felber, F. (1992) The theoretical population genetics of autopolyploidy. *Oxford Surveys of Evolutionary Biology* 8, 185–217.

Bhattacharyya, M.K., Smith, A.M., Ellis, N., Hedley, C. and Martin, C. (1990) The wrinkled-seed character of pea described by Mendel is caused by a transposon-like insertion in a gene encoding starch-branching enzyme. *Cell* 60, 115–122.

Billington, H.L., Mortimer, A.M. and McNeilly, T. (1988) Divergence and genetic structure in adjacent grass populations. I. Quantitative genetics. *Evolution* 42, 1267–1277.

Binford, L.F. (1968) Post-pleistocene adaptations. In: Binford, S.R. and Binford, L.R. (eds) *New Perspectives in Archeology*. Aldine, Chicago.

Bingham, E.T. (1980) Maximizing heterozygosity in autopolyploids. In: Lewis, W.D. (ed.) *Polyploidy: Biological Relevance*, Plenum Press, New York, pp. 471–490.

Birky, C.W. (1983) Relaxed cellular controls and organelle heredity. *Science* 222, 468–475.

Bisht, M.S. and Mukai, Y. (2001) Genomic *in situ* hybridization identifies genome donor of finger millet (*Eleusine coracana*). *Theoretical and Applied Genetics* 102, 825–832.

Blake, N.K., Lehfeldt, B.R., Lavin, M. and Talbert, L.E. (1999) Phylogenetic reconstruction based on low copy DNA sequence data in an allopolyploid: the B genome of wheat. *Genome* 42, 351–360.

Blanc, G., Barakat, A., Guyot, R., Cooke, R. and Delseny, M. (2000) Extensive duplication and reshuffling in the *Arabidopsis* genome. *Plant Cell* 12, 1093–1101.

Blattner, F.R. and Bandani Mendez, A.G. (2001) RAPD data do not support a second centre of barley domestication in Morocco. *Genetic Resources and Crop Evolution* 48, 13–19.

Bloom, W.L. (1976) Multivariate analysis of the introgressive replacement of *Clarkia nitans* by *Clarkia speciosa polyantha* (Onagraceae). *Evolution* 30, 412–424.

Boffey, S.A. and Leech, R.M. (1982) Chloroplast DNA levels and the control of chloroplast division in light grown wheat leaves. *Plant Physiology* 69, 1387–1391.

Bond, D.A. (1995) Faba bean: *Vicia faba* (Leguminosae – Papilionoide. In: Smartt, J. and Simmonds, N.W. (eds) *Evolution of Crop Plants*. Longman Scientific & Technical, Harlow, UK, pp. 312–316.

Bordes, F. (1968) *The Old Stone Age*. New York, McGraw-Hill.

Boster, J.S. (1985) Selection for perceptual distinctiveness: evidence from Aguaruna cultivars of *Manihot esculenta*. *Economic Botany* 39, 310–325.

Botstein, D., White, R.L., Skolnick, M.H. and Davis, R.W. (1980) Construction of a genetic map in humans using restriction fragment length polymorphisms. *American Journal of Human Genetics* 32, 137–144.

Brace, C.L., Nelson, H. and Korn, N. (1979) *Atlas of Human Evolution*. Holt, Reinhart and Winston, New York.

Bradshaw, H.D., Otto, K.G., Frewen, B.G., McKay, J.K. and Schemske, D.W. (1998) Quantitative trait loci affecting differences in floral morphology between two species of monkeyflower (*Mimulus*). *Genetics* 149, 367–382.

Breese, E.L. (1989) *Regeneration and Multiplication of Germplasm Resources in Seed Genebanks: The Scientific Background*. International Board for Plant Genetic Resources, Rome.

Bres-Patry, C., Lorieux, M., Clément, G., Bangratz, M. and Ghesquière, A. (2001) Heredity and genetic mapping of domestication-related traits in a temperate *japonica* weedy race. *Theoretical and Applied Genetics* 102, 118–126.

Bretagnolle, F. and Thompson, J.D. (1995) Gametes with the somatic chromosome number: mechanisms of their formation and role in the evolution of autopolyploid plants. *New Phytologist* 129, 1–22.

Bretting, P.K. and Goodman, M.M. (1989) Karyotypic variation in Mesoamerican races of maize and its systematic significance. *Economic Botany* 43, 109–124.

Briggs, D. and Walters, S.M. (1984) *Plant Variation and Evolution*. Cambridge University Press, Cambridge.

Briggs, F.N. and Knowles, P.F. (1967) *Introduction to Plant Breeding*. Reinhold Publishing Corporation, New York.

Bringhurst, R.S. (1990) Cytogenetics and evolution in American *Fragaria*. *HortScience* 25, 879–881.

Bringhurst, R.S. and Senanayaka, Y.D.A. (1966) The evolutionary significance of natural *Fragaria chiloensis* × *F. vesca* hybrids resulting from unreduced gametes. *American Journal of Botany* 53, 1000–1006.

Bringhurst, R.S. and Voth, V. (1984) Breeding octoploid strawberries. *Iowa State Journal of Research* 58, 371–382.

Brooks, H.J. and Vest, G. (1988) Public programs on genetics and breeding of horticultural crops in the US. *HortScience* 20, 826–830.

Brown, A.H.D. (1988) Isozymes, plant population, genetic structure and genetic evolution. *Theoretical and Applied Genetics* 52, 145–157.

Brown, A.H.D. and Marshall, D.R. (1981) Evolutionary changes accompanying colonization in plants. In: Scudder, G.C.E. and Reveal, J.L. (eds) *Evolution Today*. Hunt Institute for Botanical Documentation, Carnegie–Mellon University, Pittsburgh, pp. 351–363.

Brown, A.H.D. and Marshall, D.R. (1995) A basic sampling strategy: theory and practice. In: Guarino, L., Rao, V.R. and Reid, R. (eds) *Collecting Plant Genetic Diversity*. CAB International, Wallingford, UK, pp. 75–92.

Brown, A.H.D., Doyle, J.L., Grace, J.P. and Doyle, J.J. (2002) Molecular phylogenetic relationships within and among diploid races of *Glycine tomentella* (Leguminosae). *Australian Systematics and Botany* 15, 37–47.

Brubaker, C.L. and Wendel, J.F. (1994) Reevaluating the origin of domesticated cotton (*Gossypium hirsutum* Malvaceae) using nuclear restriction fragment length polymorphisms (RFLPs). *American Journal of Botany* 81, 1309–1326.

Brücher, H. (1988) *Useful Plants of Neotropical Origin and their Wild Relatives*. Springer-Verlag, New York.

Brush, S.B., Carney, H.J. and Human, Z. (1981) Dynamics of Andean potato agriculture. *Economic Botany* 35, 70–88.

Buckler, E.S. and Holtsford, T.P. (1996) *Zea* systematics: ribosomal ITS evidence. *Molecular Biology and Evolution* 13, 612–622.

Buerkle, C.A., Morris, R.J., Asmussen, M.A. and Rieseberg, L.H. (2000) The likelihood of homoploid hybrid speciation. *Heredity* 84, 441–451.

Burke, J.M. and Arnold, M.L. (2001) Genetics and the fitness of hybrids. *Annual Review of Genetics* 35, 31–52.

Burke, J.M., Tang, S., Knapp, S.J. and Rieseberg, L.H. (2002) Genetic analysis of sunflower domestication. *Genetics* 161, 1257–1267.

Burnham, C. (1962) *Discussions in Cytogenetics*. Burgress, Minneapolis.

Busbice, T.H. (1968) Effects of inbreeding on fertility in *Medicago sativa* L. *Crop Science* 8, 231–234.

Busbice, T.H. and Wilsie, C.P. (1966) Inbreeding depression and heterosis in autotetraploids with application to *Medicago sativa* L. *Euphytica* 15, 53–67.

Bush, G.L. (1975) Modes of animal speciation. *Annual Review of Ecology and Systematics* 6, 339–364.

Bush, M.B., Piperno, D.R. and Colinvaux, P.A. (1989) A 6,000 year history of Amazonian maize cultivation. *Nature* 340, 303–305.

Butterfass, T. (1979) *Patterns of Chloroplast Reproduction*. Springer-Verlag, New York.

Butzer, K.W. (1995) Biological transfer, agricultural change and environmental implications of 1492. In: Duncan, R.R. (ed.) *International Germplasm Transfer: Past and Present*. Special Publication No. 23, CSSA, Madison, Wisconsin.

Byers, D.S. (ed.) (1967) *The Prehistory of the Tehuacan Valley*, Vol. 1. Andover Foundation, University of Texas Press, Austin.

Caicedo, A.L., Gaitan, E., Duque, M.C., Toro, O., Debouck, D.G. and Tohme, J. (1999) Analyzing *Phaseolus lunatus* L. and related wild species of South America by AFLP fingerprinting. *Crop Science* 39, 1497–1507.

Camargo, L.E.A. and Osborn, T.C. (1996) Mapping loci controlling flowering time in *Brassica oleraceae*. *Theoretical and Applied Genetics* 92, 610–616.

Campbell, B.G. (1982) *Humankind Emerging*, 3rd edn. Little and Brown, Boston.

Campbell, D.R. and Waser, N.M. (2001) Genotype-by-environment interaction and the fitness of plant hybrids in the wild. *Evolution* 55, 669–676.

Carlson, P.S. (1972) Locating genetic loci with aneuploids. *Molecular and General Genetics* 114, 273–280.

Carney, S.E. and Arnold, M.L. (1997) Differences in pollen-tube growth rate and reproductive isolation between Louisiana irises. *Journal of Heredity* 88, 545–549.

Carney, S.E., Cruzen, M.B. and Arnold, M.L. (1994) Reproductive interactions between hybridizing irises – analysis of pollen-tube growth and fertilization success. *American Journal of Botany* 81, 1169–1175.

Carr, D.E. and Dudash, M.R. (1996) Inbreeding depression in two species of *Mimulus* (Scrophulariaceae) with contrasting mating system. *American Journal of Botany* 83, 586–593.

Carr, D.E. and Dudash, M.R. (1997) The effects of five generations of enforced selfing on potential male and female function in *Mimulus guttatus*. *Evolution* 51, 1797–1807.

Carr, G.D. and Kyhos, D.W. (1981) Adaptive radiation in the Hawaiian silversword alliance (Compositae–Madiinae). I. Cytogenetics of spontaneous hybrids. *Evolution* 35, 543–556.

Carson, H.L. (1971) Speciation and the founder principle. *University of Missouri Stadler Genetics Symposium* 3, 51–70.

Carson, H.L. and Templeton, A.R. (1984) Genetic revolutions in relation to speciation phenomena: the founding of new populations. *Annual Review of Ecology and Systematics* 15, 97–131.

Cavalli-Sforza, L.L. and Bodner, W.F. (1971) *The Genetics of Human Populations*. W.H. Freeman, San Francisco.

Cavalli-Sforza, L.L. and Cavalli-Sforza, F. (1995) *The Great Human Diasporas*. Addison-Wesley, Reading, Massachusetts.

Chang, T.T. (1975) Exploration and survey in rice. In: Frankel, O.H. and Hawkes, J.G. (eds) *Crop Genetic Resources for Today and Tomorrow*. Cambridge University Press, Cambridge, pp. 159–165.

Chang, T.T. (1995) Rice, *Oryza sativa* and *Oryza glaberrima* (Gramineae – Oryzeae). In: Smartt, J. and Simmonds, N.W. (eds) *Evolution of Crop Plants*. Longman Scientific and Technical, Harlow, UK.

Charlesworth, D. and Charlesworth, B. (1987) Inbreeding depression and its evolutionary consequences. *Annual Review of Ecology and Systematics* 18, 237–268.

Charlesworth, D. and Charlesworth, B. (1989) Inbreeding depression and its evolutionary consequences. *Annual Review of Systematics and Ecology* 18, 237–268.

Chen, J., Isshiki, S., Tashiro, Y. and Miyazaki, S. (1997) Biochemical affinities between *Cucumis hystrix* Chakr. and two cultivated *Cucumis* species (*C. sativus* L. and *C. melo* L.) based on isozyme analysis. *Euphytica* 97, 139–141.

Childe, V.G. (1952) *New Light on the Most Ancient East*. Routledge and Paul, London.

Choi, I.Y., Lee, J.K., Goo, H.J., Park, J.H., Kim, N.S., Park, C.H. and Chang, K.J. (2002) Genetic diversity and relationships among *Dioscorea alata* L. and related *Dioscorea* species revealed by AFLP analysis. *Korean Journal of Genetics* 24, 305–312.

Chomkos, S.A. and Crawford, G.N. (1978) Plant husbandry in prehistoric eastern North America: new evidence for its development. *American Antiquity* 43, 405–408.

Chooi, W.Y. (1971) Variation in nuclear DNA content in the genus *Vicia*. *Genetics* 68, 195–211.

Choumane, W. and Heizmann, P. (1988) Structure and variability of nuclear ribosomal genes in the genus *Helianthus*. *Theoretical and Applied Genetics* 76, 481–489.

Chyi, Y.S. and Weeden, N.F. (1984) Relative isozyme band intensities permit the identification of the 2n gamete parent for triploid apple cultivars. *HortScience* 19, 258–260.

Clark, G. (1967) *The Stone Age Hunters*. McGraw-Hill, New York.

Clausen, J. (1922) Studies on the collective species *Viola tricolor* L. II. *Botanisk Tidsskrift* 37, 363–416.

Clausen, J. (1926) Genetical and cytological investigations on *Viola tricolor* L. and *V. arvensis* Murr. *Hereditas* 8, 1–156.

Clausen, J. (1951) *Stages in the Evolution of Plant Species*. Cornell University Press, Ithaca, New York.

Clausen, J. and Heisey, W.M. (1958) *Experimental Studies on the Nature of Species. IV. Genetic Structure of Ecological Races*. Publication 615, Carnegie Institute of Washington, Washington, DC.

Clausen, J., Keck, D.D. and Heisey, W.M. (1940) *Experimental Studies on the Nature of Species. I. Effect of Varied Environments on Western North American Plants*. Publication No. 520, Carnegie Institute of Washington, Washington, DC.

Clegg, M., Allard, R. and Kahler, A. (1972) Is the gene the unit of selection? Evidence from two experimental plant populations. *Proceedings of the National Academy of Sciences USA* 69, 2474–2478.

Clegg, M.T. (1990) Molecular diversity in plant populations. In: Brown, A.H.D. (ed.) *Molecular Diversity in Plant Populations*. Sinauer, Sunderland, Massachusetts.

Clegg, M.T. and Brown, A.H.D. (1983) The founding of plant populations. In: Schonewald-Cox, C.M., Chambers, S.M., MacBryde, B. and Thomas, W.L. (eds) *Conservation Genetics*. Benjamin/Cummins, Menlo Park, California.

Clegg, M.T. and Durbin, M.L. (2000) Flower color variation: a model for the experimental study of evolution. *Proceedings of the National Academy of Sciences USA* 97, 7016–7023.

Clegg, M.T., Rawson, R.Y. and Thomas, K. (1984) Chloroplast DNA variation in pearl millet and related species. *Genetics* 106, 449–461.

Cleland, R.E. (1972) *Oenothera: Cytogenetics and Evolution*. Academic Press, New York.

Clutton-Brock, J. (1999) *A Natural History of Domesticated Mammals*. Cambridge University Press, Cambridge.

Coates, D.J., Yen, D.E. and Gaffey, P.M. (1988) Chromosome variation in taro, *Colancasis esculenta*: implications for origin in the Pacific. *Cytologia* 53, 551–560.

Cock, J.H. (1982) Cassava, a basic energy source in the tropics. *Science* 218, 755–762.

Cohen, M.N. (1977) *The Food Crisis in Prehistory: Overpopulation and the Origins of Agriculure*. Yale University Press, New Haven, Connecticut.

Colwell, R. (1951) The use of radioactive isotopes in determining spore distribution patterns. *American Journal of Botany* 38, 511–523.

Colwell, R.K., Norse, E.A., Pimentel, D., Sharples, F.E. and Simberloff, D. (1985) Genetic engineering in agriculture. *Science* 229, 111–112.

Comai, L. (2000) Genetic and epigenetic interactions in allopolyploid plants. *Plant Molecular Biology* 43, 387–399.

Comai, L., Tyagi, A.P., Winter, K., Holmes-Davis, R., Reynolds, S.H., Stevens, Y. and Byers, B. (2000) Phenotypic instability and rapid gene silencing in newly formed *Arabidopsis* allotetraploids. *The Plant Cell* 12, 1551–1567.

Constable, G. (1973) *The Neanderthals*. Time-Life, New York.

Cook, L.M. and Soltis, P.M. (1999) Mating systems of diploid and allotetraploid populations of *Tragapogon* (Asteraceae). I. Natural populations. *Heredity* 82, 237–244.

Cook, L.M. and Soltis, P.S. (2000) Mating systems of diploid and allotetraploid populations of *Tragopogon* (Asteraceae). *Heredity* 84, 410–415.

Coons, G.H. (1975) Interspecific hybrids between *Beta vulgaris* and the wild species of *Beta*. *Journal of the American Society of Sugar Beet Technology* 18, 281–386.

Coulibaly, S., Pasquet, R.S., Papa, R. and Gepts, P. (2002) AFLP analysis of the phenetic organization and genetic diversity of *Vigna unguiculata* L. Walp. reveals extensive gene flow between wild and domesticated types. *Theoretical and Applied Genetics* 104, 358–366.

Coursey, D.G. (1967) *Yams*. Longman Group UK, London.

Coursey, D.G. (1972) The civilization of the yam: interrelationships of man and yams in Africa and the Indo-Pacific region. *Archaeology and Physical Anthropology of Oceana* 7, 215–233.

Coursey, D.G. (1975) The origin and domestication of yams in Africa. In: Harlan, J.R. (ed.) *Origins of African Plant Domestication*. Mouton, The Hague.

Coursey, D.G. (1979) Yams. In: Simmonds, N.E. (ed.) *Evolution of Crop Plants*. Longmans, London, pp. 70–74.

Coyne, J.A., Barton, N.H. and Turelli, M. (1997) Perspective: a critique of Wright's shifting balance theory of evolution. *Evolution* 51, 643–671.

Coyne, J.A., Barton, N.H. and Turelli, M. (2000) Is Wright's shifting balance process important in nature? *Evolution* 54, 306–317.

Crawford, D.J. (1990) *Plant Molecular Systematics: Macromolecular Approaches*. John Wiley & Sons, New York.

Crawford, D.J. and Ornduff, R. (1989) Enzyme electrophoresis and evolutionary relationships among 3 species of *Lasthenia* (Asteraceae, Heliantheae). *American Journal of Botany* 76, 289–296.

Crawford, G.W. (1992) Prehistoric plant domestication in East Asia. In: Cowan, C.W. and Watson, P.J. (eds) *The Origins of Agriculture*. Smithsonian Institution Press, Washington and London, pp. 7–38.

Cronn, R., Brothers, M., Klier, K., Bretting, P.K. and Wendel, J.F. (1997) Allozyme variation in domesticated annual sunflower and its wild relatives. *Theoretical and Applied Genetics* 95, 532–545.

Crouch, H.K., Crouch, J.H., Madsen, S., Vuylstake, D.R. and Ortiz, R. (2000) Comparative analysis of phenotypic and genotypic diversity among plantain landraces (*Musa* spp., AAB group). *Theoretical and Applied Genetics* 101, 1056–1065.

Crow, J.F. (1945) A chart of the χ^2 and t distributions. *Journal of the American Statistical Association* 40, 376.

Crow, J.F. and Kimura, M. (1970) *An Introduction to Population Genetics Theory*. Harper and Row, New York.

Crowe, L.K. (1964) The evolution of outbreeding in plants I. The angiosperms. *Heredity* 19, 435–457.

Crowe, L.K. (1971) The polygenic control of outbreeding in *Borago officinalis*. *Heredity* 27, 111–118.

Cubero, J.I. (1974) On the origin of *Vicia faba*. *Theoretical and Applied Genetics* 45, 45–47.

Cullis, C.A. (1987) The generation of somatic and heritable variation in response to stress. *American Naturalist* 130, 562–573.

Dale, P.J. (1992) Spread of engineered genes to wild relatives. *Plant Physiology* 100, 13–15.

D'Andrea, A.C., Klee, M. and Casey, J. (2001) Archaeobotanical evidence for pearl millet (*Pennisetum glaucum*) in sub-Saharan Africa. *Antiquity* 75, 341–348.

Daniels, J. and Roach, B.T. (1987) Taxonomy and evolution. In: Heinz, D.J. (ed.) *Sugarcane Improvement through Breeding*. Elsevier Press, Amsterdam.

Daniels, J., Roach, B.T., Daniels, C. and Paton, N.H. (1991) The taxonomic status of *Saccharum barberi* Jeswiet and *S. sinense* Roxb. *Sugarcane* 3, 11–16.

Dark, S.O.S. (1971) Experiments in the cross pollination of sugar beet in the field. *National Institute of Agricultural Biology* 12, 242–266.

Darlington, C.D. and Mather, K. (1949) *The Elements of Genetics*. G. Allen, London.

Darrow, G.M. (1966) *The Strawberry, History, Breeding and Physiology*. Holt, Rinehart and Winston, New York.

Da Silva, J.A.G., Sorrels, M.E., Burnquist, W.L. and Tanksley, S.D. (1993) RFLP linkage map and genome analysis of *Saccharum spontaneum*. *Genome* 36, 782–791.

Davies, D.R. (1995) Peas: *Pisum sativum* (Leguminosae – Papilionoideae). In: Smartt, J. and Simmonds, N.W. (eds) *Evolution of Crop Plants*. Longman Scientific and Technical, Harlow, UK, pp. 294–296.

Dean, C. and Leech, R.M. (1982) Genome expression during normal leaf development. *Plant Physiology* 70, 1605–1608.

Dearing, J.A. and Guarino, L. (1995) Bibliographic databases for plant germplasm collectors. In: Guarino, L., Rao, V.R. and Reid, R. (eds) *Collecting Plant Genetic Diversity*. CAB International, Wallingford, UK.

Debener, T., Salamini, F. and Gebhardt, C. (1990) Phylogeny of wild and cultivated *Solanum* species based on nuclear restriction fragment length polymorphisms (RFLPs). *Theoretical and Applied Genetics* 79, 360–368.

Debouck, D.G. (1991) Systematics and morphology. In: van Schoonhoven, A. and Voysest, O. (eds) *Common beans: research for crop improvement*. CAB International, Wallingford, UK, pp. 55–118.

Debouck, D.G. (2000) Genetic resources of *Phaseolus* beans: patterns in time, space and people. In: Fueyo, M.A., Gonzalez, A.J., Ferreira, J.J. and Giraldez, R. (eds) *II. Seminario de Judía de la Península Ibérica. Actas de la Asociación Espanola de Leguminosas*. Villaviciosa, Asturias, pp. 17–39.

Debouck, D.G. and Smartt, J. (1995) Beans *Phaseolus* spp. (Leguminosae–Papilionideae). In: Smartt, J. and Simmonds, N.W. (eds) *Evolution of Crop Plants*. Longman Scientific and Technical, Harlow, pp. 287–294.

Decker, D.S. (1988) Origin(s), evolution and systematics of *Cucurbita pepo* (Cucurbitaceae). *Economic Botany* 42, 4–15.

Decker-Walters, D.S. (1990) Evidence for multiple domestications of *Curcurbita pepo*. In: Bates, D.M. (ed.) *Biology and Utilization of the Cucurbitaceae*. Cornell University Press, Ithaca, New York.

Decker-Walters, D., Staub, J., López-Sesé, A. and Nakata, E. (2001) Diversity in landraces and cultivars of bottle gourd (*Lagenaria siceraria*; Cucurbitaceae) as assessed by random amplified polymorphic DNA. *Genetic Resources and Crop Evolution* 48, 369–380.

Deerr, N. (1949) *The History of Sugar*. Wiley, New York.

Delgado Salinas, A., Turley, T., Richman, A. and Lavin, M. (1999) Phylogenetic analysis of the cultivated and wild species of *Phaseolus* (Fabaceae). *Systematic Botany* 24, 438–460.

de Nettancourt, D. (1977) *Incompatibility in Angiosperms*. Springer Verlag, Berlin.

de Nettancourt, D. (2001) *Incompatibility and Incongruity in Wild and Cultivated Plants*. Springer-Verlag, Berlin.

den Nijs, A.P.M. and Visser, D.L. (1985) Relationships between African species of the genus *Cucumis* L. estimated by the production, vigor and fertility of F_1 hybrids. *Euphytica* 34, 279–290.

de Pamphilis, C.W. and Palmer, J.D. (1990) Loss of photosynthetic and chlororespiratory genes from the plastid genome of a parasitic flowering plant. *Nature* 348, 337–339.

Deu, M., DeLeon, D.G., Glaszmann, J.C., DeGremont, I., Chantereau, J., Lanaud, C. and Hamon, P. (1994) RFLP diversity in cultivated *Sorghum* in relation to racial differentiation. *Theoretical and Applied Genetics* 88, 838–844.

Devlin, B. and Ellstrand, N.C. (1990) The development and application of a refined method for estimating gene flow from an angiosperm paternity analysis. *Evolution* 44, 248–258.

Devos, K.M. and Gale, M.D. (2000) Genome relationships: the grass model in current research. *Plant Cell* 12, 637–646.

Devos, K.M., Chinoy, M.D., Liu, C.J. and Gale, M.D. (1992a) RFLP-based genetic map of the homologous group 3 chromosomes of wheat and rye. *Theoretical and Applied Genetics* 83, 931–939.

Devos, K.M., Atkinson, M.D., Chinoy, C.N., Liu, C.J. and Gale, M.D. (1992b) Chromosomal rearrangements in rye genome relative to that of wheat. *Theoretical and Applied Genetics* 85, 673–680.

deWet, J.M.J. (1968) Diploid–tetraploid–haploid cycles and the origin of variability in *Dichanthium* agamospecies. *Evolution* 22, 394–397.

deWet, J.M.J. (1978) Systematics and evolution of *Sorghum* section *Sorghum* (Gramineae). *American Journal of Botany* 65, 477–484.

deWet, J.M.J. (1995) Minor cereals: various genera (Gramineae). In: Smartt, J. and Simmonds, N.W. (eds) *Evolution of Crop Plants*. Longman Scientific & Technical, Harlow, UK, pp. 202–207.

deWet, J.M.J. and Huckabay, J.P. (1967) The origin of *Sorghum bicolor*. II. Distribution and domestication. *Evolution* 21, 787–802.

deWet, J.M.J., Oestry-Stidd, L.L. and Cubero, J.I. (1979) Origins and evolution of foxtail millets *Setaria italica*. *Journal of Agriculture and Traditional Botanical Applications* 26, 53–64.

deWet, J.M.J., Newell, C.A. and Brink, D.E. (1984a) Counterfeit hybrids between *Tripsicum* and *Zea* (*Gramineae*). *American Journal of Botany* 71, 245–251.

deWet, J.M.J., Rao, K.E.P., Brink, D.E. and Mengesha, M.H. (1984b) Systematics and evolution of *Eleusine coracana* (Gramineae). *American Journal of Botany* 71, 550–557.

Dewey, D.R. (1966) Inbreeding depression in diploid, tetraploid and hexaploid crested wheatgrass. *Crop Science* 6, 144–147.

Dewey, D. (1980) Some applications and misapplications of induced polyploidy in plant breeding. In: Lewis, W.D. (ed.) *Polyploidy: Biological Relevance*. Plenum Press, New York, pp. 445–476.

Diamond, J. (1998) *Guns, Germs, and Steel: the Fates of Human Societies*. W.W. Norton, New York.

Diaz, J., Schmiediche, P. and Austin, D.F. (1996) Polygon of crossibility between eleven species of *Ipomoea*: Section Batatas (Convolvulaceae). *Euphytica* 88, 189–200.

Dilcher, D. (2000) Toward a new synthesis: major evolutionary trends in the angiosperm fossil record. *Proceedings of the National Academy of Sciences USA* 97, 7030–7036.

Dillehay, T.D. (2000) *The Settlement of the Americas*. Basic Books, New York.

Djé, Y., Heuertz, M., Lefebvre, C. and Vekemans, X. (2000) Assessment of genetic diversity within and among germplasm accessions in cultivated sorghum using microsatellite markers *Theoretical and Applied Genetics* 100, 918–925.

Dobzhansky, T. (1970) *Genetics of the Evolutionary Process*. Columbia University Press, New York.

Dobzhansky, T. and Pavlovsky, O. (1953) Indeterminate outcome of certain experiments on *Drosophila* populations. *Evolution* 7, 198–210.

Dobzhansky, T., Ayala, F.J., Stebbins, G.L. and Valentine, J.W. (1977) *Evolution*. W.H. Freeman and Company, San Francisco.

Dodds, K.S. (1965) The history and relationships of cultivated potatoes. In: Hutchinson, B. (ed.) *Essays in Crop Evolution*. Longman, London.

Doebley, J. (1989) Isozyme evidence and the evolution of crop plants. In: Soltis, D.E. and Soltis, P.S. (eds) *Isozymes in Plant Biology*. Dioscorides Press, Portland, Oregon, pp. 165–191.

Doebley, J. (1990a) Molecular evidence for gene flow among *Zea* species. *Bioscience* 40, 443–448.

Doebley, J. (1990b) Molecular evidence and the evolution of maize. *Economic Botany* 44, 6–27.

Doebley, J. and Stec, A. (1993) Inheritance of the morphological differences between maize and teosinte: comparison of results for two F_2 populations. *Genetics* 134, 559–570.

Doebley, J., Goodman, M.M. and Stuber, C.W. (1984) Isozyme variation in maize from the southwestern United States: taxonomic and anthropological implications. *Maydica* 28, 97–120.

Doebley, J., Goodman, M.M. and Stuber, C.W. (1985) Isozyme variation in the races of maize from Mexico. *American Journal of Botany* 72, 629–639.

Doebley, J., Goodman, M.M. and Stuber, C.W. (1986) Exceptional genetic divergence of northern flint corn. *American Journal of Botany* 73, 64–69.

Doebley, J., Goodman, M.M. and Stuber, C.W. (1987) Patterns of isozyme variation between maize and Mexican annual teosinte. *Economic Botany* 41, 234–246.

Doebley, J., Stec, A., Wendel, J. and Edwards, M. (1990) Genetic and morphological analysis of a maize–teosinte F_2 population: implications for the origin of maize. *Proceedings of the National Academy of Sciences USA* 87, 9888–9892.

Doebley, J., Stec, A. and Gustus, C. (1995) *Teosinte branched1* and the origin of maize: evidence for epistasis and the evolution of dominance. *Genetics* 141, 333–346.

Doganlar, S., Frary, A., Daunay, M.-C., Lester, R.N. and Tanksley, S.D. (2002) Conservation of gene function in the *Solanaceae* as revealed by comparative mapping of domestication traits in eggplant. *Genetics* 161, 1713–1726.

Doggett, H. (1988) *Sorghum*. Longman, London.

Doggett, H. and Rao, K.E.P. (1995) Sorghum, *Sorghum bicolor* (Gramineae–Androponeae). In: Smartt, J. and Simmonds, N.W. (eds) *Evolution of Crop Plants*. Longman Scientific and Technical, Harlow, UK, pp. 173–180.

Doolittle, W.F. and Sapienza, C. (1980) Selfish genes, the phenotype paradigm and genome evolution. *Nature* 284, 601–603.

Dooner, H.K. and Martinez-Ferez, I.M. (1997) Recombination occurs uniformly within the *bronze* gene, a meiotic hotspot in the maize genome. *Plant Cell* 9, 1633–1646.

Dorado, O., Rieseberg, L.H. and Arias, D.M. (1992) Chloroplast DNA introgression in southern California sunflowers. *Evolution* 46, 566–572.

Dorweiler, J.E. and Doebley, J. (1997) Developmental analysis of *teosinte glume architecture 1*: a key locus in the evolution of maize (Poaceae). *American Journal of Botany* 84, 1313–1322.

Dorweiler, J., Stec, A., Kermicle, J. and Doebley, J. (1993) *Teosinte glume architecture 1*: a genetic locus controlling a key step in maize evolution. *Science* 262, 233–235.

Douches, D.S. and Quiros, C.F. (1988a) Genetic strategies to determine the mode of 2n gamete production in diploid potatoes. *Euphytica* 38, 247–260.

Douches, D.S. and Quiros, C.F. (1988b) Additional isozyme loci in tuber-bearing solanums: inheritance and linkage relationships. *Journal of Heredity* 60, 183–191.

Dowrick, V.P.J. (1956) Heterostyly and homostyly in *Primula obconica*. *Heredity* 10, 219–236.

Doyle, J. (1985) *Altered Harvest: Agriculture, Genetics and the Fate of the World's Food Supply*. Viking Press, New York.

Doyle, J.J. and Beachy, R.N. (1990) Ribosomal gene variation in soybean (*Glycine*) and its relatives. *Theoretical and Applied Genetics* 70, 369–376.

Doyle, J.J., Doyle, J.L. and Brown, A.H.D. (1990a). A chloroplast-DNA phylogeny of the wild perennial relatives of soybean (*Glycine* subgenus *Glycine*): congruence with morphological and crossing groups. *Evolution* 44, 371–389.

Doyle, J.J., Doyle, J.L., Brown, A.H.D. and Grace, J.P. (1990b). Multiple origins of polyploids in the *Glycine tabacina* complex inferred from chloroplast DNA polymorphism. *Proceedings of the National Academy of Sciences USA* 87, 714–717.

Doyle, J.J., Doyle, J.L. and Brown, A.H.D. (1999) Origins, colonization, and linkage recombination in a widespread perennial soybean polyploid complex. *Proceedings of the National Academy of Sciences USA* 96, 10741–10745.

Doyle, J.J., Doyle, J.L., Brown, A.H.D. and Palmer, R.G. (2002) Genomes, multiple origins, and linkage recombination in the *Glycine tomentella* (Leguminosae) polyploid complex: histone H3-D sequences. *Evolution* 56, 1388–1402.

Dudash, M.R., Carr, D.E. and Fenster, C.B. (1997) Five generations of enforced selfing and outcrossing in *Mimulus guttatus*: inbreeding depression variation at the population and family level. *Evolution* 51, 54–65.

Dunbier, M.W. and Bingham, E.T. (1975) Maximum heterozygosity in alfalfa: results using haploid-derived autotetraploids. *Crop Science* 15, 527–531.

Durbin, M.L., Learn, G.H., Huttey, G.A. and Clegg, M.L. (1995) Evolution of the chalcone synthase gene family in the genus *Ipomoea*. *Proceedings of the National Academy of Sciences USA* 92, 3338–3342.

Durbin, M.L., McCaig, B. and Clegg, M.T. (2000) Molecular evolution of chalcone synthase multigene family in the morning glory family. *Plant Molecular Biology* 42, 79–92.

Durbin, M.L., Denton, A.L. and Clegg, M.T. (2001) Dynamics of mobile element activity in chalcone synthase loci in the common morning glory (*Ipomoea purpurea*). *Proceedings of the National Academy of Sciences USA* 98, 5084–5089.

Dvorak, J. and Zhang, H.B. (1990) Variation in repeated nucleotide sequences sheds light on the origin of the wheat B and G genomes. *Proceedings of the National Academy of Sciences USA* 87, 9640–9644.

Dvorak, J., Luo, M.-C., Yang, Z.-L. and Zhang, H.-B. (1998) The structure of the *Aegilops tauschii* genepool and the evolution of hexaploid wheat. *Theoretical and Applied Genetics* 97, 657–670.

East, E.M. (1916) Studies on size inheritance in *Nicotiana*. *Genetics* 1, 164–176.

Eckardt, N.A. (2001) A sense of self: the role of DNA sequence elimination in allopolyploidization. *The Plant Cell* 13, 1699–1704.

Edwards, A.W.F. and Cavalli-Sforza, L.L. (1964) Reconstruction of evolutionary trees. In: Heywood, V.E. and McNeill, J. (eds) *Phenetic and Phylogenetic Classification*. Systematics Association, London.

Ehrendorfer, F. (1979) Polyploidy and distribution. In: Lewis, W.D. (ed.) *Polyploidy : Biological Relevance*. Plenum Press, New York, pp. 45–60.

Ehrlich, P.R. and Raven, P.H. (1969) Differentiation of populations. *Science* 165, 1228–1232.

Eldridge, N. and Gould, S.J. (1972) Punctuated equilibria: an alternative to phyletic gradualism. In: Schopf, T.J.M. (ed.) *Models in Paleobiology*. Freeman, Cooper, San Francisco, pp. 82–115.

Ellstrand, N.C. (1988) Pollen as a vehicle for escape of engineered genes? In: Hodgson, J. and Sugden, A.M. (eds) *Planned Release of Genetically Engineered Organisms*, Biotechnology/Trends in Ecology and Evolution Special Publication, Elsevier Publications, Cambridge, UK, pp. 30–32.

Ellstrand, N.C. (2001) When transgenes wander, should we worry? *Plant Physiology* 125, 1543–1545.

Ellstrand, N.C. and Hoffman, C.A. (1990) Hybridization as an avenue of escape for engineered genes. *BioScience* 40, 438–442.

Ellstrand, N.C. and Marshall, P.L. (1985) Interpopulational gene flow by pollen in wild radish, *Raphanus sativus*. *American Naturalist* 126, 606–616.

Ellstrand, N.C. and Schierenbeck, K.A. (2000) Hybridization as a stimulus for the evolution of invasiveness in plants? *Proceedings of the National Academy of Sciences USA* 97, 7043–7050.

Ellstrand, N.C., Whitkus, R. and Rieseberg, L.H. (1996) Distribution of spontaneous plant hybrids. *Proceedings of the National Academy of Sciences USA* 93, 5090–5093.

Ellstrand, N.C., Prentice, H.C. and Hancock, J.F. (1999) Gene flow and introgression from domesticated plants into their wild relatives. *Annual Review of Ecology and Systematics* 30, 539–563.

Emms, S.K. and Arnold, M.L. (1997) The effect of habitat on parental and hybrid fitness: transplant experiments with Louisiana irises. *Evolution* 51, 1112–1119.

Emms, S.K. and Arnold, M.L. (2000) Site-to-site differences in pollinator visitation patterns in a Louisiana iris hybrid zone. *Oikos* 91, 568–578.

Emory, K.P. and Sinoto, Y.H. (1964) Préhistoire de la polynésie. *J. Soc. Océanistes* 20, 39–41.

Endrizzi, J.E., Turcotte, E.L. and Kohel, R.J. (1985) Genetics, cytology and evolution of *Gossypium*. *Advances in Genetics* 23, 271–375.

Engles, J.M.M., Ramanatha, V., Brown, A.H.D. and Jackson, M.T. (eds) (2002) *Managing Plant Genetic Diversity*. CAB International, Wallingford, UK.

Engle, L.M., Chang, T.T. and Ramirez, D.A. (1969) The cytogenetics of sterility in F_1 hybrids and *indica* × *javanica* varieties of rice (*Oryza sativa*). *Philippinian Agriculture* 53, 289–307.

Eshbaugh, W.H. (1975) Genetic and biochemical systematics of chili peppers (*Capsicum*-Solanaceae). *Bulletin of the Torrey Botanical Club* 102, 396–403.

Eshbaugh, W.H. (1980) The taxonomy of the genus *Capsicum* (Solanaceae) – 1980. *Phytologia* 47, 153–166.

Eshbaugh, W.H., Guttman, S.I. and McLeod, M.J. (1983) The origin and evolution of domesticated *Capsicum* species. *Journal of Ethnobiology* 3, 49–54.

Eubanks, M.W. (2001a) The mysterious origin of maize. *Economic Botany* 55, 492–514.

Eubanks, M.W. (2001b) An interdisciplinary perspective on the origin of maize. *Latin American Antiquity* 12, 91–98.

Evans, G.M. (1995) Rye, *Secale cereale* (Gramineae–Tricinae). In: Smartt, J. and Simmonds, N.W. (eds) *Evolution of Crop Plants*. Longman Scientific & Technical, Harlow, UK, pp. 166–170.

Eyre-Walker, A., Gaut, R.G., Hilton, H., Feldman, D.L. and Gaut, B.S. (1998) Investigation of the bottleneck leading to the domestication of maize. *Proceedings of the National Academy of Sciences USA* 95, 4441–4446.

Fabijanski, S., Fedak, G., Armstrong, K. and Altosaar, I. (1990) A repeated sequence probe for the C genome in *Avena* (oats). *Theoretical and Applied Genetics* 79, 1–7.

Faegri, K.L. and van der Pijl, V. (1979) *The Principles of Pollination Ecology*. Pergamon Press, Oxford.

Falconer, D.S. (1981) *Introduction to Quantitative Genetics*, 2nd edn. Longman, London.

Falconer, D.S. and Mackay, T.F.C. (1996) *Introduction to Quantitative Genetics*. Burgess Press, New York.

Fang, D., Krueger, R.R. and Roose, M.L. (1998) Phylogenetic relationships among selected *Citrus* germplasm accessions revealed by inter-simple sequence repeat (ISSR) markers. *Journal of the American Society for Horticultural Science* 123, 612–617.

Federici, C.T., Fang, D.Q., Scora, R.W. and Roose, M.L. (1998) Phylogentic relationships within the genus *Citrus* (Rutaceae) and related genera as revealed by RFLP and RAPD analysis. *Theoretical and Applied Genetics* 96, 812–822.

Fedoroff, N. (2000) Transposons and genome evolution in plants. *Proceedings of the National Academy of Sciences USA* 97, 7002–7007.

Fehr, W.R. (1987) *Principles of Cultivar Development*, Vol. I. *Theory and Technique*. Macmillan, New York.

Feldman, M. (1963) Evolutionary studies in the *Aegilops-Triticum* group with special emphasis on causes of variability in the polyploid species of section Pleinoathera. PhD Thesis, Hebrew University, Jerusalem.

Feldman, M., Galili, G. and Levy, A.A. (1986) Genetic and evolutionary aspects of allopolyploidy in wheat. In: Barigozzi, C. (ed.) *The Origin and Domestication of Cultivated Crops*. Elsevier, Amsterdam, pp. 83–101.

Feldman, M., Liu, B., Segal, G., Addo, S., Levy, A.A. and Vega, J.M. (1997) Rapid elimination of low-copy DNA sequences in polyploid wheat: a possible mechanism for differentiation of homologous chromosomes. *Genetics* 147, 1381–1387.

Fisher, R.A. (1930) *The Genetical Theory of Natural Selection*. Clarendon, Oxford.

Flannery, K.V. (1968) Archeological systems theory and early Mesoamerica. In: Meggers, B.J. (ed.) *Anthropological Archeology in the Americas*. Anthropological Society of Washington, Washington, DC.

Flor, H.H. (1954) *Identification of Races of Flax Rust by Lines with Single Rust-conditioning Genes*. Technical Bulletin No. 1087, US Department of Agriculture, Washington, DC.

Fominaya, A., Vega, C. and Ferrer, E. (1988) Giemsa C-banded karotypes of *Avena* species. *Genome* 30, 627–632.

Ford, E.B. (1975) *Ecological Genetics: The Evolution of Super-genes*. Chapman & Hall, London.

Ford, R.E. (ed.) (1985) *Prehistoric Food Production in North America*. Museum of Anthropology, University of Michigan, Ann Arbor.

Ford-Lloyd, B.V. and Williams, J.T. (1975) A revision of *Beta* section *Vulgaris* (Chenopodiaceae) with new light on the origin of cultivated beets. *Botanical Journal of the Linnean Society* 71, 89–102.

Forsline, P.L., Dickson, E.E. and Djangalieu, A.D. (1994) Collection of wild *Malus*, *Vitus* and other fruit species genetic resources in Kazakstan and neighboring republics. *HortScience* 29, 433.

Forsline, P.L., Towill, L.E., Waddell, J.W., Sushnoff, C., Lamboy, W.F. and McFerson, J.R. (1998) Recovery and longevity of cryopreserved dormant apple buds. *Journal of the American Society for Horticultural Science* 123, 365–370.

Fowler, N.L. and Levin, D.A. (1984) Ecological constraints on the establishment of a novel polyploid in competition with its diploid progenitor. *American Naturalist* 124, 703–711.

Frankel, O. and Munday, A. (1962) The evolution of wheat. In: *The Evolution of Living Organisms*. Symposium of the Royal Society, Victoria, Australia.

Frary, A., Nesbitt, T.C., Frary, A., Grandillo, S., van der Knapp, E., Cong, B., Liu, J.P., Meller, J., Elber, R., Alpert, K.B. and Tanksley, S.D. (2000). fw2.2: a quantitative trait locus key to the evolution of tomato fruit size. *Science* 289, 85–88.

Freeling, M. (1973) Simultaneous induction by anaerobiosis or 2,4-D of multiple enzymes specified by two unlinked genes: differential *Adh-1–Adh-2* expression in maize. *Molecular and General Genetics* 127, 215–227.

Fregene, M., Angel, F., Gomez, R., Rodriquez, F., Chavarriaga, P., Roca, W., Tohme, J. and Bonierbale, M. (1997) A molecular genetic map of cassava (*Manihot esculenta* Cranz). *Theoretical and Applied Genetics* 95, 431–441.

Fregene, M.A., Vargas, J., Ikae, J., Angel, F., Tohme, J., Asiedu, R.A., Akoroda, M.O. and Roca, W.M. (1994) Variability of chloroplast DNA and nuclear ribosomal DNA in cassava (*Manihot esculenta* Crantz) and its wild relatives. *Theoretical and Applied Genetics* 89, 719–727.

Friebe, B., Badaeva, E.D., Kammer, K. and Gill, B.S. (1996) Standard karyotypes of *Aegilops uniaristata, Ae. mutica, Ae. comosa* subspecies *comosa* and *heldreichii*. *Plant Systematics and Evolution* 202, 199–210.

Friesen, L.F., Jones, T.L., Acker, R.C.V. and Morrison, I.N. (2000) Identification of *Avena fatua* populations resistant to imazamethabenz, flamprop and fenoxaprop-P. *Weed Science* 48, 532–540.

Fryxell, P.A. (1957) Mode of reproduction of higher plants. *Botanical Review* 23, 135–233.

Fryxell, P.A., Craven, L.A. and Stewert, J.M. (1992) A revision of *Gossypium* section *Grandicalyx* (Malvaceae) including the description of six new species. *Systematic Botany* 17, 91–114.

Futuyma, D.J. (1979) *Evolutionary Biology*. Sinauer Associates, Sunderland, Massachusetts.

Galinat, W.C. (1973) Intergenomic mapping of maize, teosinte and *Tripsicum*. *Evolution* 27, 644–655.

Gallez, G.P. and Gottleib, L.D. (1982) Genetic evidence for the hybrid origin of the diploid plant *Stephanomeria diegensis*. *Evolution* 36, 1158–1167.

Gaut, B.S. and Doebley, J.F. (1997) DNA sequence evidence for the segmental allopolyploid origin of maize. *Proceedings of the National Academy of Sciences USA* 94, 6809–6814.

Ge, S., Sang, T., Lu, B.-R. and Hong, D.-Y. (1999) Phylogeny of rice genomes with emphasis on origins of allopolyploid species. *Proceedings of the National Academy of Sciences USA* 96, 14400–14405.

Ge, S., Sang, T., Lu, B.-R. and Hong, D.-Y. (2001) Rapid and reliable identification of rice genomes by RFLP analysis of PCR-amplified Adh genes. *Genome* 44, 1136–1142.

Gepts, P. (1988) A middle American and an Andean common gene pool. In: Gepts, P. (ed.) *Genetic Resources of Phaseolus Beans*. Kluwer Academic, Dordrecht.

Gepts, P. (1990) Biochemical evidence bearing on the domestication of *Phaseolus* (Fabaceae) beans. *Economic Botany* 44, 28–38.

Gepts, P. (1998) Origin and evolution of common bean: past events and recent trends. *HortScience* 33, 1124–1130.

Gepts, P. and Bliss, F.A. (1985) F_1 weakness in the common bean. *Journal of Heredity* 76, 447–450.

Gibbons, A. (2001a) The riddle of coexistence. *Science* 291, 1725–1729.

Gibbons, A. (2001b) The peopling of the Pacific. *Science* 291, 1735–1737.

Gibson, G. and Wagner, G. (2000) Canalization in evolutionary genetics: a stabilizing theory? *BioEssays* 22, 372–380.

Gilbert, N.E. (1967) Additive combining abilities fitted to plant breeding data. *Biometrics* 23, 45–49.

Glaszmann, J.C., Benoit, H. and Arnaud, M. (1984) Classification des riz cultivés (*Oryza sativa* L.): utilisation de la variabilité isoenzymatique. *L'Agronomie Tropicale* 39, 51–66.

Golenberg, E.M. (1989) Estimation of gene flow and genetic neighborhood size by indirect methods in a selfing annual *Triticum dicoccoides*. *Evolution* 41, 1326–1334.

Golson, J. (1984) New Guinea agricultural history: a case study. In: Denoon, D. and Snowden, C. (eds) *A Time to Plant and a Time to Uproot: a History of Agriculture in Papua New Guinea*, Institute of Papua New Guinea Studies, Pappa, pp. 55–64.

Gomes, J., Jadric, S., Winterhatter, M. and Brkic, S. (1982) Alcohol dehydrogenase isoenzymes in chickpea cotyledons. *Phytochemistry* 21, 1219–1224.

Goodman, M.M. (1990) Genetic and germ plasm stocks worth saving. *Journal of Heredity* 81, 11–16.

Goodnight, C.J. and Wade, M.J. (2000) The ongoing synthesis: a reply to Coyne, Barton and Turelli. *Evolution* 54, 317–324.

Goodrich, C.E. (1863) The potato. Its diseases – with incidental remarks on its sorts and culture. *Transactions of the New York State Agricultural Society* 23, 103–134.

Goodwin, T.W. and Mercer, E.I. (1972) *Introduction to Plant Biochemistry*. Pergamon Press, London.

Gordon, H. and Gordon, M. (1957) Maintenance of polymorphism by potentially injurious genes in eight natural populations of the platyfish, *Xiphophorous maculatus*. *Journal of Genetics* 5, 44–51.

Gorman, C. (1969) Hoabinhian: a pebble-tool complex with early plant associations in Southeast Asia. *Science* 163, 671–673.

Gornall, R.J. (1983) Recombination systems and plant domestication. *Biological Journal of the Linnean Society* 20, 375–383.

Gottleib, L.B. (1974) Genetic stability in a peripheral isolate of *Stephanomeria exigua* ssp. *coronaria* that fluctuates in population size. *Genetics* 76, 551–556.

Gottleib, L.B. (1977a) Evidence for duplication and divergence of the structural gene for PGI in diploid species of *Clarkia*. *Genetics* 86, 289–307.

Gottleib, L.B. (1977b) Phenotypic variation in *Stephanomeria exigua* ssp. *coronaria* (Compositae) and its recent derivative species 'malheurensis'. *American Journal of Botany* 64, 873–880.

Gottleib, L.D. (1981) Electrophoretic evidence and plant populations. In: Reinhold, L. and Harborne, J.B. (eds) *Progress in Phytochemistry*. Pergamon Press, Oxford, pp. 1–46.

Gottleib, L.D. (1982) Conservation and duplication of isozymes in plants. *Science* 216, 373–380.

Gottleib, L.D. (1984) Genetics and morphological evolution in plants. *American Naturalist* 123, 681–709.

Gottleib, L.D. and Ford, V.S. (1997) A recently silenced, duplicate *PgiC* locus in *Clarkia*. *Molecular Biology and Evolution* 14, 125–132.

Gottleib, L.D. and Greve, L.C. (1981) Biochemical properties of duplicated isozymes of phosphoglycose isomerase in the plant *Clarkia xantiana*. *Biochemical Genetics* 19, 155–172.

Gottleib, L.D. and Higgins, R.C. (1984) Phosphoglucose isomerase expression in species of *Clarkia* with and without a duplication of the coding gene. *Genetics* 107, 131–140.

Gould, S.J. and Eldridge, N. (1977) Punctuated equilibria: the tempo and mode of evolution reconsidered. *Paleobiology* 3, 115–151.

Grandillo, S. and Tanksley, S.D. (1996) QTL analysis of horticultural traits differentiating the cultivated tomato from the closely related species *Lycopersicon pimpinellifolium*. *Theoretical and Applied Genetics* 92, 935–951.

Grandillo, S., Ku, H.M. and Tanksley, S.D. (1999) Identifying the loci responsible for natural variation in fruit size and shape in tomato. *Theoretical and Applied Genetics* 99, 978–987.

Grant, V. (1952) Genetic and taxonomic studies in *Gilia*. II. *Gilia capitate abrotanifolia*. *Aliso* 2, 363–373.

Grant, V. (1958) The regulation of recombination in plants. *Cold Spring Harbor Symposium of Quantitative Biology* 23, 337–363.

Grant, V. (1963) *The Origin of Adaptations*. Columbia University Press, New York.

Grant, V. (1966) Linkage between viability and fertility in a species cross in *Gilia*. *Genetics* 54, 867–880.

Grant, V. (1971) *Plant Speciation*. Colombia University Press, New York.

Grant, V. (1975) *The Genetics of Flowering Plants*. Columbia University Press, New York.

Grant, V. (1977) *Organismic Evolution*. W.H. Freeman, San Francisco.

Grant, V. (1981) *Plant Speciation*, 2nd edn. Columbia University Press, New York.

Grant, V. (1985) *The Evolutionary Process: a Critical Review of Evolutionary History*. Columbia University Press, New York.

Grant, V. and Grant, A. (1954) Genetic and taxonomic studies in *Gilia*. VII. The woodland *Gilias*. *Aliso* 3, 59–91.

Grant, V. and Grant, A. (1960) Genetic and taxonomic studies in *Gilia*. XI. Fertility relationships of the diploid Cobwebby *Gilias*. *Aliso* 3, 203–287.

Gray, A.R. (1982) Taxonomy and evolution of broccoli (*Brasssica oleracea* var. *italica*). *Economic Botany* 36, 397–410.

Green, R., Vardi, A. and Galun, E. (1986) Chloroplast DNA in citrus. The plastome of *Citrus*. Physical map, variation among *Citrus* cultivars and species and comparison with related genera. *Theoretical and Applied Genetics* 72, 170–177.

Gregory, W.C., Krapovickas, A. and Gregory, M.P. (1980) Structure, variation, evolution and classification in *Arachis*. In: Summerfield, R.J. and Bunting, A.H. (eds) *Advances in Legume Science*. Royal Botanical Gardens, Kew, pp. 469–481.

Griffing, B. (1956) A generalized treatment of the use of diallel crosses in quantitative inheritance. *Heredity* 10, 31–50.

Grime, J.P. and Hunt, R. (1975) Relative growth rate: its range and adaptive significance in a local flora. *Journal of Ecology* 63, 393–422.

Grun, P. (1979) Evolution of the cultivated potato: a cytoplasmic analysis. In: Hawkes, J.G., Lester, R.N. and Skelding, A.D. (eds) *The Biology and Taxonomy of the Solanaceae*. Symposium Series No. 7, Linnean Society, London pp. 655–665.

Grun, P. (1990) The evolution of the cultivated potato. *Economic Botany* 44, 39–55.

Grun, P. and Kao, T.H. (1989) Contrast of mitochondrial DNA restriction endonuclease fragments of *Solanum tuberosum* ssp. *tuberosum* and its putative progenitor *Solanum tuberosum* ssp. *andigena*. *American Journal of Botany* 76 (Suppl.), 146.

Grun, P., Ochoa, C. and Capage, D. (1977) Evolution of cytoplasmic factors in tetraploid cultivated potatoes (Solanaceae). *American Journal of Botany* 64, 412–420.

Guarino, L., Rao, V.R. and Reid, R. (1995) *Collecting Plant Genetic Diversity*. CAB International, Wallingford, UK.

Gulson, O. and Roose, M.L. (2001a) Chloroplast and nuclear genome analysis of the parentage of lemons. *Journal of the American Society for Horticultural Science* 126, 210–215.

Gulson, O. and Roose, M.L. (2001b) Lemons: diversity and relationships with selected *Citrus* genotypes as measured with nuclear genome markers. *Journal of the American Society for Horticultural Science* 126, 309–317.

Gustafsson, O. (1948) Polyploidy, life form and vegetative reproduction. *Hereditas* 34, 1–22.

Haghighi, K.R. and Ascher, P.D. (1988) Fertile, intermediate hybrids between *Phaseolus vulgaris* and *P. acutifolius* from congruity backcrossing. *Sexual Plant Reproduction* 1, 51–58.

Hahn, S.K. (1909) *Die Entstehurg den Pfuzkultun*. C. Winter, Heidelberg.

Hahn, S., Bai, K.V. and Asiedu, R.A. (1990) Tetraploids, triploids and 2n pollen from diploid interspecies crosses with *Cassava*. *Theoretical and Applied Genetics* 79, 433–439.

Hails, R.S. (2000) Genetically modified plants: the debate continues. *Trends in Ecology and Evolution* 15, 14–18.

Haldane, J.B.S. (1957) The cost of natural selection. *Journal of Genetics* 55, 511–524.

Haldane, J.B.S. (1960) More precise expressions for the cost of natural selection. *Journal of Genetics* 57, 351–360.

Hammer, K., Fitatenko, A.A. and Korzun, V. (2000) Microsatellite markers – a new tool for distinguishing diploid wheat species. *Genetic Resources and Crop Evolution* 47, 497–505.

Hamon, P. and Bakary, T. (1990) The classification of the cultivated yams (*Dioscorea cayenensis-rotundata*) of West Africa. *Euphytica* 47, 179–187.

Hamrick, J.L. and Godt, M.J.W. (1997) Allozyme diversity in cultivated crops. *Crop Science* 37, 26–30.

Hamrick, J.R. and Allard, R.W. (1972) Microgeographic variation in allozyme frequencies in *Avena barbata*. *Proceedings of the National Academy of Sciences USA* 69, 2100–2104.

Hancock, J.F. (1999) *Strawberries*. CAB International, Wallingford, UK.

Hancock, J.F. (2003) A framework for assessing the risk of transgenic crops. *BioScience* 53, 512–519.

Hancock, J.F. and Bringhurst., R.S. (1981) Evolution in California populations of diploid and octoploid *Fragaria* (Rosaceae): a comparison. *American Journal of Botany* 68, 1–5.

Hancock, J.F. and Hokanson, K.E. (2001) Invasiveness of transgenic vs. exotic plant species: How useful is the analogy? In: Strauss, S.H. and Bradshaw, H.D. (eds) *Proceedings of the First International Symposium on Ecological and Social Aspects of Transgenic Populations*. Oregon State University, Corvallis, pp. 187–192.

Hancock, J.F., Grumet, R. and Hokanson, S.C. (1996) The opportunity for escape of engineered genes from transgenic crops. *HortScience* 31, 1080–1085.

Hancock, J.F., Lavin, A. and Retamales, J. (1999) Our southern strawberry heritage: *Fragaria chiloensis* of Chile. *HortScience* 34, 814–816.

Handel, S.N. (1983a) Contrasting gene flow patterns and genetic subdivision in adjacent populations of *Cucumis sativus* (Cucurbitaceae). *Evolution* 37, 760–771.

Handel, S.N. (1983b) Pollination ecology, plant population structure and gene flow. In: Real, L. (ed.) *Pollination Biology*. Academic Press, New York.

Hanelt, P. (1972) Zur Geschriche des Anbaues von *Vicia faba* und ihre Gliederung. *Kulturpflanze* 20, 209–223.

Hanelt, P., Schafer, H. and von Schultze-Motel, J. (1972) Die Stellung von *Vicia faba* in der Gettung Vicia und Betrachtungen zur Entstehung dieser Kulurart. *Kulturpflanze* 20, 263–275.

Hansche, P.E., Bringhurst, R.S. and Voth, V. (1968) Estimates of genetic and environmental parameters in the strawberry. *Proceedings of the American Society for Horticultural Science* 92, 338–345.

Harborne, J.B. (1982) *Introduction to Ecological Biochemistry*. Academic Press, London and New York.

Hardy, G.H. (1908) Mendelian proportions in a mixed population. *Science* 28, 49–50.

Harlan, J.R. (1965) The possible role of weed races in the evolution of cultivated plants. *Euphytica* 14, 173–176.

Harlan, J.R. (1967) Agricultural origins: centers and non-centers. *Science* 174, 468–474.

Harlan, J.R. (1969) Evolutionary dynamics of plant domestication. *Japanese Journal of Genetics* 44, 337–343.

Harlan, J.R. (1975) Geographic patterns of variation in some cultivated crops. *Journal of Heredity* 66, 182–191.

Harlan, J.R. (1976) Plant and animal distribution in relation to domestication. *Philosophical Transactions of the Royal Botanical Society, London* 275, 13–25.

Harlan, J.R. (1992a) *Crops and Man*. American Society of Agronomy, Madison, Wisconsin.

Harlan, J.R. (1992b) Indigenous African agriculture. In: Cowen, C.W. and Watson, P.J. (eds) *The Origins of Agriculture*. Smithsonian Institute Press, Washington, DC, pp. 59–70.

Harlan, J.R. and deWet, J.M.J. (1971) Toward a rational classification of cultivated plants. *Taxonomy* 20, 509–517.

Harlan, J.R. and deWet, J.M.J. (1972) A simplified classification of cultivated sorghum. *Crop Science* 12, 172–176.

Harlan, J.R. and deWet, J.M.J. (1975) On Ö. Winge and a prayer: the origins of polyploidy. *Botanical Review* 41, 361–390.

Harlan, J.R. and Zohary, D. (1966) Distribution of wild wheats and barley. *Science* 153, 1074–1080.

Harlan, J.R., deWet, J.M.J. and Price, E.G. (1973) Comparative evolution of cereals. *Evolution* 27, 311–325.

Harris, S.A., Robinson, J.P. and Junipae, B.E. (2002) Genetic clues to the origin of apples. *Trends in Genetics* 18, 426–430.

Harrison, R.E., Luby, J.J., Furnier, G.R. and Hancock, J.F. (1997) Morphological and molecular variation among populations of octoploid *Fragaria virginiana* and *F. chiloensis* (Rosaceae) from North America. *American Journal of Botany* 84, 612–620.

Hart, G.E. (1988) Genetics and evolution of multilocus isozymes in hexaploid wheat. In: Rattazzi, M.C., Scandalios, J.G. and Whitt, G.S. (eds) *Isozymes*, Vol. 10: *Genetics and Evolution*. A.R. Liss, New York.

Hartl, D.L. (1980) *Principles of Population Genetics*. Sinauer Associates, Sunderland, Massachusetts.

Hatchett, J.H. and Gallum, R. (1970) Genetics of the ability of the Hessian fly, *Mayetiola destructor*, to survive on wheats having different genes for resistance. *Annual Report of the Entomological Society of America* 63, 1400–1407.

Hauber, D.P. (1986) Autotetraploidy in *Haplopappus spinulosus* hybrids: evidence from natural and synthetic tetraploids. *American Journal of Botany* 73, 1595–1606.

Hautea, R.A., Coffman, W.R., Sorrells, M.E. and Bergstrom, G.C. (1987) Inheritance of partial resistance to powdery mildew in spring wheat. *Theoretical and Applied Genetics* 73, 609–615.

Haviland, W.A. (1996) *Human Evolution and Prehistory*, Vol. 8. Holt, Rinehard and Winston, New York.

Hawkes, J.G. (1967) History of the potato. In: Harris, P.M. (ed.) *The Potato Crop*. Chapman & Hall, London.

Hawkes, J.G. (1979) Evolution and polyploidy in potato species. In: Hawkes, J.G., Lester, R.N. and Skelding, D. (eds) *The Biology and Taxonomy of the Solanaceae*. Seminar Series No. 7, Linnean Society, London, pp. 637–646.

Hawkes, J.G. (1990) *The Potato: Evolution, Biodiversity and Genetic Resources*. Belhaven Press, Washington, DC.

Hawtin, G.C., Singh, K.B. and Sexena, M.C. (1980) Some recent developments in the understanding and improvement of *Cicer* and *Lens*. In: Summerfield, R.J. and Bunting, A.H. (eds) *Advances in Legume Science*. Royal Botanical Gardens, Kew.

Hedrick, P.W. (1980) Hitchhiking: a comparison of linkage and partial selfing. *Genetics* 94, 791–808.

Hedrick, P.W. (1983) *Genetics of Populations*. Jones and Bartlett Publishers, Boston.

Hedrick, P.W. and Holden, L. (1979) Hitch-hiking: an alternative to coadaptation for the barley and slender wild oat examples. *Heredity* 43, 79–86.

Hedrick, U.P. (1917) *The Peaches of New York*. Report of the New York Agricultural Station, Albany, New York.

Heiser, C. (1979) *The Gourd Book*. University of Oklahoma Press, Norman.

Heiser, C. (1990) *Seed to Civilization: the Story of Food*. Harvard University Press, Cambridge, Massachusetts.

Heiser, C.B. (1947) Hybridization between the sunflower species *Helianthus annuus* and *H. petiolaris*. *Evolution* 1, 249–262.

Heiser, C.B. (1949) Study in the evolution of the sunflower species *Helianthus annuus* and *H. bolanderi*. *University of California Publications in Botany* 23, 157–208.

Heiser, C.B. (1951) Hybridization in the annual sunflowers: *Helianthus annuus* × *H. debilis* var. *cucumerifolius*. *Evolution* 5, 42–51.

Heiser, C.B. (1958) Three new annual sunflowers (*Helianthus*) from the southwestern United States. *Rhodora* 60, 272–283.

Heiser, C.B. (1973) Introgression re-examined. *Botanical Review* 39, 347–366.

Heiser, C.B. (1976) *The Sunflower*. University of Oklahoma Press, Norman.

Heiser, C.B. (1995a) Peppers: *Capsicum* (Solanaceae). In: Smartt, J. and Simmonds, N.W. (eds) *Evolution of Crop Plants*. Longman Scientific & Technical, Harlow, UK, pp. 449–451.

Heiser, C.B. (1995b) Sunflowers: *Helianthus* (Compositae). In: Smartt, J. and Simmonds, N.W. (eds) *Evolution of Crop Plants*. Longman Scientific and Technical, Harlow, UK, pp. 51–53.

Heiser, C.B. and Smith, P.G. (1953) The cultivated *Capsicum* peppers. *Economic Botany* 7, 214–227.

Helbaek, H. (1959) Domestication of food plants in the Old World. *Science* 130, 365–372.

Helbaek, H. (1966) Commentary on the phylogenesis of *Triticum* and *Hordeum*. *Economic Botany* 20, 350–360.

Helentjaris, T.D., Weber, D.F. and Wright, S. (1988) Identification of the genomic locations of genes of duplicated nucleotide sequences in maize by analysis of restriction fragment length polymorphisms. *Genetics* 118, 353–363.

Helm, J. (1963) Morphologisch-taxonomische Gliederung der Kultursippen von *Brassica oleraceae*. *Kulturpflanze* 11, 92–210.

Hemingway, J.S. (1995) Mustards: *Brassica* ssp. and *Sinus alpa*. In: Smartt, J. and Simmonds, N. (eds) *Evolution of Crop Plants*. Longman Scientific & Technical, Harlow, UK, pp. 82–86.

Hermsen, J.G.T. (1984) Nature, evolution and breeding of polyploids. *Iowa State Journal of Research* 58, 421–436.

Heslop-Harrison, J.S. (2000) Comparative genome organization in plants: from sequence and markers to chromatin and chromosomes. *The Plant Cell* 12, 617–635.

Hesse, C.O. (1975) Peaches. In: Janick, J. and Moore, J.N. (eds) *Advances in Fruit Breeding*. Purdue University Press, West Lafayette, Indiana, pp. 285–335.

Heun, M., Schafer-Pregl, R., Klawan, D., Castagna, R., Accerbi, M., Borghi, B. and Salamini, F. (1997) Site of Einkorn wheat domestication identified by DNA fingerprinting. *Science* 278, 1312–1314.

Higham, C.F.W. (1984) Prehistoric rice cultivation in southeast Asia. *Scientific American* 250, 138–146.

Hillman, G. (1975) The plant remains from Tell Abu Hureyra: a preliminary report. *Proceedings of the Prehistoric Society* 41, 70–73.

Hillman, G. (1978) On the origin of domesticated rye – *Secale cereale*: the finds from a ceramic Can Hasan III in Turkey. *Anatolian Studies* 28, 157–174.

Hilu, K.W. (1983) The role of single-gene mutations in the evolution of flowering plants. *Evolutionary Biology* 16, 97–128.

Hilu, K.W. and Johnson, J.L. (1992) Ribosomal DNA variation in finger millet and wild species of *Eleusine* (Poaceae). *Theoretical and Applied Genetics* 83, 895–902.

Ho, P.T. (1975) *The Cradle of the East*. University of Chicago Press, Chicago.

Ho, P.T. (1969) The origin of Chinese agriculture. *American Historical Review* 75, 1–36.

Ho, P.T. (1977) The indigenous origins of Chinese agriculture. In: Reed, C.A. (ed.) *Origins of Agriculture*. Mouton, The Hague.

Hoang-Tang, D. and Liang, G.H. (1988) The genomic relationship between cultivated sorghum [*Sorghum bicolor* (L.) Moench] and Johnsongrass [*S. halepense* (L.) Pers.]: a re-evaluation. *Theoretical and Applied Genetics* 76, 277–284.

Hoang-Tang, D., Dube, S.K., Liang, G.H. and Kung, S.D. (1991) Possible repetitive DNA markers for Eusorghum and Parasorghum and their potential use in examining phylogenetic hypotheses on the origin of *Sorghum* species. *Genome* 34, 241–250.

Hodgson, R.W. (1961) Taxonomy and nomenclature in *Citrus*. In: Price, W.C. (ed.) *Proceedings of the Second Conference of the International Organization of* Citrus. University of Florida Press, Gainesville, Florida, pp. 1–7.

Hogenboom, N.G. (1973) A model for incongruity in intimate partner relationships. *Euphytica* 22, 219–233.

Hogenboom, N.G. (1975) Incompatibility and incongruity: two different mechanisms for the non-functioning of intimate partner relationships. *Proceedings of the Royal Society of London Ser. B* 188, 361–375.

Hoisington, A.M.R. and Hancock, J.F. (1981) Effect of allopolyploidy on the activity of selected enzymes in *Hibiscus*. *Plant Systematics and Evolution* 138, 189–198.

Hokanson, K. and Hancock, J. (2000) Early-acting inbreeding depression in three species of *Vaccinium*. *Sexual Plant Reproduction* 13, 145–150.

Hokanson, S.C., Grumet, R. and Hancock, J.F. (1997a) Effect of border rows and trap/donor ratios on pollen-mediated gene movement. *Ecological Applications* 7, 1075–1081.

Hokanson, S.C., McFerson, J.R., Forsline, P.L., Lamboy, W.F., Luby, J.J., Djangaliev, A.D. and Aldwinckle, H.S. (1997b) Collecting and managing wild *Malus* germplasm in its center of diversity. *HortScience* 32, 173–176.

Holden, C. (1998) No last word on language origins. *Science* 282, 1455–1458.

Holden, J.H.W. (1976) Oats. In: Simmonds, N.W. (ed.) *Evolution of Crop Species*. Longman, London.

Holliday, R.J. and Putwain, P.D. (1980) Evolution of herbicide resistance in *Senecio vulgaris*: variation in susceptibility to simazine between and within populations. *Journal of Applied Ecology* 17, 779–792.

Holsinger, K.E. and Ellstrand, N.C. (1984) The evolution and ecology of permanent translocation heterozygotes. *American Naturalist* 12, 448–471.

Hopf, M. (1969) Plant remains and early farming at Jericho. In: Ucko, P.J. and Dimbleby, G.W. (eds) *Domestication and Exploitation of Plants and Animals*. Aldine, Chicago.

Hopf, M. (1983) Jericho plant remains. In: Kenyon, K.M. and Holland, T.A. (eds) *Excavations at Jerico*, Vol. 5. British School of Archaeology in Jerusalem, London, pp. 576–621.

Hopf, M. (1986) Archaeological evidence of the spread and use of some members of the Leguminosae family. In: *The Origin and Domestication of Cultivated Crops*, Elsevier Science Publishers, New York.

Hosaka, K. (1986) Who is the mother of potato? Restriction endonuclease analysis of chloroplast DNA of cultivated potatoes. *Theoretical and Applied Genetics* 72, 606–618.

Hosaka, K. (1995) Successive domestication and evolution of the Andean potatoes as revealed by chloroplast DNA restriction-endonuclease analysis. *Theoretical and Applied Genetics* 90, 356–363.

Hosaka, K. and Hanneman, R.E. (1988) Origin of chloroplast diversity in the Andean potato. *Theoretical and Applied Genetics* 76, 333–340.

Howard, D.J. and Berlocher, S.H.B. (eds) (1998) *Endless Forms: Species and Speciation*. Oxford University Press, New York.

Huang, J.C. and Sun, M. (2000) Genetic diversity and relationships of sweet potato and its wild relatives in *Ipomoea* series *Batatas* (Convolvulaceae) as revealed by inter-simple sequence repeat (ISSR) and restriction analysis of chloroplast DNA. *Theoretical and Applied Genetics* 100, 1050–1060.

Huang, S., Sirikhachornkit, A., Sun, X., Faris, J., Gill, B., Haselkorn, R. and Gornichi, P. (2002) Genes encoding plastid acetyl-CoA carboxylase and 3-phosphoglycerate kinase of the *Triticum/Aegilops* polyploid wheat. *Proceedings of the National Academy of Sciences USA* 99, 8133–8138.

Hucl, P. and Scoles, G.J. (1985) Interspecific hybridization in the common bean: a review. *HortScience* 20, 352–357.

Huether, C.A. (1968) Exposure of natural genetic variability underlying the pentamerous corolla constancy in *Linanthus androsaceus* ssp. *androsaceus*. *Genetics* 60, 123–146.

Huether, C.A. (1969) Constancy of the pentamerous corolla phenotype in natural populations of *Linanthus*. *Evolution* 23, 572–588.

Human, Z. and Spooner, D.M. (2002) Reclassification of landrace populations of cultivated potatoes (*Solanum* sect. *Petota*). *American Journal of Botany* 89, 947–965.

Husband, B.C. and Schemske, D.W. (1996) Evolution of the magnitude and timing of inbreeding depression in plants. *Evolution* 50, 54–70.

Husband, B.C. and Schemske, D.W. (1997) The effect of inbreeding in diploid and tetraploid populations of *Epilobium angustifolium* (Onagraceae): implications for the genetic basis of inbreeding depression. *Evolution* 51, 737–746.

Husband, B.C. and Schemske, D.W. (1998) Cytotype distribution at a diploid–tetraploid contact zone in *Chamerion* (*Epilobium*) *angustifolium* (Onagraceae) *American Journal of Botany* 85, 1688–1694.

Husted, L. (1936) Cytological studies of the peanut *Arachis* II. Chromosome number, morphology and behavior and their application to the origin of the cultivated forms. *Cytologia* 7, 396–423.

Hutchinson, E.S., Price, S.C., Kahler, A.L., Morris, M.I. and Allard, R.W. (1983) An experimental verification of segregation theory in a diploidized tetraploid: esterase loci in *Avena barbata*. *Journal of Heredity* 74, 381–383.

Huttley, G.A., Durbin, M.L., Glover, D.E. and Clegg, M.T. (1997) Nucleotide polymorphism in the chalcone synthase-A locus and evolution of the chalcone synthase multigene family of common morning glory *Ipomoea purpurea*. *Molecular Evolution* 6, 549–558.

Hymowitz, T. (1995) Soybean: *Glycine max* (Leguminosae – Papilionoidae). In: Smartt, J. and Simmonds, N.W. (eds) *Evolution of Crop Plants*. Longman Scientific & Technical, Harlow, UK, pp. 261–266.

Hymowitz, T. and Harlan, J.R. (1983) Introduction of soybean to North America by Samuel Bowen in 1765. *Economic Botany* 37, 371–379.

Hymowitz, T. and Singh, R.J. (1987) Taxonomy and speciation. In: Wilcox, J.R. (ed.) *Soybeans: Improvement, Production and Uses*. American Society of Agronomy, Madison, Wisconsin, pp. 23–48.

Iltis, H.H. (1983) From teosinte to maize: the catastrophic sexual transmutation. *Science* 222, 886–894.

Iltis, H.H. (2000) Homeotic sexual translocations and the origin of maize (*Zea mays*, Poaceae): a new look at an old problem. *Economic Botany* 54, 7–42.

Iltis, H.H. and Doebley, J.F. (1980) Taxonomy of *Zea* (Gramineae). II. Subspecific categories in the *Zea mays* complex and genetic synopsis. *American Journal of Botany* 67, 994–1004.

Ingram, G.B. and Williams, J.T. (1984) *In situ* conservation of wild relatives of crops. In: Holden, J.H. and Williams, J.T. (eds) *Crop Genetic Resources: Conservation and Evaluation*. Allen & Unwin, London.

Iruela, M., Rubio, J., Cubero, J.I., Gil, J. and Millán, T. (2002) Phylogenetic analysis in the genus *Cicer* and cultivated chickpea using RAPD and ISSR markers. *Theoretical and Applied Genetics* 104, 643–651.

Irwin, S.V., Kaufusi, P., Banks, K., de la Pena, R., and Cho, J.J. (1998) Molecular characterization of taro (*Colocasia esculenta*) using RAPD markers. *Euphytica* 99, 183–189.

Ishii, T., Xu, Y. and McCouch, S.R. (2001) Nuclear- and chloroplast-microsatellite variation in A-genome species of rice. *Genome* 44, 658–666.

Iwanaga, M. and Peloquin, S.J. (1982) Origin and evolution of cultivated tetraploid potatoes via 2*n* gametes. *Theoretical and Applied Genetics* 61, 161–169.

Jaaska, V. (1983) *Secale* and *Triticale*. In: Tanksley, S.D. and Orton, T.J. (eds) *Isozyme in Plant Breeding and Evolution*. Elsevier, Amsterdam, pp. 79–101.

Jaccard, P. (1908) Nouvelles recherches sur la distribution florale. *Bull. Soc. Vaudoise Sci. Nat.* 44, 223–270.

Jackson, R.C. and Murry, B.G. (1983) Colchicine induced quadrivalent formation in *Helianthus*: evidence of ancient polyploidy. *Theoretical and Applied Genetics* 64, 219–222.

Jain, S.K. and Bradshaw, A.D. (1966) Evolutionary divergence among adjacent plant populations. I. The evidence and its theoretical basis. *Heredity* 21, 407–441.

James, J. (1979) New maize-×-*Tripsicum* hybrids. *Euphytica* 28, 239–247.

James, S.H. (1965) Complex hybridity of *Isotoma petraea*. I. The occurrence of interchange heterozygosity, autogamy and a balanced lethal system. *Heredity* 20, 341–353.

Janick, J., Cummins, J.N., Brown, S.K. and Hemmat, M. (1996) Apples. In: Janick, J. and Moore, J.N. (eds) *Fruit Breeding,* Vol. 1, *Tree and Nut Fruits*. John Wiley & Sons, New York, pp. 1–79.

Jannoo, N., Grivet, L., Sequin, M., Paulet, F., Domaingue, R., Rao, P.R., Dookun, A., D'Hont, A. and Glaszmann, J.C. (1999) Molecular investigation of the genetic base of sugarcane cultivars. *Theoretical and Applied Genetics* 99, 171–184.

Janushevich, Z.V. (1984) The specific composition of wheat finds from ancient agricultural centers in the USSR. In: Zeist, W.V. and Casparie, W.A. (eds) *Plants and Ancient Man*. Balkema, Rotterdam, pp. 267–276.

Janzen, D.H. (1966) Coevolution of mutualism between ants and acacias in Central America. *Evolution* 20, 249–275.

Jarret, R.L. and Litz, R.E. (1986) Enzyme polymorphism in *Musa acuminata* Colla. *Journal of Heredity* 77, 183–188.

Jarvis, D.I. and Hodgkin, T. (1999) Wild relatives and crop cultivars: detecting natural introgression and farmer selection of new genetic combinations in agroecosystems. *Molecular Ecology* 8, S159–S173.

Jasieniuk, M. and Maxwell, B.D. (1994) Population genetics and the evolution of herbicide resistance in weeds. *Phytoprotection* 75, 25–35.

Jellen, E.N. and Ladizinsky, G. (2000) Giemsa C-banding in *Avena insularis* Ladizinsky. *Genetic Resources and Crop Evolution* 47, 227–230.

Jenkins, J.A. (1948) The origin of the cultivated tomato. *Economic Botany* 2, 379–392.

Jennings, D.L. (1995) Cassava: *Manihot esculenta* (Euphorbiaceae). In: Smartt, J. and Simmonds, N.W. (eds) *Evolution of Crop Plants*. Longman Scientific & Technical, Harlow, UK, pp. 129–132.

Jensen, R.A. (1976) Enzyme recruitment in evolution of new function. *Annual Review of Microbiology* 30, 409.

Jiang, C.-X., Chee, P.W., Draye, X., Morrell, P.L., Smith, C.W. and Paterson, A.H. (2000) Multilocus interactions restrict gene expression in interspecific populations of polyploid *Gossypium* (cotton). *Evolution* 54, 798–814.

Jobst, J., King, K. and Hemleben, V. (1998) Molecular evolution of the internal transcribed spacers (ITS1 and ITS2) and phylogenetic relationships among species of the family Cucurbitaceae. *Molecular Phylogenetics and Evolution* 9, 204–219.

Johannsen, W. (1903) *Uber Enblichkeit in Populationen und in reinen Linien*. Gustav Fischer, Jena.

Johanson, D.C. and Eday, M.A. (1981) *Lucy: the Beginnings of Humankind*. Simon and Schuster, New York.

Johanson, D.C. and White, T.D. (1979) A systematic assessment of early African hominids. *Science* 203, 321–330.

Johns, T.A. and Keen, S.L. (1986) Ongoing evolution of the potato on the Altiplano of western Bolivia. *Economic Botany* 40, 409–424.

Johnson, A.W. and Packer, J.G. (1965) Polyploidy and environment in arctic Alaska. *Science* 148, 237–239.

Johnson, B.L. (1972) Protein electrophoretic profiles and the origin of the B genome of wheat. *Proceedings of the National Academy of Sciences USA* 69, 1398–1402.

Johnston, J.A., Wesselingh, R.A., Bouck, A.C., Donovan, L.A. and Arnold, M.L. (2001) Intimately linked or hardly speaking? The relationship between genotype and environmental gradients in a Louisiana iris hybrid population. *Molecular Ecology* 10, 673–681.

Johnston, M.O. and Schoen, D.J. (1996) Correlated evolution of self-fertilization and inbreeding depression: an experimental study of nine populations of *Amsinckia* (Boraginaceae). *Evolution* 50, 1478–1491.

Jones, A. (1967) Should Nishiyama's K123 (*Ipomoea trifida*) be designated *I. batatas*? *Economic Botany* 21, 163–166.

Jones, A. (1990) Unreduced pollen from a wild tetraploid relative of sweet potato. *Journal of the American Society for Horticultural Science* 115, 512–516.

Jones, N., Ougham, H. and Thomas, H. (1997) Markers and mapping: we are all geneticists now. *New Phytologist* 137, 165–177.

Josefsson, E. (1967) Distribution of thioglucosides in different parts of *Brassica* plants. *Phytochemistry* 32, 151–159.

Joseph, M.C., Randall, D.D. and Nelson, C.J. (1981) Photosynthesis in polyploid tall fescue. II. Photosynthesis and ribulose-1,5-bisphosphate carboxylase of polyploid tall fescue. *Plant Physiology* 68, 894–898.

Joshi, S.P., Gupta, V.S., Aggarwal, R.K., Ranjekar, P.K. and Brar, D.S. (2000) Genetic diversity and phylogenetic relationship as revealed by inter simple sequence repeat (ISSR) polymorphism in the genus *Oryza*. *Theoretical and Applied Genetics* 100, 1311–1320.

Jusuf, M. and Pernes, J. (1985) Genetic variability of foxtail millet (*Setaria italica* P. Beauv.). *Theoretical and Applied Genetics* 71, 385–391.

Kahler, A.L., Allard, R.W. and Miller, R.D. (1984) Mutation rates for enzyme and morphological loci in barley (*Hordeum vulgare* L.). *Genetics* 106, 729–734.

Kalton, R.R., Smit, A.G. and Leffel, R.C. (1952) Parent–inbred progeny relationships of selected orchardgrass clones. *Agronomy Journal* 44, 481–486.

Kaplan, L. and Lynch, T.F. (1999) *Phaseolus* (Fabaceae) in archaeology: AMS radioactive carbon dates and their significance in pre-Columbian agriculture. *Economic Botany* 53, 261–272.

Kayser, M., Brauer, S., Weiss, G., Schiefenhovel, W., Underhill, P.A. and Stoneking, M. (2001) Independent histories of human Y chromosomes from Melanesia and Australia. *American Journal of Human Genetics* 68, 173–190.

Kennedy-O'Byrne, J. (1957) Notes on African grasses: XXIX: a new species of *Eleusine* from tropical South Africa. *Kew Bulletin* 1, 65–72.

Kerby, K. and Kuspira, J. (1988) Cytological evidence bearing on the origin of the B genome in polyploid wheats. *Genome* 30, 36–43.

Kerster, H. and Levin, D. (1968) Neighborhood size in *Lithospermum caroliniense*. *Genetics* 60, 577–587.

Khairallah, M., Adams, M.W. and Sears, B.B. (1990) Mitochondrial DNA polymorphisms of Malawian bean lines: further evidence for two gene pools. *Theoretical and Applied Genetics* 80, 753–761.

Khush, G.S. (1962) Cytogenetics and evolutionary studies in *Secale*. II. Interrelationships of the wild species. *Evolution* 16, 484–496.

Khush, G.S. and Stebbins, G.L. (1961) Cytogenetic and evolutionary studies in *Secale*. I. Some new data on the ancestry of *S. cereale*. *American Journal of Botany* 48, 721–730.

Kiang, Y.T., Antonovics, J. and Wu, L. (1979) The extinction of wild rice (*Oryza perennis formosana*) in Taiwan. *Journal of Asian Ecology* 1, 1–9.

Kihara, H. (1954) Considerations on the evolution and distribution of *Aegilops* species based on the analyzer method. *Cytologia* 19, 336–357.

Kihara, H., Yamashita, H. and Tanaka, M. (1959) Genomes of 6 species of *Aegilops*. *Wheat Information Service* 8, 3–5.

Kim, S.-C. and Rieseberg, L.H. (2001) The contribution of epistasis to species differences in annual sunflowers. *Molecular Ecology* 10, 683–690.

Kishima, Y., Mikami, T., Hirai, A., Sugiura, M. and Kinoshita, T. (1987) *Beta* chloroplast genomes: analysis of fraction I protein and chloroplast DNA variation. *Theoretical and Applied Genetics* 73, 330–336.

Kislev, M.E. (1985) Early neolithic horsebean from Yiftah'el, Israel. *Science* 228, 319–320.

Klinger, T. and Ellstrand, N.C. (1994) Engineered genes in wild populations – fitness of weed-crop hybrids of *Raphanus sativus*. *Ecological Applications* 4, 117–120.

Kloppenburg, J.R. (1988) *First the Seed*. Cambridge University Press, Cambridge.

Knight, J. (2003) A dying breed. *Nature* 421, 568–570.

Knight, R.L. (1948) The role of major genes in the evolution of economic characters. *Journal of Genetics* 48, 370–387.

Kobayashi, K. (1984) Proposed polyploid complex of *Ipomoea trifida*. In: *Proceedings of the Sixth Symposium of the International Society for Tropical Fruit Crops*. International Potato Center, Lima, Peru, pp. 561–568.

Kochert, G., Stalker, H.T., Gimenes, M., Galgaro, L., Lopes, C.R. and Moore, K. (1996) RFLP and cytogenetic evidence on the origin and evolution of allopolyploid domesticated peanut, *Arachis hypogaea* (Leguminoseae). *American Journal of Botany* 83, 1282–1291.

Koinange, E.M.K., Singh, S.P. and Gepts, P. (1996) Genetic control of the domestication syndrome in common bean. *Crop Science* 36, 1037–1045.

Kollipara, K.P., Sing, R.J., Hymowitz, T. (1997) Phylogenetic and genomic relationships in the genus *Glycine* Willd based on sequences from the ITS region of nuclear rDNA. *Genome* 40, 57–68.

Konishi, T. and Linde-Laursen, I. (1988) Spontaneous chromosomal rearrangements in cultivated and wild barleys. *Theoretical and Applied Genetics* 75, 237–243.

Korban, S.S. (1986) Interspecific hybridization in *Malus*. *HortScience* 21, 41–48.

Korban, S.S. and Skirvin, R.M. (1984) Nomenclature of the cultivated apple. *HortScience* 19, 177–180.

Krebs, S.L. and Hancock, J.F. (1989) Tetrasomic inheritance of isozyme markers in the highbush blueberry, *Vaccinium corymbosum* L. *Heredity* 63, 11–18.

Krebs, S.L. and Hancock, J.F. (1991) Embryonic genetic load in the highbush blueberry, *Vaccinium corymbosum* (Ericaceae). *American Journal of Botany* 78, 1427–1437.

Ku, H.-M., Vision, T., Liu, J. and Tanksley, S.D. (2000) Comparing sequenced segments of the tomato and *Arabidopsis* genomes: large-scale duplication followed by selective gene loss creates a network of synteny. *Proceedings of the National Academy of Sciences USA* 2000, 9121–9126.

Kumar, O.A., Panda, R.C. and Rao, K.G.R. (1987) Cytogenetic studies of the F_1 hybrids of *Capsicum annum* with *C. chinense* and *C. baccatum*. *Theoretical and Applied Genetics* 74, 242–246.

Kumar, P.S. and Hymowitz, T. (1989) Where are the diploid ($2n = 2x = 20$) genome donors of *Glycine* Willd. (Leguminosae, Papilionoideae)? *Euphytica* 40, 221–226.

Kuruvilla, K. and Singh, A. (1981) Karotypic and electrophoretic studies on taro and its origin. *Euphytica* 30, 405–413.

Labdi, M., Robertson, L.D., Singh, K.B. and Charrier, A. (1996) Genetic diversity and phylogenetic relationships among the annual *Cicer* species as revealed by isozyme polymorphisms. *Euphytica* 88, 181–188.

Ladizinsky, G. (1985) Founder effect in crop evolution. *Economic Botany* 39, 191–199.

Ladizinsky, G. (1999) Identification of lentil's wild genetic stock. *Genetic Resources and Crop Evolution* 46, 115–118.

Ladizinsky, G., Braun, D., Goshen, D. and Muehlbauer, F.J. (1984) The biological species of the genus *Lens*. *Botanical Gazette* 145, 253–261.

Ladizinsky, G. (1975) On the origin of the broad bean, *Vicia faba* L. *Israel Journal of Botany* 24, 80–88.

Ladizinsky, G. (1995) Chickpea, *Cicer arietinum* (Leguminosae – Papilionoideae). In: Smartt, J. and Simmonds, N.W. (eds) *Evolution of Crop Plants*. Longman Scientific & Technical, Harlow, UK, pp. 258–261.

Ladizinsky, G. (1998) A new species of *Avena* from Sicily, possibly the tetraploid progenitor of hexaploid oats. *Genetic Resources and Crop Evolution* 45, 263–269.

Ladizinsky, G. (1999) Cytogenetic relationships between *Avena insularis* (2n = 28) and both *A. strigosa* (2n = 14) and *A. murphyi* (2n = 28). *Genetic Resources and Crop Evolution* 46, 501–504.

Ladizinsky, G. and Adler, A. (1976a) Genetic relationships among the annual species of *Cicer*. *Theoretical and Applied Genetics* 48, 197–203.

Ladizinsky, G. and Adler, A. (1976b) The origin of chickpea *Cicer arietinum*. *Euphytica* 88, 181–188.

Ladizinsky, G. and Zohary, D. (1971) Notes on species delimination, species relationships and polyploidy in *Avena*. *Euphytica* 20, 380–395.

Ladizinsky, G. and Genizi, A. (2001) Could early gene flow have created similar allozyme-gene frequencies in cultivated and wild barley? *Genetic Resources and Crop Evolution* 48, 101–104.

Lamboy, W.F. and Alpha, C.G. (1998) Using simple sequence repeats (SSRs) for DNA fingerprinting germplasm accessions of grape (*Vitis* L.) species. *Journal of the American Society for Horticultural Science* 123, 182–188.

Lampa, G.K. and Bendich, A.J. (1984) Changes in mitochondrial DNA levels during development of pea. *Planta* 162, 463–468.

Lander, E.S., Green, P., Abrahamson, J., Barlow, A., Daly, M.J., Lincoln, S.E. and Newberg, L. (1987) MAP-MAKER: an interactive computer package for constructing primary genetic maps of experimental and natural populations. *Genomics* 1, 174–181.

Langer, R.H.M. and Hall, C.D. (1982) *Agricultural Plants*, Cambridge University Press, New York.

Langercrantz, U. and Lydiate, D.J. (1996) Comparative genome mapping in *Brassica*. *Genetics* 4, 1903–1910.

Larter, E.N. (1995) *Triticale*: Triticosecale spp. (Gramineae – Triticinae). In: Smartt, J. and Simmonds, N.W. (eds) *Evolution of Crop Plants*. Longman Scientific & Technical, Harlow, UK, pp. 181–183.

Lasker, G.W. and Tyzzer, R. (1982) *Physical Anthropology*. Holt, Rinehart and Winston, New York.

Latimer, H. (1958) *A Study of the Breeding Barrier between* Gilia australis *and* Gilia splendens. Claremont University College Press, Claremont, California.

Leakey, M.D. (1971) Discovery of postcranial remains of *Homo erectus* and associated artefacts in Bed IV at Olduvai Gorge, Tanzania. *Nature* 231, 380–383.

Leakey, R. and Lewin, R. (1992) *Origins Reconsidered: In Search of What Makes Us Human*. Anchor Books, New York.

Leath, S. and Pederson, W.D. (1986) Comparison of near-isogenic maize lines with and without the *Ht1* gene for resistance to four foliar pathogens. *Phytopathology* 76, 108–111.

Lebot, V. and Aradhya, K.M. (1991) Isozyme variation in taro (*Colocasia esculenta* (L.) Schott.) from Asia and Oceana. *Euphytica* 56, 55–66.

Lee, J.M., Grant, D., Vallejos, C.E. and Shoemaker, R.C. (2001) Genome organization in dicots. II. *Arabidopsis* as a 'bridging species' to resolve genome evolution events among legumes. *Theoretical and Applied Genetics* 103, 765–773.

Lee, R.E. (1968) What hunters do for a living, or how to make out on scarce resources. In: Lee, R.B. and DeVore, I. (eds) *Man the Hunter*, Aldine, Chicago, pp. 30–48.

Leggett, J.M. (1992) Classification and speciation in *Avena*. In: Marshall, H.G. and Sorrells, M.E. (eds) *Oat Science and Technology*. Agronomy Monograph 33, ASA and CSSA, Madison, Wisconsin, pp. 29–53.

Leggett, J.M. (1996) Using and conserving *Avena* genetic resources. In: Scoles, G.J. and Rossnagel, B.G. (eds) *Barley Chromosome Coordinators Workshop at the V International Oat Conference and VII International Barley Genetics Symposium*. University of Saskatchewan, Saskatchewan, Canada, pp. 128–132.

Leggett, J.M. and Thomas, H. (1995) Oat evolution and cytogenetics. In: Welch, W. (ed.) *The Oat Crop. Production and Utilization*. Chapman & Hall, London, pp. 121–149.

Levin, D.A. (1983) Polyploidy and novelty in flowering plants. *American Naturalist* 122, 1–25.

Levin, D.A. (2000) *The Origin, Expansion, and Demise of Plant Species*. Oxford University Press, New York.

Levin, D.A. and Kerster, H.W. (1967) Natural selection for reproductive isolation in *Phlox*. *Evolution* 21, 679–687.

Levin, D.A. and Kerster, H.W. (1969) Density-dependent gene dispersal in *Liatris*. *American Naturalist* 103, 61–74.

Levin, D.A. and Schmidt, K.P. (1985) Dynamics of a hybrid zone in *Phlox*: an experimental demographic investigation. *American Journal of Botany* 72, 1404–1409.

Levin, D.A., Torres, A.M. and Levy, M. (1979) Alcohol dehydrogenase activity in diploid and autotetraploid *Phlox*. *Biochemical Genetics* 17, 35–42.

Levin, D.A., Francisco-Ortega, J. and Jansen, R.K. (1995) Hybridization and the extinction of rare plant species. *Conservation Biology* 10, 10–16.

Levinson, C. and Gutman, G.A. (1987) Slipped-strand mispairing: a major mechanism for DNA sequence evolution. *Molecular Biology and Evolution* 4, 203–221.

Levy, M. and Levin, D.A. (1975) Genetic heterozygosity and variation in permanent translocation heterozygotes of the *Oenothera biennis* complex. *Genetics* 79, 493–512.

Levy, M. and Winternheimer, P.L. (1977) Allozyme linkage disequilibria among chromosome complexes in the permanent translocation heterozygote *Oenothera biennis*. *Evolution* 31, 465–476.

Levy, M., Steiner, E.E. and Levin, D.A. (1975) Allozyme genetics in permanent translocation heterozygotes of the *Oenothera biennis* complex. *Biochemical Genetics* 13, 487–500.

Lewis, D. (1943) Physiology of incompatibility in plants. III. Autopolyploids. *Journal of Genetics* 45, 171–185.

Lewis, E.B. (1951) Pseudoallelism and gene evolution. *Cold Spring Harbor Symposium of Quantitative Biology* 16, 159–172.

Lewis, H. (1962) Catastrophic selection as a factor in speciation. *Evolution* 16, 257–271.

Lewis, H. (1966) Speciation in flowering plants. *Science* 152, 167–172.

Lewis, W.H. and Lewis, M. (1955) The genus *Clarkia*. *University of California Publications in Botany* 20, 241–392.

Lewontin, R.C. and Kojima, K. (1960) The evolutionary dynamics of complex polymorphisms. *Evolution* 14, 458–472.

Li, C.-D., Rossnagel, B.G. and Scoles, G.J. (2000) Tracing the phylogeny of the hexaploid oat *Avena sativa* with satellite DNAs. *Crop Science* 40, 1755–1763.

Li, H.-L. (1970) The origin of cultivated plants in Southeast Asia. *Economic Botany* 24, 3–19.

Linares, C., Ferrer, E. and Fominaya, A. (1998) Discrimination of the closely related A and D genomes of the hexaploid oat *Avena sativa* L. *Proceedings of the National Academy of Sciences USA* 95, 12450–12455.

Linder, C.R., Taha, I., Seiler, G.L., Snow, A.A. and Rieseberg, L.H. (1998) Long-term introgression of crop genes into sunflower populations. *Theoretical and Applied Genetics* 96, 339–347.

LingHwa, T. and Morishima, H. (1997) Genetic characterization of weedy rices and the inference on their origins. *Breeding Science* 47, 153–160.

Liu, B., Vega, J.M., Segal, G., Abbo, S., Rodova, M. and Feldman, M. (1998a) Rapid genomic changes in newly synthesized amphiploids of *Triticum* and *Aegilops*. I. Changes in low copy non-coding sequences. *Genome* 41, 272–277.

Liu, B., Vega, J.M. and Feldman, M. (1998b) Rapid genomic changes in newly synthesized amphiploids of *Triticum* and *Aegilops*. II. Changes in low-copy coding DNA sequences. *Genome* 41, 535–542.

Livingstone, K.D., Lackney, V.K., Blauth, J.R., van Wijk, R. and Jahn, M.K. (1999) Genome mapping in *Capsicum* and the evolution of genome structure in the Solanaceae. *Genetics* 152, 1183–1202.

Loomis, W.F. and Gilpin, M.E. (1986) Multigene families and vestigial sequences. *Proceedings of the National Academy of Sciences USA* 83, 2143–2147.

Lubbers, E.L., Gill, K.S., Cox, T.S. and Gill, B.S. (1991) Variation of molecular markers among geographically diverse accessions of *Triticum tauschii*. *Genome* 34, 354–361.

Lukens, L.N. and Doebley, J. (1999) Epistatic and environmental interactions for quantitative trait loci involved in maize evolution. *Genetical Research* 74, 291–302.

Lush, W.M. and Evans, L.T. (1981) The domestication and improvement of cowpeas (*Vigna unguiculata* (L.) Walp.). *Euphytica* 30, 579–587.

Lyman, J.C. and Ellstrand, N.C. (1984) Clonal diversity in *Taraxacum officinale* (Compositae), an apomict. *Heredity* 53, 1–10.

MacKey, J. (1954) Neutron and x-ray experiments in wheat and a revision of the speltoid problem. *Hereditas* 40, 65–180.

MacKey, J. (1970) Significance of mating systems for chromosomes and gametes in polyploids. *Hereditas* 66, 165–176.

MacNeisch, R.S., Nelken-Terner, A. and Johnson, I.W. (1967) *The Prehistory of the Tehuacan Valley*. University of Texas Press, Austin.

MacLeod, M.J., Eshbaugh, W.H. and Guttman, S.I. (1979) An electrophoretic study of *Capsicum* (Solanaceae): the purple flowered taxa. *Economic Botany* 106, 326–333.

MacLeod, M.J., Eshbaugh, W.H. and Guttman, S.I. (1982) Early evolution of chili peppers (*Capsicum*). *Economic Botany* 36, 361–368.

McClintock, B. (1953) Induction of instability at selected loci in maize. *Genetics* 38, 579–599.

McClintock, B. (1956) Controlling elements and the gene. *Cold Spring Harbor Symposium on Quantitative Biology* 21, 197–216.

McClung de Tapia, E. (1992) *The Origins of Agriculture in Mesoamerica and Central America*. In: Cowan, C.W. and Watson, P.J. (eds) *The Origins of Agriculture.* Smithsonian Institute Press, Washington, DC.

McCubbin, A.G. and Kao, T.-H. (2000) Molecular recognition and response in pollen and pistil interactions. *Annual Review of Cellular and Developmental Biology* 16, 333–364.

McFadden, E.S. and Sears, E.R. (1946) The origin of *Triticum spelta* and its free-threshing hexaploid relatives. *Journal of Heredity* 37, 81–91.

McNaughton, I.H. and Harper, J.L. (1960) The comparative biology of closely related species living in the same area. I. External breeding barriers between *Papaver* species. *New Phytologist* 59, 15–26.

McNeilly, T. and Antonovics, J. (1968) Evolution in closely adjacent plant populations. IV. Barriers to gene flow. *Heredity* 23, 205–218.

Magoon, M.L., Krishnan, R. and Bai, K.V. (1969) Morphology of pachytene chromosomes and meiosis in *Manihot esculenta*. *Cytologia* 34, 612–624.

Maloney, B.K., Higham, C.F.W. and Bannanurag, R. (1989) Early rice cultivation in Southeast Asia: archaeological and palynological evidence from the Bang Pakong Valley, Thailand. *Antiquity* 63, 363–370.

Mangelsdorf, P.C. (1953) *Wheat*. Scientific American, New York.

Mangelsdorf, P.C. (1958) Reconstructing the ancestor of corn. *Proceedings of the American Philosophic Society* 102, 454–463.

Mangelsdorf, P.C. (1974) *Corn: Its Origin, Evolution and Improvement*. Harvard University Press, Cambridge, Massachusetts.

Mangelsdorf, P.C. (1986) The origin of corn. *Scientific American* 255, 80–86.

Mangelsdorf, P.C. and Reeves, R.G. (1939) *The Origin of Indian Corn and its Relatives*. Bulletin 549, Texas Agricultural Experiment Station, College Station.

Manwell, C. and Baker, C.M. (1970) *Molecular Biology and the Origin of Species*. University of Washington Press, Seattle.

Marshack, A. (1976) Implications of the paleolithic symbolic evidence for the origin of language. *American Scientist* 64, 136–145.

Marshall, D. and Allard, R. (1970) Isozyme polymorphisms in natural populations of *Avena fatua* and *A. barbata*. *Heredity* 25, 373–382.

Marshall, D.R. and Brown, A.D.H. (1975) Optimum sampling strategies in genetic conservation. In: Frankel, O.H. and Hawkes, J.G. (eds) *Crop Genetic Resources for Today and Tomorrow*. Cambridge University Press, New York, pp. 53–80.

Marshall, E. (2001) Pre-Clovis sites fight for acceptance. *Science* 291, 1730–1732.

Martin, F.W. and Jones, A. (1972) The species of *Ipomoea* closely related to the sweet potato. *Economic Botany* 26, 201–215.

Martin, G.B. and Adams, M.W. (1987a) Landraces of *Phaseolus vulgaris* (Fabaceae) in Northern Malawi. I. Regional variation. *Economic Botany* 41, 190–203.

Martin, G.B. and Adams, M.W. (1987b) Landraces of *Phaseolus vulgaris* (Fabaceae) in Northern Malawi. II. Generation and maintenance of variability. *Economic Botany* 41, 204–215.

Martin, T.J. and Ellingboe, A.H. (1976) Differences between compatible parasite host genotypes involving the Pm4 locus of wheat and the corresponding genes in *Erysiphe graminis* f. sp. *Tritici*. *Phytopathology* 66, 1435–1438.

Martinez-Zapater, J.M. and Oliver, J.L. (1984) Genetic analysis of isozyme loci in tetraploid potaoes (*Solanum tuberosum* L.). *Genetics* 108, 669–679.

Masterson, J. (1994) Stomatal size in fossil plants – evidence for polyploidy in the majority of angiosperms. *Science* 264, 421–424.

Mather, K. (1943) Polygenic balance in the canalization of development. *Nature* 151, 68–71.

Mather, K. (1973) *Genetic Structure of Populations*. Chapman & Hall.

Mather, K. and Jinks, J.L. (1977) *Introduction to Biometrical Genetics*. Cornell University Press, Ithaca.

Matsuoka, Y., Vigouroux, Y., Goodman, M.M., Sanches, J., Buckler, E. and Doebley, J. (2002) A single domestication for maize shown by multilocus microsatellite genotyping. *Proceedings of the National Academy of Sciences USA* 99, 6080–6084.

Mayer, M.S. and Bagga, S.K. (2002) The phylogeny of *Lens* (Leguminosae): new insight from ITS sequence analysis. *Plant Systematics and Evolution* 232, 145–154.

Mayr, E. (1942) *Systematics and the Origin of Species*. Columbia University Press, New York.

Mayr, E. (1954) Change of genetic environment and evolution. In: Huxley, J., Hardy, A.C. and Ford, E.B. (eds) *Evolution as a Process*. Allen & Unwin, London, pp. 157–180.

Mayr, E. (1963) *Animal Species and Evolution*. Harvard University Press, Cambridge, Massachusetts.

Mehra, K.L. (1963a) Differentiation of cultivated and wild *Eleusine* species. *Phyton* 20, 189–198.

Mehra, K.L. (1963b) Considerations of the African origin of *Eleusine coracana*. *Current Science* 32, 300–301.

Menacio-Hautea, D., Fatokum, C.A., Kumar, L., Danesh, D. and Young, N.D. (1993) Comparative genome analysis of mungbean (*Vigna radiata* L. Wilczek) and cowpea (*V. unguiculata*) using RFLP analysis. *Theoretical and Applied Genetics* 86, 797–810.

Mendel, G. (1866) Versuche uber Pflanzen-Hybridan. *Verhandlugen des Naturforschenden Vereins in Brünn* 4, 1–47.

Mendiburu, A.O. and Peloquin, S.J. (1977) The significance of 2N gametes in potato breeding. *Theoretical and Applied Genetics* 49, 53–61.

Mendoza, H.A. and Haynes, F.L. (1974) Genetic basis of heterosis for yield in the autotetraploid potato. *Theoretical and Applied Genetics* 45, 21–26.

Menkir, A., Goldsbrough, P. and Ejeta, G. (1997) RAPD based assessment of genetic diversity in cultivated races of sorghum. *Crop Science* 37, 564–569.

Merrick, L.C. (1995) Squashes, pumpkins and gourds: *Cucurbita* (Cucurbitaceae). In: Smartt, J. and Simmonds, N.W. (eds) *Evolution of Crop Species*. Longman Scientific & Technical, Harlow, UK, pp. 57–105.

Michaelis, P. (1954) Cytoplasmic inheritance in *Epilobium* and its theoretical significance. *Advances in Genetics* 6, 287–401.

Michaelis, P. (1967) IV. The investigation of plastome segregation by the pattern-analysis. *Nucleus* 10, 1–14.

Mignouna, H.D., Ellis, N.T.H., Knox, M.R., Asiedu, R. and Ng, Q.N. (1998) Analysis of genetic diversity in Guinea yams (*Dioscorea* spp.) using AFLP fingerprinting. *Tropical Agriculture* 75, 224–229.

Mikami, T., Kishima, Y., Sugiura, M. and Kinoshita, T. (1984) Chloroplast DNA diversity in the cytoplasms of sugarbeet and its related species. *Plant Science Letters* 36, 231–235.

Miller, J.S. and Venable, D.L. (2000) Polyploidy and evolution of gender in plants. *Science* 289, 2335–2338.

Mindzie, C.M., Doutrelepont, H., Vrydaghs, L., Swennen, R.J., Beeckman, H. and de Langhe, E. (2001) First archeological evidence of banana cultivation in cental Africa during the third millennium before present. *Vegetation History and Archaebotany* 10, 1–6.

Mitchell-Olds, T. (1996) Genetic constraints on life history evolution – quantitative trait loci influencing growth and flowering in *Arabidopsis thaliana*. *Evolution* 50, 140–145.

Mitton, J.B. and Grant, M.C. (1984) Associations among protein heterozygosity, growth rate and developmental homeostasis. *Annual Review of Ecology and Systematics* 15, 479–499.

Mok, D.W.S. and Peloquin, S.J. (1975) Three mechanisms of 2n pollen formation in diploid potatoes. *Canadian Journal of Genetics and Cytology* 17, 217–225.

Molina, M.D. and Naranjo, C.A. (1987) Cytogenetic studies in the genus *Zea*: 1. Evidence for five as the basic chromosome number. *Theoretical and Applied Genetics* 73, 542–550.

Molina-Cano, J.L., Moralejo, M., Igartua, E. and Romagosa, I. (1999) Further evidence supporting Morocco as a centre of origin of barley. *Theoretical and Applied Genetics* 98, 913–918.

Moore, P. (1976) How far does pollen travel? *Nature* 260, 280–281.

Morden, C.W., Doebley, J.F. and Schertz, K.F. (1989) Allozyme variation in Old World races of *Sorghum bicolor*. *American Journal of Botany* 76, 247–255.

Moringa, T. (1965) Cytogenetical investigations on *Oryza* species. In: International Rice Research Institute (ed.) *Cytogenetical Investigations on* Oryza *Species*. Elsevier Press, New York.

Morishima, H. and Oka, H.I. (1970) The dynamics of plant domestication: Cultivation experiments with wild *Oryza* populations. *Evolution* 25, 356–364.

Mortensen, D.A., Bastiaans, L. and Sattin, M. (2000) The role of ecology in the development of weed management systems: an outlook. *Weed Research* 40, 49–62.

Moscone, E.A., Lambrou, M. and Ehrendorfer, F. (1996) Fluorescent chromosome banding in the cultivated species of *Capsicum* (Solanaceae). *Plant Systematics and Evolution* 202, 37–63.

Mulcahy, D.L. (1979) The rise of the angiosperms: a genecological factor. *Science* 206, 20–30.

Müntzing, A. (1932) Cyto-genetic investigations on synthetic *Galeopsis tetrahit*. *Hereditas* 16, 105–164.

Müntzing, A. (1936) The evolutionary significance of autopolyploidy. *Hereditas* 21, 263–378.

Murai, K., Xu, N.Y. and Tsunewaki, K. (1989) Studies on the origin of crop species by restriction endonuclease analysis of genomic DNA. III. Chloroplast DNA variation and interspecific relationships in the genus *Secale*. *Japanese Journal of Genetics* 64, 35–47.

Nair, N.V., Nair, S., Sreenivasan, T.V. and Mohan, M. (1999) Analysis of genetic diversity and phylogeny in *Saccharum* and related genera using RAPD markers. *Genetic Resources and Crop Evolution* 46, 73–79.

Nasrallah, J.B. (2002) Recognition and rejection of self in plant reproduction. *Science* 296, 305–308.

Nassar, H.N., Nassar, N.M.A., Vieira, C. and Saraiva, L.S. (1995) Cytogenetic behavior of the interspecific hybrid of *Manihot neusana* Nassar and cassava *Manihot esculenta* Crantz, and its backcross progeny. *Canadian Journal of Plant Science* 75, 675–678.

National Research Council, N.R. (1989) *Lost Crops of the Incas: Little Known Plants of the Andes with Promise for Worldwide Cultivation*. National Academy Press, Washington, DC.

Nayar, N.W. (1973) Origin and cytogenetics of rice. *Advances in Genetics* 17, 153–292.

Neale, D.B., Saghai-Maroof, M.A., Allard, R.W. and Jorgensen, R.A. (1988) Chloroplast DNA diversity in populations of wild and cultivated barley. *Genetics* 120, 195–213.

Nee, M. (1990) The domestication of the *Cucurbita* (Cucurbitaceae). *Economic Botany* 44, 56–68.

Nei, M. (1972) Genetic distance between populations. *American Naturalist* 106, 283–292.

Nei, M. (1987) *Molecular Evolutionary Genetics*. Columbia University Press, New York.

Nei, M. and Li, W.-H. (1975) Probability of identical monomorphism in related species. *Genetic Research* 26, 31–43.

Nei, M. and Li, W.-H. (1979) Mathematical model for studying genetic variation in terms of restriction endonucleases. *Proceedings of the National Academy of Sciences USA* 76, 5269–5273.

Newell, C.A. and DeWet, J.M.J. (1973) A cytological survey of *Zea–Tripsicum* hybrids. *Canadian Journal of Genetics and Cytology* 15, 763–778.

Newell, C.J. and Hymowitz, T. (1978) A reappraisal of the subgenus *Glycine*. *American Journal of Botany* 65, 168–179.

Newton, K.J. (1983) Genetics of mitochondrial isozymes. In: Tansley, S.D. and Orton, T.J. (eds) *Isozymes in Plant Genetics and Breeding*. Elsevier, Amsterdam.

Ng, N.Q. (1995) Cowpea, *Vigna unguiculata* (Leguminoseae – Papilionoidaea). In: Smartt, J. and Simmonds, N. (eds) *Evolution of Crop Species*. Longman Scientific & Technical, London, pp. 326–333.

Ng, N.Q. and Maréchal, R. (1985) Cowpea taxonomy, origin and germplasm. In: Singh, S. and Rachie, K. (eds) *Cowpea Research, Production and Utilization*. Wiley, Chichester, UK.

Nicolosi, E., Deng, Z.N., Gentile, A., Malfa, S.L., Continella, G. and Tribulato, E. (2000) Citrus phylogeny and genetic origin of important species as investigated by molecular markers. *Theoretical and Applied Genetics* 100, 1155–1166.

Nilen, R.A. (1971) *Barley Genetics II*. Pullman, Washington, DC.

Nilsson-Ehle, H. (1909) Kreuzungsuntersuchungen an Hafer and Weizen. *Acta Universitatis Lundensis* 5, 1–122.

Nishiyama, I. and Temamura, T. (1962) Mexican wild form of sweet potato. *Economic Botany* 16, 304–314.

Nishiyama, I., Miyasaki, T. and Sakamoto, S. (1975) Evolutionary autopolyploidy of the sweet potato (*Ipomoea batatas* (L.) Lam.) and its progenitors. *Euphytica* 24, 197–208.

Noggle, G.R. (1946) The physiology of polyploidy in plants I. Review of literature. *Lloydia* 9, 153–173.

Noguti, Y., Oka, H. and Otuka, T. (1940) Studies on the polyploidy of *Nicotiana* induced by the treatment with colchicine. II. Growth rate and chemical analysis of diploid and its autotetraploid in *Nicotiana rustica* and *N. tabacum*. *Japanese Journal of Botany* 10, 343–364.

Normile, D. (1997) Yangtze seen as earliest rice site. *Science* 275, 309–310.

Ochiai, T., Nguyen, V.X., Tahara, M. and Yoshino, H. (2001) Geographical differentiation of Asian taro, *Colocasia esculenta* (L.) Schott, detected by RAPD and isozyme analysis. *Euphytica* 122, 219–234.

Ochoa, C.M. (1990) *The Potatoes of South America: Bolivia*. Cambridge University Press, New York.

O'Hanlon, P.C., Peakall, R. and Briese, D.T. (1999) Amplified fragment length polymosphism (AFLP) reveals introgression in weedy *Onopordum* thistles: hybridization and invasion. *Molecular Ecology* 8, 1239–1246.

Ohno, S. (1970) *Evolution by Gene Duplication*. Springer-Verlag, New York.

Oka, H.I. (1974) Experimental studies on the origin of cultivated rice. *Genetics* 78, 475–486.

Oka, H.I. (1975) The origin of cultivated rice and its adaptive evolution. In: *Rice in Asia*. University of Tokyo Press, Tokyo.

Oka, H.I. (1988) *Origin of Cultivated Rice*. Japan Societies Press, Tokyo.

Oliver, J.L. and Martinez-Zapater, J.M. (1984) Allozyme variability and phylogenetic relationships in the cultivated potato (*Solanum tuberosum*) and related species. *Plant Systematics and Evolution* 148, 1–18.

Olmo, H.P. (1995) Grapes: *Vitis, Muscadinia* (Vitaceae). In: Smartt, J. and Simmonds, N.W. (eds) *Evolution of Crop Species*. Longman Scientific and Technical, Harlow, UK, pp. 485–490.

Olsen, K.M. (2002) Population history of *Manihot esculenta* (Euphorbiaceae) inferred from nuclear DNA. *Molecular Ecology* 11, 901–911.

Olsen, K.M. and Schaal, B.A. (1999) Evidence on the origin of cassava: phylogeography of *Manihot esculenta*. *Proceedings of the National Academy of Sciences USA* 96, 5586–5591.

Olsen, K.M. and Schaal, B.A. (2001) Microsatellite variation in cassava (*Manihot esculenta*, Euphorbiaceae) and its wild relatives: further evidence for a southern Amazonian origin of domestication. *American Journal of Botany* 88, 131–142.

Orgel, L.E. and Crick, F.H.C. (1980) Selfish DNA: the ultimate parasite. *Nature* 284, 604–607.

Orjeda, G., Freyre, R. and Iwanaga, M. (1990) Production of 2*n* pollen in diploid *Ipomoea trifida* a putative ancestor of sweet potato. *Journal of Heredity* 81, 462–467.

Ortiz, R. (1997) Morphological variation in *Musa* germplasm. *Genetic Resources and Crop Evolution* 44, 393–404.

Osuji, J.O., Harrison, G., Crouch, J. and Heslop-Harrison, J.S. (1997) Identification of the genomic constitution of *Musa* lines (bananas, plantains and hybrids) using molecular cytogenetics. *Annals of Botany* 80, 787–793.

Otto, S.P. and Whitton, J. (2000) Polyploid incidence and evolution. *Annual Review of Genetics* 34, 401–437.

Ownbey, M. (1950) Natural hybridization and amphiploidy in the genus *Tragapogon*. *American Journal of Botany* 37, 487–499.

Ozkan, H., Levy, A.A. and Feldman, M. (2001) Allopolyploidy-induced rapid genome evolution in the wheat (*Aegilops–Triticum*) group. *The Plant Cell* 13, 1735–1747.

Palmer, J.D. (1985) Comparative organization of chloroplast genomes. *Annual Review of Genetics* 19, 325–354.

Palmer, J.D., Shields, C.R., Cohen, D.B. and Orton, T.J. (1983) Chloroplast DNA evolution and the origin of amphidiploid *Brassica* species. *Theoretical and Applied Genetics* 65, 181–189.

Palmer, J.R. (1987) Chloroplast DNA evolution and biosystematic uses of chloroplast DNA variation. *American Naturalist* 130, S6–S29.

Parfitt, D.E. and Arulsekar, S. (1989) Inheritance and isozyme diversity for GOT and PGM among grape cultivars. *Journal of the American Society of Horticultural Science* 114, 486–491.

Parfitt, D.E., Arulsekar, S. and Ramming, D.W. (1985) Identification of plum × peach hybrids by isoenzyme analysis. *HortScience* 20, 246–248.

Pasquet, R.S. (1997) A new subspecies of *Vigna unguiculata* (*Leguminosae – Papilionideae*). *Kew Bulletin* 52, 840.

Pasquet, R.S. (1998) Morphological study of cultivated cowpea *Vigna unguiculata* (L.) Walp. Importance of ovule number and definition of cv gr *melanophthalmus*. *Agronomie* 18, 61–70.

Pasquet, R.S. (1999) Genetic relationships among subspecies of *Vigna unguiculata* (L.) Walp. based on allozyme variation. *Theoretical and Applied Genetics* 98, 1104–1119.

Patel, G.I. and Olmo, H.P. (1955) Cytogenetics of *Vitis*, I. The hybrid *V. vinifera* × *V. rotundifolia*. *American Journal of Botany* 42, 141–159.

Paterniani, E. (1969) Selection for reproductive isolation between two species of maize, *Zea mays* L. *Evolution* 23, 534–547.

Paterson, A.H. (1995) Molecular dissection of quantitative traits: progress and prospects. *Genome Research* 5, 321–333.

Paterson, A.H., Schertz, K.F., Lin, Y.-R., Liu, S.-C. and Chang, Y.-L. (1995) The weediness of wild plants: molecular analysis of genes influencing dispersal and persistence of johnsongrass, *Sorghum halepense* (L.) Pers. *Proceedings of the National Academy of Sciences USA* 9, 6127–6131.

Paterson, A.H., Schertz, K.F., Lin, Y. and Li, Z. (1998) *Case History in Plant Domestication: Sorghum, an Example of Cereal Evolution*. CRC Press, London.

Paterson, A.H., Bowers, J.E., Burlow, M.D., Draye, X., Elsik, C.G., Jiang, C.-X., Katsar, C.S., Lan, T.-H., Lin, Y.-R., Ming, R. and Wright, R.J. (2000) Comparative genomics of plant chromosomes. *The Plant Cell* 12, 1523–1539.

Pearsall, D.M. (1992) *The Origins of Plant Cultivation in South America*. In: Cowen, C.W. and Watson, P.J. (eds) *Origins of Agriculture*. Smithsonian Institute Press, Washington, DC, pp. 173–206.

Peck, S.L., Ellner, S.P. and Gould, F. (1998) A spacially explicit stochastic model demonstrates the feasibility of Wright's shifting balance theory. *Evolution* 52, 1834–1839.

Pederson, W.L. and Leath, S. (1988) Pyramiding major genes for resistance to maintain residual effects. *Annual Review of Phytopathology* 26, 369–378.

Pedrosa, A., Schweizer, D. and Guerra, M. (2000) Cytological heterozygosity and the hybrid origin of sweet orange [*Citrus sinensis* (L.) Oseck]. *Theoretical and Applied Genetics* 100, 361–367.

Pejic, I., Ajmone-Marsan, P., Morgante, M., Kozumplick, V., Castiglioni, P., Taramino, G. and Motto, M. (1998) Comparative analysis of genetic similarity among maize inbred lines detected by RFLPs, RAPDs, SSRs and AFLPs. *Theoretical and Applied Genetics* 97, 1248–1255.

Pendergast, H.D.V. (1995) *Published Sources of Information on Wild Plant Species*. CAB International, Wallingford, UK.

Perl-Treves, R. and Galun, E. (1985) Phylogeny of *Cucumis* based on isozyme variability and its comparison with plastome phylogeny. *Theoretical and Applied Genetics* 71, 430–436.

Pfeiffer, J.E. (1978) *The Emergence of Man*. Harper and Row, New York.

Phillips, L.L. (1966) The cytology and phytogenetics of the diploid species of *Gossypium*. *American Journal of Botany* 53, 328–335.

Phillips, L.L. (1979) Cotton. In: Simmonds, N.E. (ed.) *Evolution of Crop Species*. Longman, London.

Phillipson, D.W. (1977) The excavation of Godedra rock shelter, Axum: an early occurrence of cultivated finger millet in northern Ethiopia. *Azania* 12, 53–82.

Pickersgill, B. (1969) The archeological record of chili peppers (*Capsicum* spp.). *American Antiquity* 35, 54–61.

Pickersgill, B. (1989) Cytological and genetical evidence on the domestication and diffusion of crops within the Americas. In: Harris, D. and Hillman, G. (eds) *Farming and Foraging*. Unwin Hyman, London.

Pillay, M., Nwakanma, D.C. and Tenkouano, A. (2000) Identification of RAPD markers linked to A and B genome sequences in *Musa* L. *Genome* 43, 763–767.

Piperno, D.R. and Flannery, K.V. (2001) The earliest archeological maize (*Zea mays* L.) from highland Mexico: new accelerator mass spectrometry dates and their implications. *Proceedings of the National Academy of Sciences USA* 98, 2101–2103.

Piperno, D.R. and Stothert, K.E. (2003) Phytolith evidence for early Holocene *Cucurbita* domestication in southwest Ecuador. *Science* 299, 1054–1055.

Piperno, D.R., Ranere, A.J., Holst, I. and Hansel, P. (2000) Starch grains reveal early root crop horticulture in the Panamanian tropical forest. *Nature* 407, 894–897.

Plucknett, D.L. (1979) Edible aroids. In: Simmonds, N.W. (ed.) *Evolution of Crop Species*. Longman, London, pp. 10–12.

Plucknett, D.L., Pena, R.S. and Obero, F. (1970) Taro (*Colacasia esculenta*). *Field Crop Abstracts* 23, 413–426.

Plucknett, D.L., Smith, N.J.H., Williams, J.T. and Anishetty, N.M. (1987) *Gene Banks and the World's Food*. Princeton University Press, Princeton, New Jersey.

Poncet, V., Lamy, F., Enjalbert, J., Joly, H., Sarr, A. and Robert, T. (1998) Genetic analysis of the domestication syndrome in pearl millet (*Pennisetum glaucum* L., Poaceae): inheritance of the major characters. *Heredity* 81, 648–658.

Powell, J.R. (1975) Protein variation in natural populations of animals. *Evolutionary Biology* 8, 79–119.

Powell, W., Morgante, M., Andre, C., Hanafey, M., Vogel, J., Tingey, S. and Rafalski, A. (1996) The comparison of RFLP, RAPD, AFLP, and SSR (microsatellite) markers for germplasm analysis. *Molecular Breeding* 2, 225–238.

Prakash, S. and Lewontin, R.C. (1968) A molecular approach to the study of genic heterozygosity in natural populations. III. Direct evidence of coadaptation in gene arrangements of *Drosophila*. *Proceedings of the National Academy of Sciences USA* 59, 398–405.

Prakash, S. and Lewontin, R.C. (1971) A molecular approach to the study of genic heterozygosity. V. Further direct evidence of co-adaptation in inversions of *Drosophila*. *Genetics* 69, 405–408.

Pratt, R.C. and Nabham, G.P. (1988) Evolution and diversity of *Phaseolus acutifolius* genetic resources. In: Gepts, P. (ed.) *Genetic Resources of Phaseolus Beans*. Kluwer, Dordrecht, The Netherlands, pp. 409–440.

Price, H.J. (1988) DNA content variation among higher plants. *Annals Missouri Botanical Garden* 75, 1248–1257.

Price, H.J., Chambers, K.L. and Bachmann, K. (1981) Geographic and ecological distribution of genomic DNA content variation in *Microseris douglasii* (Asteraceae). *Botanical Gazette* 142, 415–426.

Price, H.J., Chambers, K.L., Bachmann, K. and Riggs, J. (1986) Patterns of mean nuclear DNA content in *Microseris douglasii* (Asteraceae) populations. *Botanical Gazette* 147, 496–507.

Prideaux, T. (1973) *Cro-Magnon Man*. Time–Life, New York.

Prince, J.P., Loaiza-Figueroa, F. and Tanksley, S.D. (1992) Restriction fragment length polymorphisms and genetic distance among Mexican accessions of *Capsicum*. *Genome* 35, 726–732.

Prince, J.P., Lackney, V.K., Angeles, C., Blauth, J.R. and Kyle, M.M. (1995) A survey of DNA polymorphism within the genus *Capsicum* and the fingerprinting of pepper cultivars. *Genome* 38, 224–231.

Pringle, H. (1998) The slow birth of agriculture. *Science* 282, 1446–1450.

Proctor, M.C.F. and Yeo, P.F. (1973) *The Pollination of Flowers*. Collins, London.

Purseglove, J.W. (1972) *Tropical Crops: Monocotyledons*. Longman, London.

Putz, F.E. and Mooney, H.A. (1992) Green in leaf and tendril – the biology of vines. *Science* 256, 1339–1343.

Qu, L., Hancock, J.F. and Whallon, J.H. (1998) Evolution in an autopolyploid group displaying predominantly bivalent pairing at meiosis: genomic similarity of diploid *Vaccinium darrowi* and autotetraploid *V. corymbosum* (Ericaceae). *American Journal of Botany* 85, 698–703.

Quiros, C.F. (1982) Tetrasomic inheritance for multiple alleles in alfalfa. *Genetics* 101, 117–127.

Quiros, C.F. and McHale, N. (1985) Genetic analysis of isozyme variation in diploid and tetraploid potatoes. *Genetics* 111, 131–145.

Raina, S.N. and Mukai, Y. (1999) Genomic *in situ* hybridization in *Arachis* (Fabaceae) identifies the diploid wild progenitors of cultivated (*A. hypogeae*) and related wild (*A. monticola*) peanut species. *Plant Systematics and Evolution* 214, 251–262.

Rajhathy, T. and Thomas, H. (1974) *Cytogenetics of Oats* (*Avena*). Miscellaneous Publication 2, Genetic Society of Canada, Ottawa.

Raker, C.M. and Spooner, D.M. (2002) Chilean tetraploid cultivated potato, *Solanum tuberosum*, is distinct from the Andean populations: microsatallite data. *Crop Science* 42, 1451–1458.

Ramachandran, C. and Narayan, R.K.J. (1985) Chromosomal DNA variation in *Cucumis*. *Theoretical and Applied Genetics* 69, 497–502.

Ramser, J., Weising, K., Lopez-Peralta, C., Terhalle, W., Terauchi, R. and Kahl, G. (1997) Molecular marker based taxonomy and phylogeny of Guinea yam (*Dioscorea rotundata*). *Genome* 40, 903–915.

Ramsey, J. and Schemske, D.W. (1998) Pathways, mechanisms, and rates of polyploid formation in flowering plants. *Annual Review of Ecology and Systematics* 29, 467–501.

Rana, R.S. and Jain, H.K. (1965) Adaptive role of interchange heterozygosity in the annual *Chrysanthemum*. *Heredity* 20, 21–29.

Randall, D.D., Nelson, C.J. and Asay, K.H. (1977) Ribulose bisphosphate carboxylase: altered genetic expression in tall fescue. *Plant Physiology* 59, 38–41.

Randolph, L.F. (1966) *Iris nelsonii*, a new species of Louisiana iris of hybrid origin. *Baileya* 14, 143–169.

Rao, K.E.P., deWet, J.M.J., Brink, D.E. and Mengesha, M.H. (1987) Infraspecific variation and systematics of cultivated *Setaria italica*, foxtail millet (Poaceae). *Economic Botany* 41, 108–116.

Reed, B.M. (2001) Implementing cryogenic storage of clonally propagated plants. *Cryo-Letters* 22, 97–104.

Renfrew, J.M. (1973) *Palaeoethnobotany. The prehistoric food plants of the Near East and Europe*. Methuen, London.

Rhodes, F.H.T. (1983) Gradualism, punctuated equilibrium and the origin of species. *Nature* 305, 269–272.

Rhodes, M.M. and Dempsey, E. (1966) Induction of chromosome doubling at meiosis by the elongate gene in maize. *Genetics* 54, 505–522.

Rick, C.M. (1947) Partial suppression of hair development indirectly affecting fruitfulness and the proportion of cross pollination in a tomato mutant. *American Naturalist* 81, 185–202.

Rick, C.M. (1995) Tomato: *Lycopersion esculentum* (Solanaceae). In: Smartt, J. and Simmonds, N.W. (eds) *Evolution of Crop Plants*. Longman Scientific & Technical, Harlow, UK, pp. 452–457.

Rick, C.M. and Fobes, J.R. (1977) Allozyme variation in the cultivated tomato and closely related species. *Bulletin of the Torrey Botanical Club* 102, 376–384.

Rick, C.M. and Holle, M. (1990) Andean *Lycopersicon esculentum* var. *cerasiformae*: genetic variation and evolutionary significance. *Economic Botany* 44, 69–78.

Rick, C.M., Fobes, J.F. and Holle, M. (1977) Genetic variation in *Lycopersicon pimpinellifolium*: evidence of evolutionary change in mating system. *Plant Systematics and Evolution* 127, 139–170.

Rieseberg, L.H. (1991) Homoploid reticulate evolution in *Helianthus* (Asteraceae): evidence from ribosomal genes. *American Journal of Botany* 78, 1218–1237.

Rieseberg, L.H. (1995) The role of hybridization: old wine in new skins. *American Journal of Botany* 82, 944–953.

Rieseberg, L.H. (2000) Genetic mapping as a tool for studying speciation. In: Soltis, D.E., Soltis, P.S. and Doyle, J.J. (eds) *Molecular Systematics of Plants II. DNA Sequencing*. Kluwer Academic Publishers, Boston, Massachusetts.

Rieseberg, L.H. and Ellstrand, N.C. (1993) What can molecular and morphological markers tell us about plant hybridization? *Critical Reviews in Plant Science* 12, 213–241.

Rieseberg, L.H. and Gerber, D. (1995) Hybridization in the Catalina Island mountain mahogany (*Cerocarpus traskiae*): RAPD evidence. *Conservation Biology* 9, 199–203.

Rieseberg, L.H. and Seiler, G.J. (1990) Molecular evidence and the origin and development of the domesticated sunflower, *Helianthus annuus* (Asteraceae). *Economic Botany* 44, 79–91.

Rieseberg, L.H. and Warner, D.A. (1987) Electrophoretic evidence for hybridization between *Tragopogon mirus* and *T. miscellus* (Compositae). *Systematic Botany* 12, 281–285.

Rieseberg, L.H., Soltis, D.E. and Palmer, J.D. (1988) A molecular examination of introgression between *Helianthus annuus* and *H. bolanderi* (Compositae). Evolution 42, 227–238.

Rieseberg, L.H., Beckstrom-Sternberg, S. and Doan, K. (1990a) *Helianthus annuus* ssp. *texanus* has chloroplast DNA and nuclear RNA genes of *Helianthus debilis* ssp. *cucumerifolus*. *Proceedings of the National Academy of Sciences USA* 87, 593–597.

Rieseberg, L.H., Carter, R. and Zona, S. (1990b) Molecular tests of the hypothesized hybrid origin of two diploid *Helianthus* species (Asteraceae). *Evolution* 44, 1498–1511.

Rieseberg, L.H., van Fossen, C. and Desrochers, A. (1995) Hybrid speciation accompanied by genomic reorganization in wild sunflowers. *Nature* 375, 313–316.

Rieseberg, L.H., Sinervo, B., Linder, C.R., Ungerer, M. and Arias, D.M. (1996) Role of gene interactions in hybrid speciation: evidence from ancient and experimental hybrids. *Science* 272, 741–745.

Rieseberg, L.H., Whitton, J. and Gardner, K. (1999) Hybrid zones and genetic architecture of a barrier to gene flow between two sunflower species. *Genetics* 152, 713–727.

Rieseberg, L.H., Baird, S.J.E. and Gardner, K.A. (2000) Hybridization, introgression and linkage evolution. *Plant Molecular Biology* 42, 205–224.

Riley, H.P. (1938) A character analysis of colonies of *Iris hexagona* var. *giganticaerulea* and natural hybrids. *American Journal of Botany* 25, 727–738.

Riley, R. (1955) The cytogenetics of the differences between some *Secale* species. *Journal of Agricultural Science* 46, 277–283.

Riley, R., Unrau, J. and Chapman, V. (1958) Evidence on the origin of the B genome of wheat. *Journal of Heredity* 49, 91–98.

Rissler, J. and Mellon, M. (1996) *The Ecological Risks of Engineered Plants*. MIT Press, Cambridge, Massachusetts.

Ritland, K. and Clegg, M.L. (1987) Evolutionary analysis of plant DNA sequences. *The American Naturalist* 130, S74–S100.

Rivera, M. (1991) The prehistory of northern Chile: a synthesis. *Journal of World Prehistory* 5, 1–47.

Roa, A.C., Maya, M.M., Duque, M.C., Tohme, J., Allem, A.C. and Bonierbale, M.W. (1997) AFLP analysis of relationships among cassava and other *Manihot* species. *Theoretical and Applied Genetics* 95, 741–750.

Rodriguez, J.M., Berke, T., Engle, L. and Nienhuis, J. (1999) Variation among and within *Capsicum* species revealed by RAPD markers. *Theoretical and Applied Genetics* 99, 147–156.

Rogers, D.J. and Appan, S.G. (1973) Manihot *and* Manihoides *(Euphorbiacea)*. Neotropical Monograph No. 13, Hafner Press, New York.

Rohlf, F.J. (1998) *NTSYS-pc Numerical Taxonomy and Multivariate Analysis System, Version 2.0*. Exeter Publishing, Setauket, New York.

Rohlf, F.J. and Schnell, G.D. (1971) An investigation of the isolation-by-distance model. *American Naturalist* 105, 295–324.

Rohweder, H. (1937) Versuch zur Erfassung der mengemmassigen Bedeckung des Dors und Zingst mit polyploiden Pflanzen. *Planta* 27, 501–549.

Roose, M.L. and Gottleib, L.D. (1976) Genetic and biochemical consequences of polyploidy in *Tragopogon*. *Evolution* 30, 818–830.

Roose, M.L. and Gottleib, L.D. (1978) Stability of structural gene number in diploid species with different amounts of nuclear DNA and different chromosome numbers. *Heredity* 40, 159–163.

Roose, M.L. and Gottleib, L.D. (1980) Biochemical properties and level of expression of alcohol dehydrogenases in the allotetraploid plant *Tragopogon miscellus* and its diploid progenitors. *Biochemical Genetics* 18, 1065–1085.

Roose, M.L., Soost, R.K. and Cameron, J.W. (1995) Citrus (*Rutaceae*). In: Smartt, J. and Simmonds, N. (eds) *Evolution of Crop Species*. Longman Scientific and Technical, Harlow, UK, pp. 443–449.

Roughgarden, J. (1976) Resource partitioning among competing species – a coevolutionary approach. *Theoretical and Population Biology* 9, 388–424.

Rowe, P. and Rosales, F.E. (1996) Bananas and plantains. In: Janick, J. and Moore, J.N. (eds) *Fruit Breeding* Vol. 1: *Tree and Tropical Fruits*. John Wiley & Sons, New York, pp. 167–212.

Rowewal, S.S., Ramanujam, S. and Mehra, K.L. (1969) Plant type in bengalgrain. *Indian Journal of Genetics* 26, 255–261.

Rowlands, D.C. (1959) A case of mimicry in plants of *Vicia sativa* in lentil crops. *Genetica* 30, 435–446.

Russell, J.R., Fuller, J.D., Macaulay, M., Hatz, B.G., Jahoor, A., Powell, W. and Waugh, R. (1997) Direct comparison of levels of genetic variation among barley accesssions detected by RFLPs, AFLPs, SSRs and RAPDs. *Theoretical and Applied Genetics* 95, 714–722.

Sage, T.L., Bertin, R.I. and Williams, E.G. (1994) Ovarian and other late-acting self-incompatibility systems. In: Williams, E.G., Knox, R.B. and Clark, A.E. (eds) *Genetic Control of Self-incompatibility and Reproductive Development in Flowering Plants*. Kluwer Academic, Amsterdam, pp. 116–140.

Sage, T.L., Strumas, F., Cole, W.W. and Barrett, S.C.H. (1999) Differential ovule development following self- and cross-pollination: the basis of self-sterility in *Narcissus triandrus* (Amaryllidaceae). *American Journal of Botany* 86, 855–870.

Salaman, R.N. (1949) *The History and Social Influence of the Potato*. Cambridge University Press, Cambridge.

Salick, J., Cellinese, N. and Knapp, S. (1997) Indigenous diversity of cassava: generation, maintenance, use and loss among the Amuesha Peruvian Upper Amazon. *Economic Botany* 51, 6–19.

Salimath, S.S., Deoliveira, A.C., Godwin, I.D. and Bennetzen, J.L. (1995) Assessment of genome origins and genetic diversity in the genus *Eleusine* with DNA markers. *Genome* 38, 757–763.

Sampson, D.R. and Tarumoto, I. (1976) Genetic variances in an eight parent half diallel of oats. *Canadian Journal of Genetics and Cytology* 18, 419–427.

Sanderson, J.A. and Hulbert, E.O. (1955) Sunlight as a source of radiation. In: Hollaender, A. (ed.) *Radiation Biology II. Ultraviolet and Related Radiations*. McGraw-Hill, New York.

Sanjur, O.I., Piperno, D., Andres, T.C. and Wessel-Beaver, L. (2002) Phylogenetic relationships among domesticated and wild species of *Cucurbita* (Cucurbitaceae) inferred from a mitochondrial gene: implications for crop plant evolution and areas of origin. *Proceedings of the National Academy of Sciences USA* 99, 535–540.

SanMiguel, P. and Bennetzen, J.L. (1998) Evidence that a recent increase in maize genome size was caused by the massive amplification of intergene retrotranspositions. *Annals of Botany* 82, 37–44.

Sasaki, K. (1986) Development and type of East Asian agriculture. In: Hanihara, K. (ed.) *The Origin of Japanese*. Shogakukan, Tokyo.

Sasanuma, T., Miyashita, N.T. and Tsunewaki, K. (1996) Wheat phylogeny determined by RFLP analysis of nuclear DNA 3: intra- and interspecific variations of five *Aegilops* species. *Theoretical and Applied Genetics* 92, 928–934.

Sauer, C.O. (1952) *Agricultural Origins and Dispersals*. MIT Press, Cambridge, Massachusetts.

Sauer, J.D. (1967) The grain amaranths and their relatives – a revised taxonomic and geographic survey. *Annals of the Missouri Botanical Garden* 54, 103–137.

Sauer, J.D. (1993) *Historical Geography of Crop Plants: a Select Roster*. CRC Press, Boca Raton, Florida.

Scandalios, J.G. (1969) Genetic control of multiple molecular forms of enzymes in plants: a review. *Biochemical Genetics* 3, 37–79.

Schaal, B.A. (1975) Population structure and local differentiation in *Liatris cylindracea*. *American Naturalist* 109, 511–528.

Schäfer, H.I. (1973) Zur Taxonomie der *Vicia narbonensis*-Gruppe. *Kulturpflanze* 21, 211–273.

Schemske, D.W. (2000) Understanding the origin of species. *Evolution* 54, 1069–1073.

Schemske, D.W. and Bierzychudek, P. (2001) Perspective: evolution of flower color in the desert annual *Linanthus parryae*: Wright revisited. *Evolution* 55, 1269–1282.

Schemske, D.W. and Bradshaw, H.D., Jr. (1999) Pollinator preference and the evolution of floral traits in monkeyflowers (*Mimulus*). *Proceedings of the National Academy of Sciences USA* 96, 11910–11915.

Schilling, E. and Heiser, C. (1981) Infrageneric classification of *Helianthus* (Compositae). *Taxon* 30, 393–403.

Schmidt, T. and Heslop-Harrison, J.S. (1998) Genomes, genes and junk: the large-scale organization of plant chromosomes. *Trends in Plant Science* 3, 195–199.

Schmit, V. and Debouck, D.G. (1991) Observations on the origin of *Phaseolus polyanthus*. *Economic Botany* 45, 345–364.

Schumann, C.M. and Hancock, J.F. (1990) Paternal inheritance of plastids in *Medicago sativa*. *Theoretical and Applied Genetics* 78, 863–866.

Scora, R.W. (1975) On the history and origin of *Citrus*. *Bulletin of the Torrey Botanical Club* 102, 369–375.

Scorza, R. and Okie, W. (1991) Peaches (*Prunus*). In: Moore, J.N. and Ballington, J.R. (eds) *Genetic Resources of Temperate Fruit and Nut Crops*. International Society of Horticultural Science, Wageningen, The Netherlands.

Scott, N.S. and Possingham, J.V. (1980) Chloroplast DNA in expanding spinach leaves. *Journal of Experimental Botany* 31, 1082–1092.

Sears, B. (1980) Elimination of plastids during spermatogenesis and fertilization in the plant kingdom. *Plasmid* 4, 233–255.

Sears, E.R. (1944) Cytogenetic studies with polyploid species of wheat. II. Additional chromosomal aberrations in *Triticum vulgare*. *Genetics* 29, 232–246.

Seavey, S.R. and Bawa, K.S. (1986) Late-acting self incompatibility in angiosperms. *Botanical Review* 52, 195–219.

Seavey, S.R. and Carter, S.K. (1994) Self-sterility in *Epilobium obcordatum* (Onagraceae). *American Journal of Botany* 81, 331–338.

Second, G. (1982) Origin of the genetic diversity of cultivated rice (*Oryza* spp.): study of the polymorphism scored at 40 isozyme loci. *Japanese Journal of Genetics* 57, 25–57.

Second, G. (1985) A new insight into the genome differentiation in *Oryza* L. through isozymic studies. In: Sharma, A.K. and Sharma, A. (eds) *Advances in Chromosome and Cell Genetics*. University Press, New York, pp. 45–78.

Sefc, K.M., Lopes, M.S., Lefort, F., Botta, R., Roubelakis-Angelakis, K.A., Ibanez, J., Pejic, I., Wagner, H.W., Gloossl, J. and Steinkellner, H. (2000) Microsatellite variability in grapevine cultivars from different European regions and evaluation of assignment testing to assess the geographic origin of cultivars. *Theoretical and Applied Genetics* 100, 498–505.

Settler, T.L., Schrader, L.E. and Bingham, E.T. (1978) Carbon dioxide exchange rates, transpiration, and leaf characteristics in genetically equivalent ploidy levels in alfalfa. *Crop Science* 18, 327–332.

Shah, D. (1986) Engineering herbicide tolerance in transgenic plants. *Science* 233, 478–481.

Shaked, H., Kashkush, K., Ozkan, H., Feldman, M. and Levy, A.A. (2001) Sequence elimination and cytosine methylation are rapid and reproducible responses of the genome to wide hybridization and allopolyploidy in wheat. *Plant Cell* 13, 1749–1759.

Shands, H.L. (1990) Plant genetic resources and conservation: the role of the gene bank in delivering useful genetic materials to the research scientist. *Journal of Heredity* 81, 7–10.

Shii, C.T., Mok, M.C. and Mok, D.W.S. (1981) Developmental controls of morphological mutants of *Phaseolus vulgaris* L.: differential expression of mutant loci in plant organs. *Developmental Genetics* 2, 279–290.

Shimamoto, Y., Fukushi, H., Abe, J., Kanazawa, A., Gai, J., Gao, Z. and Xu, D. (1998) RFLPs of chloroplast and mitochondrial DNA in wild soybean, *Glycine soja*, growing in China. *Genetic Resources and Crop Evolution* 45, 433–439.

Shimamoto, Y., Abe, J., Gao, Z., Gai, J. and Thseng, F.-S. (2000) Characterizing the cytoplasmic diversity and phyletic relationship of Chinese landraces of soybean, *Glycine max*, based on RFLPs of chloroplast and mitochondrial DNA. *Genetic Resources and Crop Evolution* 47, 611–617.

Shipman, P. (1988) *What Does it Take to be a Meateater?* Discover Publications, New York.

Shiotani, I. and Kawase, T. (1989) Genomic structure of the sweet potato and haploids in *Ipomoea trifida*. *Japanese Journal of Breeding* 39, 57–66.

Silva, N.F. and Goring, D.R. (2001) Mechanisms of self-incompatibility in flowering plants. *Cellular and Molecular Life Sciences* 58, 1988–2007.

Simmonds, N.W. (1979) *Principles of Crop Improvement*. Longman, London.

Simmonds, N.W. (1995a) Bananas: *Musa* (Musaceae). In: Smartt, J. and Simmonds, N.W. (eds) *Evolution of Crop Plants*. Longman Scientific & Technical, Harlow, UK, pp. 370–375.

Simmonds, N.W. (1995b) Potatoes: *Solanum tuberosum* (Solanaceae). In: Smartt, J. and Simmonds, N.W. (eds) *Evolution of Crop Plants*. Longman Scientific and Technical, Harlow, UK, pp. 466–471.

Simmonds, N.W. (1995c) Food crops: 500 years of travel. In:. Duncan, R.R (ed.) *International Germplasm Transfer: Past and Present*. Special Publication No. 23, CSSA, Madison, Wisconsin, pp. 31–46.

Simpson, G.G. (1961) *Principles of Animal Taxonomy*. Columbia University Press, New York.

Sims, L.E. and Price, H.J. (1985) Nuclear DNA content variation in *Helianthus* (Asteraceae). *American Journal of Botany* 72, 1213–1219.

Singh, A.K. (1995) Groundnut: *Arachis hypogaea* (Leguminosae – Papilionoideae). In: Smarrtt, J. and Simmonds, N.W. (eds) *Evolution of Crop Plants*. Longman Scientific & Technical, Harlow, UK, pp. 246–250.

Singh, K.B. and Ocampo, B. (1993) Interspecific hybridization in annual *Cicer* species. *Journal of Genetics and Breeding* 47, 199–204.

Singh, K.B. and Ocampo, B. (1997) Exploitation of wild species for yield improvement in chickpea. *Theoretical and Applied Genetics* 95, 418–423.

Singh, K.B., Malhorta, R.S., Halila, H., Knights, E.J. and Verma, M.M. (1994) Current status and future strategy in breeding chickpea for resistance to biotic and abiotic stresses. *Euphytica* 73, 137–149.

Singh, S.P. and Hymowitz, T. (1995) The genomic relationships among six perennial species of the genus *Glycine* subgenus *Glycine*. *Euphytica* 33, 337–345.

Slatkin, M. (1985) Gene flow in natural populations. *Annual Review of Ecology and Systematics* 16, 393–430.

Small, R.L. and Wendel, J.F. (1998) The mitochondrial genome of allopolyploid cotton (*Gossypium* L.). *Journal of Heredity* 90, 251–253.

Smartt, J. (1984) Gene pools in grain legumes. *Economic Botany* 38, 24–35.

Smartt, J. (1999) *Grain Legumes: Evolution and Genetic Resources*. Cambridge University Press, Cambridge.

Smartt, J. and Simmonds, N.W. (1995) *Evolution of Crop Plants*, Vol. 2. Longman Scientific & Technical, Harlow, UK.

Smartt, J., Gregory, W.C. and Gregory, M.P. (1978) The genomes of *Arachis hypogaea* I. Cytogenetic studies of putative genome donors. *Euphytica* 27, 665–675.

Smith, B. (2001) Documenting plant domestication: the consilience of biological and archeological approaches. *Proceedings of the National Academy of Sciences USA* 98, 1324–1326.

Smith, B.D. (1989) Origins of agriculture in eastern North America. *Science* 246, 1566–1571.

Smith, B.D. (1998) *The Emergence of Agriculture*. Scientific American Library, New York.

Smith, D.C. (1946) *Sedum pulchellum*: a physiological and morphological comparison of diploid, tetraploid and hexaploid races. *Bulletin of the Torrey Botanical Club* 73, 495–541.

Smith, F.H. and Clarkson, Q.D. (1956) Cytological studies of interspecific hybridization in *Iris*, subsection Californicae. *American Journal of Botany* 43, 582–588.

Snow, A.A. and Palma, P. (1997) Commercialization of transgenic plants: potential ecological risks. *BioScience* 47, 86–96.

Snow, R. (1960) Chromosomal differentiation in *Clarkia dudleyana*. *American Journal of Botany* 47, 302–309.

Snowden, J.D. (1936) *The Cultivated Races of Sorghum*. Allard and Son, London.

Snowdon, R.J., Friedrich, T., Friedt, W. and Kohler, W. (2002) Identifying the chromosomes of the A- and C-genome diploid *Brassica* species *B. rapa* (syn. *campestris*) and *B. oleraceae* in their amphidiploid *B. napus*. *Theoretical and Applied Genetics* 104, 533–538.

Sokal, R.R. and Rohlf, F.J. (1969) *Biometry*. W.H. Freeman, San Francisco.

Sokal, R.R., Oden, N.L. and Wilson, C. (1991) Genetic evidence for the spread of agriculture in Europe by demic diffusion. *Nature* 351, 143–145.

Solano, R., Hueros, G., Fominaya, A. and Ferrer, E. (1992) Organization of repeated sequences in species of the genus *Avena*. *Theoretical and Applied Genetics* 83, 602–607.

Solbrig, O.T. and Simpson, B.B. (1977) A garden experiment on competition between biotypes of the common dandelion (*Taraxacum officinale*). *Journal of Ecology* 65, 427–430.

Soltis, D.E. (1984) Autopolyploidy in *Tolmiea menziesii* (Saxifragaceae). *American Journal of Botany* 71, 1171–1174.

Soltis, D.E. and Soltis, P.S. (1988) Electrophoretic evidence for tetrasomic segregation in *Tolmiea menziesii* (Saxifragacea). *Heredity* 60, 375–382.

Soltis, D.E. and Soltis., P.S. (1989a). Allopolyploid speciation in *Tragopogan*: insights from chloroplast DNA. *American Journal of Botany* 76, 1119–1124.

Soltis, D.E. and Soltis, P.S. (1989b) Tetrasomic inheritance in *Heuchera micrantha* (Saxifragaceae). *Journal of Heredity* 80, 123–126.

Soltis, D.E. and Soltis, P.S. (1990) Isozyme evidence for ancient polyploidy in primitive angiosperms. *Systematic Botany* 15, 328–337.

Soltis, D.E. and Soltis, P.S. (1993) Molecular data and the dynamic nature of polyploidy. *Critical Reviews in Plant Sciences* 12, 243–273.

Soltis, D.E. and Soltis, P.S. (1995) The dynamic nature of polyploid genomes. *Proceedings of the National Academy of Sciences USA* 92, 8089–8091.

Soltis, D.E. and Soltis, P.S. (1999) Polyploidy: recurrent formation and genome evoluion. *Trends in Ecology and Sytematics* 14, 348–352.

Soltis, D.E. and Soltis, P.S. (2000) Choosing an approach and an appropriate gene for phylogenetic analysis. In: Soltis, D.E., Soltis, P.S. and Doyle, J.J. (eds) *Molecular Systematics of Plants II. DNA Sequencing*. Kluwer Academic, Boston, pp. 1–42.

Soltis, P.S. and Soltis, D.E. (2000) The role of genetic and genomic attributes in the success of polyploids. *Proceedings of the National Academy of Sciences USA* 97, 7051–7057.

Song, K.M., Osborn, T.C. and Williams, P.H. (1990) *Brassica* taxonomy based on nuclear restriction fragment length polymosphisms (RFLPs). 3. Genome relationships in *Brassica* and related genera and the origin of *B. oleraceae* and *B. rapa* (syn. *campestris*). *Theoretical and Applied Genetics* 79, 497–506.

Song, K.M., Lu, P., Tang, K. and Osborn, T.C. (1995) Rapid genome change in synthetic polyploids of *Brassica* and its implications for polyploid evolution. *Proceedings of the National Academy of Sciences USA* 92, 7719–7723.

Souza Machado, V., Bandeen, J.D., Taylor, W.D. and Lavigne, P. (1977) Atrazine resistant biotypes of common ragweed and birds rape. *Research Report of the Canadian Weed Commission (East Section)* 22, 305.

Spencer, H.A. and Hawkes, J.G. (1980) On the origin of cultivated rye. *Biological Journal of the Linnean Society* 13, 299–313.

Spier, R.F.G. (1951) Some notes on the origin of taro. *Southwestern Journal of Anthropology* 7, 69–76.

Spooner, D.M. and Hijmans, R.J. (2001) Potato systematics and germplasm collecting, 1989–2000. *American Journal of Potato Research* 78, 237–268.

Sporne, K.R. (1971) *The Mysterious Origin of Flowering Plants*. Oxford University Press, London.

Sreekumari, M.T. and Mathew, P.M. (1991) Karyomorphology of five morphotypes of taro (*Colocasia esculenta* (L.) Schott.). *Cytologia* 56, 215–218.

Stadler, L.J. (1942) Some observations on gene variability and spontaneous mutation. *Spragg Memorial Lectures (Michigan State University)* 3, 3–15.

Stalker, H.T., Desi, D.S., Parry, D.C. and Hahn, J.H. (1991) Cytological and interfertility relationships of *Arachis* section *Arachis*. *American Journal of Botany* 78, 238–246.

Stapf, O. and Hubbard, C.E. (1934) *Pennisetum*. In: Prain, D. (ed.) *Flora of Tropical Africa* 9. Reeve Brothers, London.

Staub, J.E., Serquen, F.C. and Gupta, M. (1996) Genetic markers, map construction and their application in plant breeding. *HortScience* 31, 729–741.

Staudt, G. (1989) The species of *Fragaria*, their taxonomy and geographical distribution. *Acta Horticulturae* 265, 23–33.

Stebbins, G.L. (1947) Types of polyploids: their classification and significance. *Advances in Genetics* 1, 403–429.
Stebbins, G.L. (1950) *Variation and Evolution in Plants*. Columbia University Press, New York.
Stebbins, G.L. (1956) Cytogenetics and evolution of the grass family. *American Journal of Botany* 43, 890–905.
Stebbins, G.L. (1957) The hybrid origin of microspecies in the *Elymus glaucus* complex. *Cytologia* 36, 336–340.
Stebbins, G.L. (1959) Genes, chromosomes and evolution. In: Turrill, W. (ed.) *Vistas in Botany*. Pergamon Press, Elmsford, New York.
Stebbins, G.L. (1971) *Chromosomal Variation in Higher Plants*. Arnold, London.
Stebbins, G.L. (1972) Research on the evolution of higher plants: problems and prospects. *Canadian Journal of Genetics and Cytology* 14, 453–462.
Stebbins, G.L. (1974) *Flowering Plants*. Harvard University Press, Cambridge.
Stebbins, G.L. (1980) Polyploidy in plants: unsolved problems and prospects. In: Lewis, W.H. (ed.) *Polyploidy – Biological Relevance*. Plenum, New York, pp. 495–520.
Stephens, S.G. (1946) The genetics of 'corky'. I. The New World alleles and their possible role as an interspecific isolating mechanism. *Journal of Genetics* 47, 150–161.
Stephens, S.G. (1951) Possible significance of duplication in evolution. *Advances in Genetics* 4, 247–265.
Stevenson, G.C. (1965) *Genetics and Breeding of Sugarcane*. Longman, London.
Story, D.A. (1985) Adaptive strategies of archaic cultures of the West Gulf Coastal Plain. In: Ford, R.I. (ed.) *Prehistoric Food Production in North America*. Museum of Anthropology, University of Michigan, Ann Arbor.
Strauss, S.H. and Libby, W.J. (1987) Allozyme heterosis in *Radiata* pine is poorly explained by overdominance. *American Naturalist* 130, 879–890.
Stubbe, W. (1960) Untersuchungen zur genetischen Analyses des Plastoms von *Oenothera*. *Zeitschrift für Botanisch* 48, 191–218.
Stubbe, W. (1964) The role of the plastome in evolution of the genus *Oenothera*. *Genetics* 35, 28–33.
Stuber, C.W. and Goodman, M.M. (1983) Inheritance, intracellular localization and genetic variation of phosphogluco-mutase isozymes in maize (*Zea mays* L.). *Biochemical Genetics* 21, 667–689.
Stutz, H.C. (1972) The origin of cultivated rye. *American Journal of Botany* 59, 59–70.
Sun, G., Dilcher, D.L., Zheng, S. and Zhou, Z. (1998) In search of the first flower: a Jurassic angiosperm, *Archaefructus*, from northeast China. *Science* 282, 1692–1694.
Swingle, W.T. (1967) The botany of citrus and its wild relatives. In: Reuther, W., Webber, H.J. and Bachelor, L.D. (eds) *The Citrus Industry*. Division of Agricultural Science, University of California, Berkeley, pp. 1–39.
Tahara, M., Nguyen, V.X. and Yoshino, H. (1999) Isozyme analysis of Asian diploid and tetraploid taro, *Colocasia esculenta* (L.) Schott. *Aroideana* 22, 72–78.
Tai, G.C.C. (1976) Estimation of general and specific combining abilities in potato. *Canadian Journal of Genetics and Cytology* 18, 463–470.
Takhtajan, A. (1969) *Flowering Plants – Origin and Dispersal*. Oliver and Boyd, Edinburgh.

Talbert, L.E., Doebley, J.F., Larson, S. and Chandler, V.L. (1990) *Tripsicum andersonii* is a natural hybrid involving *Zea* and *Tripsicum*: molecular evidence. *American Journal of Botany* 77, 722–726.

Talbert, L.E., Blake, N.K., Storie, E.W. and Lavin, M. (1995) Variability in wheat based on low-copy DNA sequence comparisons. *Genome* 38, 951–957.

Tanaka, T. (1954) *Species Problems in Citrus*. Japanese Society for the Promotion of Science, Tokyo.

Tanksley, S.D. (1993) Mapping polygenes. *Annual Review of Genetics* 27, 205–233.

Tanksley, S.D. and McCouch, S.R. (1997) Seed banks and molecular maps: unlocking the genetic potential from the wild. *Science* 277, 1063–1066.

Tanksley, S.D., Miller, J.C. and Bernatsky, R. (1988) Molecular mapping of plant chromosomes. In: Gustafson, J.P. and Appels, R. (eds) *Chromosome Structure and Function*. Plenum Press, New York, pp. 157–173.

Tanksley, S.D., Ganal, M.W., Prince, J.P., de Vicente, M.C., Bonierbale, M.W., Broun, P., Fulton, T.M., Giovannoni, J.J., Grandillo, S. and Martin, G.B. (1992) High density molecular linkage maps of the tomato and potato genomes. *Genetics* 132, 1141–1160.

Tanno, K., Taketa, S., Takeda, K. and Komatsuda, T. (2002) A DNA marker closely linked to the *vrs1* locus (row-type gene) indicates multiple origins of six-rowed cultivated barley (*Hordeum vulgare* L.). *Theoretical and Applied Genetics* 104, 54–60.

Tattersall, I. (1998) *Becoming Human*. Harcourt, Brace & Company, Orlando, Florida.

Tautz, D. (1989) Hypervariability of simple sequences as a general source for polymorphic DNA markers. *Nucleic Acids Research* 17, 6463–6471.

Tayyar, R.I. and Waines, J.G. (1996) Genetic relationships among annual species of *Cicer* (Fabaceae) using isozyme variation. *Theoretical and Applied Genetics* 92, 245–254.

Templeton, A.R. (1981) Mechanisms of speciation – a population genetics approach. *Annual Review of Ecology* 12, 23–41.

Templeton, A.R. (1989) The meaning of species and speciation: a genetic perspective. In: Otte, D. and Endler, J.A. (eds) *Speciation and its Consequences*. Sinauer Associates, Sunderland, Massachusetts, pp. 3–27.

Tenaillon, M.I., Sawkins, M.C., Long, A.D., Gaut, R.L., Doebley, J.F. and Gaut, B.S. (2001) Patterns of DNA sequence polymorphism along chromosome 1 of maize (*Zea mays* ssp. *mays* L.). *Proceedings of the National Academy of Sciences USA* 98, 9161–9166.

Terachi, T., Ogihara, Y. and Tsunewaki, K. (1990) The molecular basis of genetic diversity among cytoplasms of *Triticum* and *Aegilops*. 7. Restriction endonuclease analysis of mitochondrial DNAs from polyploid wheats and their ancestral species. *Theoretical and Applied Genetics* 80, 366–373.

Terauchi, R., Chikaleke, V.A., Thottappilly, G. and Hahn, S.K. (1992) Origin and phylogeny of guinea yams as revealed by RFLP analysis of chloroplast DNA and nuclear ribosomal DNA. *Theoretical and Applied Genetics* 83, 743–751.

Thompson, K.F. (1979) Cabbages, kales, etc.: *Brassica oleracea* (Cruciferae). In: Simmonds, N.E. (ed.) *Evolution of Crop Species*. Longman, London, pp. 49–52.

Timko, M.P. and Vasconcelos, A.C. (1981) Photosynthetic activity and chloroplast membrane polypeptides in euploid cells of *Ricinus*. *Physiologia Plantarum* 52, 192–196.

Ting, Y.C. (1985) Meiosis and fertility of anther derived maize plants. *Maydica* 30, 161–169.
Ting, Y.C. and Kehr, A.E. (1953) Meiotic studies in the sweet potato. *Journal of Heredity* 44, 207–211.
Townsend, C.E. and Remmenga, E.E. (1968) Inbreeding in tetraploid alsike clover, *Trifolium hybridum*. *Crop Science* 8, 213–217.
Tsunewaki, K. (1989) Plasmon diversity in *Triticum* and *Aegilops* and its implications in wheat evolution. *Genome* 31, 143–154.
U, N. (1935) Genome analysis of *Brassica* with special reference to experimental formation of *B. napus*. *Journal of Heredity* 73, 335–339.
Ugent, D., Pozorski, S. and Pozorski, T. (1981) Prehistoric remains of sweet potato from the Casma Valley, Peru. *Phytologia* 49, 401–415.
Ugent, D., Pozorski, S. and Pozorski, T. (1982) Archeological potato tuber remains from the Casma Valley of Peru. *Economic Botany* 36, 182–192.
Ugent, D., Pozorski, S. and Pozorski, T. (1986) Archaeological manioc (*Manihot*) from coastal Peru. *Economic Botany* 40, 78–102.
Vaillancourt, R.E. and Weeden, N.F. (1992) Chloroplast DNA polymorphism suggests a Nigerian center of domestication for the cowpea, *Vigna unguiculata*, Leguminosae. *American Journal of Botany* 79, 1194–1199.
Valenzuela, H.R. and DeFrank, J. (1995) Agroecology of tropical underground crops for small scale agriculture. *Critical Reviews in Plant Sciences* 14, 213–238.
van Bothmer, R., Yen, C. and Yang, J. (1990) Does wild, six-rowed barley, *Hordeum agriocrithon* really exist? *FAO/IBPGR Plant Genetic Resource Newsletter* 77, 17–19.
van den Berg, R.G., Miller, J.T., Ugarte, M.L., Kardolus, J.P., Villand, J., Niehuis, J. and Spooner, D.M. (1998) Collapse of morphological species in the wild potato *Solanum brevicaule* complex (Solanaceae sect. Petota). *American Journal of Botany* 85, 92–109.
Vander Kloet, S.P. and Lyrene, P.M. (1987) Self-incompatibility in diploid, tetraploid and hexaploid *Vaccinium corymbosum*. *Canadian Journal of Botany* 65, 660–665.
van der Maesen, L.J.G. (1987) Origin, history and taxonomy of chickpea. In: Saxena, M.C. and Singh, K.B. (eds) *The Chickpea*. CAB International, Wallingford, UK.
Vanderplank, J.E. (1978) *Disease Resistance in Plants*. Academic Press, New York.
van Oss, H., Aron, Y. and Ladizinsky, G. (1997) Chloroplast DNA variation and evolution in the genus *Lens* Mill. *Theoretical and Applied Genetics* 94, 452–457.
van Slageren, M.W. (1994) *Wild Wheats: a Monograph of* Aegilops *L. and* Amblyopyrum *(Jaub. & Spach) Eig (Poaceae)*. Papers 1994, Wageningen Agricultural University, Wageningen.
van Zeist, W. and Bakker-Heeres, J.A.H. (1985) Archaeological studies in the Lavant 1. Neolithic sites in the Damascus Basin: Aswad, Ghoraifé, Ramad. *Palaeohistoria* 24, 165–256.
Vaughan, D.A. (1994) *The Wild Relatives of Rice*. The International Rice Research Institute, Manila, Philippines.
Vavilov, N.I. (1926) *Studies on the Origins of Cultivated Plants*. Institute of Applied Botany and Plant Breeding, Leningrad.
Vavilov, N.I. (1949–1950) *The Origin, Variation, Immunity and Breeding of Cultivated Crops*. Chronica Botanica, Waltham, Massachusetts.

Vences, F.J., Vaquero, F., Garcia, P. and Vega, M.P.D. L. (1987) Further studies on the phylogenetic relationships in *Secale*: on the origin of its species. *Plant Breeding* 98, 281–291.

Vetukhiv, M. (1956) Fecundity of hybrids between geographic populations of *Drosophila pseudoobscura*. *Evolution* 10, 139–146.

Viard, F., Bernard, J. and Desplanque, B. (2002) Crop–weed interactions in the *Beta vulgaris* complex at a local scale: allelic diversity and gene flow within sugar beet fields. *Theoretical and Applied Genetics* 104, 688–697.

Vijendra Das, L.D. (1970) Chromosome associations in diploid and autotetraploid *Zea mays* L. *Cytologia* 35, 259–261.

Vorsa, N. and Bingham, E.T. (1979) Cytology of 2n pollen production in diploid alfalfa, *Medicago sativa*. *Canadian Journal of Genetics and Cytology* 21, 525–530.

Vos, P., Hogers, R., Bleeker, M., Reijans, M., van de Lee, T., Hornes, M., Frijters, A., Pot, J., Peleman, J., Kuiper, M. and Zabeau, M. (1995) AFLP: a new technique for DNA fingerprinting. *Nucleic Acids Research* 23, 4407–4414.

Waddington, C.H. (1942) Canalization of development and the inheritance of aquired characteristics. *Nature* 150, 563–567.

Waddington, C.H. (1953) Genetic assimilation of an acquired character. *Evolution* 7, 118–126.

Waddington, C.H. (1957) *The Strategy of the Genes*. Allen & Unwin, London.

Wade, M.J. (1992) Sewell Wright: gene action and the shifting balance theory. *Oxford Survey of Evolutionary Biology* 8, 35–62.

Wahl, I. and Segal, A. (1986) Evolution of host–parasite balance in natural indigenous populations of wild barley and oats in Israel. In: Barigozzi, C. (ed.) *Origin and Domestication of Wild Plants*. Elsevier, Amsterdam, pp. 129–142.

Wahlund, S. (1928) Zusammersetung von Populationen und Korrelation – sercheinungen von Standpunkt der Verebungslehre aus betrachtet. *Hereditas* 1, 165–206.

Walbot, V. and Cullis, C.A. (1985) Rapid genomic change in higher plants. *Annual Review Plant Physiology* 36, 367–396.

Wallace, B. (1968) *Topics in Population Genetics: Coadaptation*. Norton, New York.

Wallace, B. (1970) *Genetic Load*. Prentice-Hall, Englewood Cliffs, New Jersey.

Walters, J.L. (1952) Heteromorphic chromosome pairs in *Paeonia californica*. *American Journal of Botany* 39, 145–151.

Wang, G.-Z., Miyashita, N.T. and Tsunewaki, K. (1997) Plasmon analysis of *Triticum* (wheat) and *Aegilops*: PCR-single-strand conformational polymorphism (PCR-SSCP) analysis of organellar DNAs. *Proceedings of the National Academy of Sciences USA* 94, 14570–14577.

Wang, Z.Y., Second, G. and Tanksley, S.D. (1992) Polymorphism and phylogenetic relationships among species in the genus *Oryza* as determined by nuclear RFLPs. *Theoretical and Applied Genetics* 83, 565–581.

Warwick, S.I. and Black, L. (1980) Uniparental inheritance of atrizine resistance in *Chenopodium album*. *Canadian Journal of Plant Science* 60, 751–753.

Waser, N.M. (1993) Population structure, optimal outbreeding, and assortative mating. In: Thornhill, N.W. (ed.) *The Natural History of Inbreeding and Outbreeding: Theoretical and Emperical Perspectives*. University of Chicago Press, Chicago, Illinois, pp. 173–199.

Washburn, S.L. and Moore, R. (1980) *Ape into Human: a Study of Human Evolution*. Little and Brown, Boston, Massachusetts.

Way, R.D., Aldwincle, H.S., Lamb, R.C., Rejman, A., Sansavini, S., Shen, T., Watkins, R., Westwood, M.N. and Yoshida, Y. (1991) Apples. In: Moore, J.N. and Ballington, J.R. (eds) *Genetic Resources in Temperate Fruit and Nut Crops*. International Society of Horticultural Science, Wageningen, pp. 1–62.

Weatherwax, P. (1954) *Indian Corn in Old America*. Academic Press, New York.

Weber, J. and May, P.E. (1989) Abundant class of human DNA polymorphisms which can be typed using the polymerase chain reaction. *American Journal of Human Genetics* 44, 388–396.

Weeden, N. (1983) Evolution of plant isozymes. In: Tanksley, S.D. and Orton, T.J. (eds) *Isozymes in Plant Genetics and Breeding*. Elsevier Science Publishers, Amsterdam, pp. 175–208.

Weeden, N. and Gottleib, L.D. (1980) Isolation of cytoplasmic enzymes from pollen. *Plant Physiology* 66, 400–403.

Weeden, N.F. and Lamb, R.C. (1987) Genetics and linkage analysis of 19 isozyme loci in apple. *Journal of the American Society for Horticultural Science* 112, 865–872.

Weeden, N.F. and Robinson, R.W. (1990) Isozyme studies in *Cucurbita*. In: Bates, D.M., Robinson, R.W. and Jeffrey, C. (eds) *Biology and Utilization of the Cucurbitaceae*. Cornell University Press, Ithaca, New York.

Weeden, N.F., Reisch, B.I. and Martins, M.-H.E. (1988) Genetic analysis of isozyme polymorphism in grape. *Journal of the American Society for Horticultural Science* 113, 765–769.

Weeden, N.F., Muehlbauer, F.G. and Ladizinsky, G. (1992) Extensive conservation of linkage relationships between pea and lentil genetic maps. *Journal of Heredity* 83, 123–129.

Weinberg, W. (1908) On the demonstration of heredity in man (translated by S.H. Boyon) (1963). *Papers in Human Genetics*. Presentations Hall, Englewood, Cliffs, New Jersey.

Weins, D., Calvin, C.L., Wilson, C.A., Davern, C.I., Frank, D. and Seavey, S.R. (1987) Reproductive success, spontaneous embryo abortion, and genetic load in flowering plants. *Oecologia* 71, 501–509.

Weir, B.S., Allard, R.W. and Kahler, A.L. (1972) Analysis of complex allozyme polymorphisms in a barley population. *Genetics* 72, 505–523.

Welsh, J. and McClelland, M. (1994) Fingerprinting genomes using PCR with arbitrary primers. *Nucleic Acids Research* 18, 7213–7218.

Wendel, J.F. (1989) New world tetraploid cottons contain old world cytoplasm. *Proceedings of the National Academy of Sciences USA* 86, 4132–4136.

Wendel, J.F. (2000) Genome evolution in polyploids. *Plant Molecular Biology* 42, 225–249.

Wendel, J.F. and Albert, V.A. (1992) Phylogenetics of the cotton genus (*Gossypium*) – character state weighted parsimony analysis of chloroplast restriction site data and its systematic and biogeographic implications. *Systematic Botany* 17, 115–143.

Wendel, J.F. and Percy, R.G. (1990) Allozyme diversity and introgression in the Galapagos Islands endemic *Gossypium darwinii* and its relationship to continental *G. barbadense*. *Biochemical Ecology and Systematics* 18, 517–528.

Wessler, S.R. (1998) Transposable elements associated with normal plant genes. *Physiologia Plantarum* 103, 581–586.

Whitaker, T.W. (1944) The inheritance of certain characters in a cross of two American species of *Lactuca*. *Bulletin of the Torrey Botanical Club* 71, 347–355.

White, E. and Brown, D. (1973) *The First Men*. Time–Life, New York.

White, M.J.D. (1978) *Modes of Speciation*. Freeman, San Francisco.

Whitehouse, M.L.K. (1950) Multiple-allelomorph incompatibility of pollen and style in the evolution of the angiosperms. *Annals of Botany* 14, 199–216.

Whitman, T.G., Morrow, P.A. and Potts, B.M. (1991) Conservation of hybrid plants. *Science* 254, 779–780.

Wittwer, S., Youtai, Y., Han, S. and Lianzheng, W. (1987) *Feeding a Billion. Frontiers of Chinese Agriculture*. Michigan State University Press, East Lansing.

Wijnheijmer, E.H.M., Brandenburg, W.A. and TerBorg, S.J. (1989) Interactions between wild and cultivated carrots (*Daucus carota* L.) in the Netherlands. *Euphytica* 40, 147–154.

Wilkes, H.G. (1977) Hybridization of maize and teosinte in Mexico and Guatemala and the improvement of maize. *Economic Botany* 31, 254–293.

Williams, E.G., Clarke, A.E. and Knox, R.B. (eds) (1994) *Genetic Control of Self-incompatibility and Reproductive Development in Flowering Plants. Advances in Cellular and Molecular Biology of Plants*, Vol. 2. Kluwer Academic, Dordrecht.

Williams, J.G.K., Kubelik, A.R., Livak, A.R., Rafalski, J.A. and Tingey, S.V. (1993) DNA polymorphisms amplified by arbitrary primers are useful as genetic markers. *Nucleic Acids Research* 18, 6531–6535.

Wilson, A.C., Carlson, S.S. and White, T.J. (1975) Biochemical evolution. *Annual Review of Biochemistry* 46, 573–639.

Wilson, C.D. and Heiser, D.B. (1979) The origin and evolutionary relationships of 'Huauzontle' (*Chenopodium nutalliae* Safford), a domesticated chenopod of Mexico. *American Journal of Botany* 66, 198–206.

Wilson, H.D. (1981) Genetic variation among South American populations of tetraploid *Chenopodium* sect. Chenopodium subsect. Cellulata. *Systematic Botany* 6, 380–398.

Wilson, H.D., Barber, S.C. and Walker, T.W. (1982) Loss of duplicate gene expression in tetraploid *Chenopodium*. *Biochemical Systematics and Ecology* 11, 7–13.

Wilson, R.E. and Rice, L.L. (1968) Allelopathy as expressed by *Helianthus annuus* and its role in old field succession. *Bulletin of the Torrey Botanical Club* 95, 432–448.

Wolf, A. and Elisens, W.J. (1993) Diploid hybrid speciation in *Penstemon* (Schrophulariacea) revisited. *American Journal of Botany* 80, 1082–1094.

Wolf, D.E., Takebayashi, N. and Rieseberg, L.H. (2001) Predicting the risk of extinction through hybridization. *Conservation Biology* 5, 1039–1053.

Workman, P.L. (1969) The analysis of simple genetic polymorphism. *Human Biology* 41, 97–114.

Wright, H.E. (1968) Natural environment of early food production north of Mesopotamia. *Science* 161, 334–339.

Wright, S. (1922) Coefficients of inbreeding and relationship. *American Naturalist* 56, 330–338.

Wright, S. (1931) Evolution in Mendelian populations. *Genetics* 16, 97–159.

Wright, S. (1943) Isolation by distance. *Genetics* 28, 114–138.

Wright, S. (1946) Isolation by distance under diverse systems of matings. *Genetics* 31, 39–59.

Wright, S. (1969) *Evolution and the Genetics of Populations* Vol. 2. *The Theory of Gene Frequencies*. University of Chicago Press, Chicago.

Wu, C.-I. (2001) The genic view of the process of speciation. *Journal of Evolutionary Biology* 14, 851–865.

Xiao, J., Li, J., Grandillo, S., Ahn, S.N., Yuan, L., Tanksley, S.D. and McCouch, S.R. (1998) Identification of trait-improving quantitative trait loci alleles from a wild rice relative, *Oryza rufipogon*. *Genetics* 150, 899–909.

Xiong, L.Z., Liu, K.D., Dai, X.K., Xu, C.G. and Zhang, Q. (1999) Identification of genetic factors controlling domestication-related traits of rice using an F_2 population of a cross between *Oryza sativa* and *O. rufipogon*. *Theoretical and Applied Genetics* 98, 243–251.

Yabuno, T. (1962) Cytogenetic studies on the two cultivated species and the wild relatives in the genus *Echinochloa*. *Cytologia* 27, 296–305.

Yamamoto, T., Lin, H., Sasaki, T. and Yano, M. (2000) Identification of heading date quantitative trait locus *Hd6* and characterization of its epistatic interactions with *Hd2* in rice using advanced backcross progeny. *Genetics* 154, 885–891.

Yamane, H., Tao, R., Sugiura, A., Hauck, N.R. and Iezzoni, A.F. (2001) Identification and characterization of S-RNases in tetraploid sour cherry (*Prunus cerasus* L.) *Journal of the American Society for Horticultural Science* 126, 661–667.

Yang, Q., Hanson, L., Bennett, M.D. and Leitch, I.J. (1999) Genome structure and evolution in the allohexaploid weed *Avena fatua* L. (Poaceae). *Genome* 42, 512–518.

Yano, M., Katayose, Y., Ashikari, M., Yamamouchi, U., Monna, L., Fuse T., Baba,T., Yamamoto, K., Umehara, Y. and Sasaki, T. (2000) *Hd1* a major photoperiod sensitivity quantitative trait locus in rice is closely related to the *Arabidopsis* flowering time gene CONSTANS. *Plant Cell* 12, 2473–2483.

Yen, D.E. (1982) Sweet potato in historical perspective. In: Villareal, R. and Griggs, T. (eds) *Sweet Potato – Proceedings 1st International Symposium*. AVRDC, Tainan, Taiwan.

Yu, S.B., Li, J.X., Tan, Y.F., Gao, Y.J., Li, X.H., Zhang, Q. and Maroff, M.A.S. (1997) Importance of epistasis as the genetic basis of heterosis in an elite rice hybrid. *Proceedings of the National Academy of Sciences USA* 94, 9226–9231.

Yule, G.L. (1902) Mendel's laws and their probable relations to intra-racial heredity. *New Phytologist* 1, 193–207.

Zhang, D., Cervantes, J., Huamán, Z., Carey, E. and Ghislain, M. (2000) Assessing genetic diversity of sweet potato (*Ipomoea batatas* (L.) Lam.) cultivars from tropical America using AFLP. *Genetic Research and Crop Evolution* 47, 659–665.

Zhukovskii, P.M. (1970) Spontaneous and experimental introgression, its role in evolution and breeding. *Soviet Genetics* 6, 449–453.

Zijlstra, S., Purimahua, C. and Lindhout, P. (1991) Pollen tube growth in interspecific crosses between *Capsicum* species. *HortScience* 26, 585–587.

Zohary, D. (1965) Colonizer species in the wheat group. In: Baker, H.G. and Stebbins, G.L. (eds) *The Genetics of Colonizing Species*. Academic Press, New York.

Zohary, D. (1972) The wild progenitor and the place of origin of the cultivated lentil: *Lens culinaris*. Merik. *Economic Botany* 26, 326–332.

Zohary, D. (1973) The origin of cultivated cereals and pulses in the Near East. In: Wahrman, J. and Lewis, K.R. (eds) *Chromosomes Today*, Vol. 4. John Wiley & Sons, New York.

Zohary, D. (1986) The origin and spread of agriculture in the old world. In: Barigozzi, C. (ed.) *The Origin and Domestication of Cultivated Plants*. Elsevier, Amsterdam, pp. 3–20.

Zohary, D. (1989) Pulse domestication and cereal domestication: how different are they? *Economic Botany* 43, 31–34.

Zohary, D. (1995) Lentil: *Lens culinaris* (Leguminoseae – Papilionoideae). In: Smartt, J. and Simmonds, N.W. (eds) *Evolution of Crop Plants*. Longman Scientific & Technical, Harlow, UK, pp. 271–274.

Zohary, D. (1999) Monophyletic vs. polyphyletic origin of the crops on which agriculture was founded in the Near East. *Genetic Resources and Crop Evolution* 46, 133–142.

Zohary, D. and Feldman, M. (1962) Hybridization between amphidiploids and the evolution of polyploids in the wheat (*Aegilops–Triticum*) group. *Evolution* 16, 44–61.

Zohary, D. and Hopf, M. (1973) Domestication of pulses in the Old World. *Science* 182, 887–894.

Zohary, D. and Hopf, M. (1988) *Domestication of Plants in the Old World*, Oxford University Press, New York.

Zohary, D. and Hopf, M. (1993) *Domestication of Plants in the Old World: the Origin and Spread of Cultivated Plants in West Asia, Europe and the Nile Valley*, 2nd edn. Clarendon Press, Oxford.

Index

Acheulian hand axe 137
Achira 172
Adaptation syndromes 161–162
Additive genetic variation 18–19
Aegilops species 91–92, 97, 193
Aerial yam 223
AFLPs (amplified fragment length polymorphisms) 24, 27–29
African Center 144, 147
African rice 185
Agave 158
Agricultural centres 143–145
Ahipa 172
Alfalfa *see* Lucerne
Allopatric speciation 110–112
Allopolyploidy 12–16
Allozymes 20–21
Almond 40, 158, 231
Alocasia (taro species) 223
Amaranthus (quinoa species) 144, 172
Andean crops 144, 172–173
Aneuploidy 12
Angiosperm origins 129–130
Annona cherimola 172
Anthoxanthum odoratum 45, 46
Antirrhinum species 118
Apomixis 39, 74–75
Apple 40, 226–228
Apple maggot 115
Apricot 40, 231
Aquilegia species 106
Arabidopsis thaliana 4, 30, 75, 90
Arachis (groundnut species) 243–244
Aripithecus ramidus 133–134
Aroids 222–223
Arracacia xanthorrhiza 172
Arrowroot 145
Australopithecus species 134–136
Autoallopolyploids 16
Autopolyploidy 12–16, 72
Avena (oat species) 30, 45, 47, 70–71, 184
Avocado 13, 144, 158

Balancing selection 46
Banana 40, 209–212
Barley 40, 104, 126, 174–176
Basul 172
Batata line 222
Beta vulgaris (sugar beet) 219
Biological species 101
Bitter vetch 202
Black mustard 234–236
Black pepper 158
Blueberry 40, 105, 158
see also Vaccinium
Bottle gourd 238
Bottlenecks 53
Brassica (cole crop species) 96–97, 168, 234
Bread wheat 191–194
Breeding systems 39–42
Broccoli 40, 234–236

Bromus mango 166
Broomcorn millet 181
Brown mustard 234–236
Brussels sprouts 40, 234–236
Bulrush millet 181–183
Butter bean 203

Cabbage 40, 234–236
Camelina sativa 168
Camote line 222
Canalization 72–73
Canna edulis 172
Canola 234–236
Capsicum (pepper species) 12, 33, 106, 108, 238–240
Capuli cherry 172
Carbon dating 156
Carica (pawpaw species) 13, 172
Cassava 212–214
Catastrophic speciation 112
Cat-tail millet 181
Cauliflower 40, 234–236
Celery 40, 158
Cerocarpus 124
Character displacement 50–51
Chenopodium species 13, 145, 155, 168, 172
Cherimoya 172
Cherry 40
    see also Prunus
Chi squared 33–35
Chickpea 40, 195–196
Chilli peppers 169, 238–240
China Center 144, 147
Chinese cabbage 234
Chinese yam 223
Chloroplast genes 10, 26–29
Chromosomal numerical changes 12–16
Chromosomal repatterning 30, 95–98, 123–124
Chrysanthemum carinatum 6
Cicer (chickpea species) 9, 109
Citron 228, 229
Citrullus lanatus (water melon) 13, 236–237
Citrus species 40, 228–229
Clarkia species 88, 89, 113–114, 124
Cleistogamy 39
Clover 40
Coadaptation 64–72, 123
Cocoa 144
Coconut (*Cocos*) 13, 158
Coevolution 49–51
Coffee arabica 13, 158
Cole crops 234–236
Colocasia esculenta (yam) 223
Columbian exchange 148–149
Common bean 40, 165–166, 167, 203
    see also Phaseolus
Corn *see* Maize
Cotton 40, 98, 126, 241–243
Cowpea 40, 196–198
Cro-Magnon 139–141, 154
Crop characteristics 159–161
Crop dispersals 145–147
Crop mimicry 125–126, 167–178
Crop origins 157–159
Crop–weed hybridizations 125–126, 170
Cucumber 40, 236, 238
Cucumis (cucumber species) 236–238
Cucurbita (squash and melon species) 23, 236–238
Cultural diffusion 145
Cush-cush yam 224
Cyphomandra betacea 172
Cyrtosperma chamissonis (yam) 223

Date 13, 40
Date-palm 158
Datura stramonium 82–83
Deficiencies 10–11
Demographic swamping 124
Dendrogram 22
Desi 196
Dichogamy 40
Dioecism 40
Dioscorea (yam species) 223–224
Directional selection 45
Disomic inheritance 12, 14, 15
Disruptive selection 45–46, 163
DNA markers 24–30
DNA transposable elements 7, 9
Domestication syndromes 161–162, 163–166, 167
Dominance variation 18–19
Dosage effects 82–84
Drosophila species 67, 73, 114
Dry beans 126
    see also common bean
Duplications 10–11
Durum wheat 191, 193

Echinochloa (millet species) 168, 181–182
Ecogeographical isolation 105
Einkorn wheat 190–192
Electrophoresis 20–21

Eleusine (millet species) 13, 183
Emmer wheat 191–193
Enhancers 4
Ensete ventricosum 211
Epilobium species 65, 79
Epistasis 18–19, 60–63
Erythrina edulis 172
Ethiopian mustard 234–236
Euchlaena species 177
Evolutionary species 102
Ex situ conservation 248
Exons 4
Extinction 124–125

Faba beans 206
Festuca microstachys 94
Fig 13, 40
Fig-leaf gourd 236
Filbert 40
Finger millet 126, 181, 183
First division restitution (FDR) 86–87
Fitness 43
Fixed heterozygosity 84, 93
Flax 40, 158
Floral isolation 106, 107
Forbidden mutations 86
Founder-induced speciation 112
Foxtail millet 181, 182
Fragaria (strawberry species) 94, 109, 233
French bean 203

Gametic incompatibility 106–108
Gametic phase disequilibria 69–71
Gene duplications 11
Gene pool system 101, 103–104
Gene-for-gene relationships 51
General combining ability 62
Genetic assimilation 124
Genetic bridge 90–93
Genetic distance 21–23
Genetic diversity in crops 169–171
Genetic drift 52–56
Genetic engineering 127
Genetic maps 30–31
Genetic transilience 114
Genetically modified organisms (GMOs) 127
Genetics of plant domestication 163–166, 167
Genome evolution 30, 31, 95–98, 123–124
Genomic allopolyploids 14
Germplasm collection strategies 249
Gilia species 64, 95, 106, 109–110, 112–113
Glycine (soybean species) 23, 93, 207
Goldenberry 172
Goosefoot 145, 158
Gossypium species 64, 124, 242
    see also cotton
Gourds 236–238
Grape 40, 230–231
Grapefruit 228, 229
Grass pea 202
Green revolution 248
Groundnut 243–244
    see also Peanut
Guava 158
Guinea pig 145
Guinea-fowl 144

Haploidy 12
Hardy-Weinberg equilibrium 32–35
Haricot bean 203
Helianthus (sunflower species) 13, 155, 123–124, 243–244
Hemp 13, 40
Herbicide mimicry 168
Heritability 19, 43
Hessian fly 51
Heterostyly 66
Hibiscus species 89–90
Highland Papayas 172
Homo
    antecessor 139
    erectus 137–138
    ergaster 136
    habilis 136–139
    heidelbergensis 139, 140
    rudolfensis 136
    sapiens 139–143
    spread 141–143
Hordeum (rye species) 70, 174–176
Host–pathogen evolution 51
Hot peppers 238
Hunter–gatherer strategy 153
Hybrid breakdowns 64–65, 109
Hybrid index 121
Hybrid inviability 109
Hybrid sterility 109

In situ conservation 249–250
Inbreeding 38–39, 41, 42, 71, 79
Inca crop species 172
Incongruity 107
Indica rice 187
Instantaneous speciation 115–117

Interracial autopolyploids 14
Interspecific hybridization 119–124
Introgressive hybridization 120–121, 125–126
Introns 4
Invasive species 120
Inversions 6, 8, 9, 67
Inverted repeats 6, 10
Ipomoea (sweet potato species) 221
Iris (species) 121–122
Isozyme variability 20–21, 23
Italian millet 181
ITS (internal transcribed spacer) 29, 30
Iva annua 145, 155

Japanese barnyard millet 181
Japonica rice 187
Javanica rice 187
Jerusalem artichoke 13, 244

Kabuli 196
Kale 234–236
Kaniwa 172
Kiwicha 172
Kohlrabi 234–236

Lactuca species 13, 106
Lagenaria siceraria (gourd) 236–238
Lamium purpureum 117
Late blight fungus 216
Layia species 65
Lemon 228, 229
Lens (species) 23, 200–202
Lentil 22, 200–202
Lepidium meyenii 172
Lettuce 13, 40, 158
Liatrus cylindracea 52
Lima bean 40, 203
Lime 228, 229
Linanthus androsaceus 73–74
Linkage conservation 30–31
LTRs (long terminal repeats) 10
Lucerne 13, 40, 72, 86
Lucuma 172
Lupinus mutabilis 172
Lycopersicon (tomato species) 82, 241

Maca 172
Madagascar bean 203
Madia sativa 166
Maize
  breeding system 40
  cob size 47
  crop types 180–181
  domestication QTL 164
  duplications 11
  evolutionary history 176–178
  fruiting structure 177
  gene pools 104
  genetic variability 24–26
  hybridization 126
  inbreeding depression 79
  origins 179–180
  protein content 48
  races 22
  reproductive isolation 117
  unreduced gametes 86
Malus (apple species) 227
Mandarin 229
Mangifera indica (mango) 13, 40
Mango 13, 40
Manihot (cassava species) 213–214
Manila hemp 210
Manioc 212–214
Marginal zone hypothesis 152
Marrow 236
Mashua 172
Mauka 172
Medicago sativa 13
  see also Lucerne
Mesoamerican centre 144, 147
Microsatellites 24
Migration 35–37, 55
Mimulus species 106
Mirabilis expansa 172
Modes of speciation 110–117
Mora de Castilla 172
Mung bean 40
Musa (banana species) 209–210
Muscadinia species 230
Musk melon 236
Mutation rates 5
Mutualism 50
Myrtus ugni 172

Naranjilla (Lulo) 172
Navel orange 228, 229
Navy bean 203
Neanderthal 139–141
Near East Center 144, 145
Neighbourhood 37
North American Centre 144
Nuclear–organelle interactions 65
Nuñas 172

Oasis theory 151
Oat 104, 183–185
Oca 172
Oenothera species 6, 66, 67–68
Oil palm 141, 158
Oil-seed rape 234–236
Oldowan chopper 137
Olea europaea 13
  see also olive
Olive 40, 158
Onion 13, 40, 158
Organelle inheritance 54
Oryza species 109, 125, 186
  see also rice
Oxalis tuberosa 172

Pacay 172
Pachyrhizus ahipa 172
Panicum miliaceum 181
Papaver species 109
Papaya 13, 40
Parajubaea cocoides 172
Parapatric speciation 110–115
Parthenocarpy 211
Passiflora species 172
Passionfruit 172
Pawpaw 144
Pea 40, 198–200
Peach 40, 231–232
Peanut 126, 243–244
  see also Groundnut
Pear 40
Pearl millet 104, 126
Pecan 40
Penstemon (millet species) 106–107, 181–182
Pepino 172
Peppers 172
Permanent translocation heterozygotes 67–69
Phaseolus (bean species) 105, 108, 110, 172, 203–206
Phlox 50
Physalis peruviana 172
Phytoliths 145
Pin and thrum 66
Pineapple 13, 144, 158
Pistachio 40, 158
Pisum (pea species) 198–199
Plant domestication
  evidence 154–157
  genetic diversity 169–170, 170–171
  reasons 151–153
  stages 153–154
  traits 161–163
Plantains 210
Pleiotropy 60
Plum 40, 231
Point mutations 5
Polymnia sonchifolia 172
Polyploid complex 90–93
Polyploidy
  allelic dose span 94
  chromosomal repatterning 95–98
  definitions 12–16
  developmental rates 81
  factors influencing establishment 78–80
  fertility 79
  gene dosage effects 82–84
  genetic bridge 90–93
  genetic differentiation 93–95
  heterozygosity 84–90
  inbreeding depression 79–80
  ionizing radiation 81
  nucleotypic effects 80–81
  photosynthesis 83–84
  self-incompatibility 79
Polysomic inheritance 14, 16
Portuguese sailors/traders 148
Postzygotic RIBs 103
Potato 72, 126, 86, 172, 214–218
Potentilla glandulosa 117–118
Pouteria lucuma 172
Prezygotic RIBs 103
Primrose 66
Promoters 4
Proso millet 181
Prunus species 172, 231
Puccinia graminis tritici 51
Pulses *see* Pea, Lentil *and* Chickpea
Pummello 229
Pumpkin 237

Q factor 193
QTL (quantitative trait loci) 18, 31
Quantitative genetics 18
Quinine 158
Quinoa 144, 158, 172
Quito palm 172

Radish 40, 158
RAPDs (randomly amplified polymorphic DNA) 24–27
Rapeseed 126
Raphanus species 13, 23, 124

Raspberry 40
Reasons for agriculture 151–153
Recombinational speciation 121
Repetitive DNA 5
Reproductive isolating barriers (RIBs) 103–110
Retroelements 7, 9–10
RFLPs (restriction fragment length polymorphisms) 24–26, 27–28, 30
Rhagoletis species 115
Rhubarb 40
Rice 40, 104, 126, 185–187
RNA transposable elements 7, 9
Rubber 30
Rubus glaucus 172
Runner bean 203
Rye 40, 104, 187–188
Ryegrass 40

Saccharum species 218
Safflower 13, 40, 158
Scarlet bean 203
Secale (rye species) 187
Second division restitution (SDR) 86–87
Seed mimicry 168
Segmental allopolyploids 15
Selection 42–49, 55–56
Self-incompatibility 40–42, 79
Selfish DNA 4, 99
Senecio species 120, 168
Sequencing DNA 29–30
Sesame 13, 158
Setaria italica (millet) 181
Shaddock 228, 229
Shifting balance theory 55, 57
Sieva bean 203
Silenced genes 90
Silversword alliance 114
Single gene traits 13
Small millets 181
Snap bean 203
Solanum (potato species) 172, 215, 241
Sorghum 40, 104, 126, 165–166, 188–190
Sour cherry 231
Sour orange 228, 229
South American Centre 144, 147
Southeast Asia Centre 144
Southern blot 24, 26
Soybean 40, 126, 207–208
Spanish sailors/traders 148
Specific combining ability 62
Speltoid mutant 67
Spinach 10, 40
Squash 40, 172, 236–238
SSRs (single sequence repeats) 24, 27–28, 29
Stabilizing selection 45, 73, 86
Stasipatric speciation 113
Stephanomeria species 115–116
Strawberry 40, 232–233
Strict autopolyploids 13, 14
Sugar beet 219–220
Sugar cane 126, 218–219
Summer squash 236
Sumpweed 145
Sunflower 13, 126, 244–245
see also Helianthus
Swede 234–235
Sweet cherry 231
Sweet orange 229
Sweet peppers 238
Sweet potato 40, 216, 220–222
Sympatric speciation 115–117

Tamarillo 172
Tangerine 228
Taraxicum officinale 75
Taro 13, 216, 222–223
Tarwi 172
Taxonomic species 100
Tea 13, 158
Teff 141
Tehuacan Valley 47, 156
Temporal isolation 105, 116
Teosinte species 177
Tepary bean 203
Tetrasomic inheritance 16
TIRs (short terminal inverted repeats) 9
Tomato 40, 240–241
Tradescantia species 105
Tragopogon species 78, 82, 85
Transgene escape 127
Translocations 6, 7
Transpositions 6–10
Triangle of U 235
Tripartate hypothesis 179
Tripsacum species 177, 179
Triticum (wheat species) 13, 91, 97, 190
Tropaeolum tuberosum 172
Turnip 40, 158, 234–236

Ugni 172
Ulluco 172
Unreduced gametes 78, 86–87

Vaccinium) 13, 79, 105, 108
    see also Blueberry
Vavilov 170
Vicia (faba bean species) 13, 168, 206
Vigna (pea species) 198–200
Viola tricolor 117
Vitis (grape species) 230

Wahlund effect 37–38
Walnut 40
Water melon 13, 40, 236
Water yam 224
Weed evolution 166–168
Wheat 40, 51, 83, 97–98, 103, 126,
    190–194
White flowered gourd 236
Winter squash 236
World population numbers 247

Xanthosoma (taro species) 223

Yacon 172
Yam 216, 223–225
Year bean 203
Yellow yam 223

Zauschneria species 65, 109
Zea mays see Maize
Zea species 177

cabi-publishing.org/bookshop
Browse Read and Buy
ANIMAL & VETERINARY SCIENCES
BIODIVERSITY CROP PROTECTION
HUMAN HEALTH NATURAL RESOURCES
ENVIRONMENT PLANT SCIENCES
SOCIAL SCIENCES
CABI Publishing
A division of CAB International
Online BOOK SHOP
Search
Reading Room
Bargains
New Titles
Forthcoming
Crop Pollination by Bees
Keith S. Delaplane and Daniel F. Mayer
A DICTIONARY OF Entomology
G Gordh and D H Headrick
Order & Pay Online!
MasterCard
VISA
AMERICAN EXPRESS
Principles of CATTLE PRODUCTION
C.J.C. Phillips
Seeds
THE ECOLOGY OF REGENERATION IN PLANT COMMUNITIES
2nd EDITION
Edited by Michael Fenner
FULL DESCRIPTION
BUY THIS BOOK
BOOK OF THE MONTH
Tel: +44 (0)1491 832111 Fax: +44 (0)1491 829292